Techniques for Evaluating Insect Resistance in Crop Plants

C. Michael Smith
Zeyaur R. Khan
Mano D. Pathak

Library of Congress Cataloging-in-Publication Data

Smith, D. Michael (Charles Michael)
    Techniques for evaluating insect resistance in crop plants / by C. Michael Smith,
Zeyaur R. Khan, Mano D. Pathak.
        p.   cm.
    Includes bibliographical references and index.
    ISBN 0-87371-856-9
    1. Food crops—Insect resistance—Measurement.   I. Khan, Zeyaur R.
II. Pathak, M. D.   III. Title.
SB608.F62S58   1994
632′.7—dc20   93-6127
                                                                                                          CIP

Preface

The growing of insect-resistant crop varieties has led to major increases in food production in many tropical developing countries of the world. The varieties have played an important role in the "Green Revolution" in south and southeast Asia. During the development of these and many other varieties, several entomological techniques and methods were developed for evaluating insect resistance. These techniques were developed to measure the levels of insect resistance, as well as to elucidate the morphological and chemical bases of insect resistance in crop plants. Many of these techniques made use of existing technologies, but in several instances, new technologies have been developed to help identify and quantify insect resistance in crop plants.

In the first textbook on insect resistance in crop plants, published in 1951, R. H. Painter described the use of photographic and volumetric methods to describe or quantify plant resistance, in addition to the quantitative methods still used today, such as the insect sweep net and visual counts of insect damage. In the ensuing 40 years, quantum changes have occurred in the ability of researchers to measure both plant and insect characteristics. Our ability to determine quantitative and qualitative differences in plant chemical constituents has been tremendously increased with the development of gas chromatography and high pressure liquid chromatography. Plant tissue extraction and distillation techniques have improved greatly from the destructive technique of steam distillation to the current nondestructive practices of molecular distillation and chromatographic head space sampling. Analytical research dealing with the chemical composition of plants has led to the development of artificial diets and rearing methods for many of the major pest insects of the world. As a result, a constant source of these insects has greatly accelerated the rate of progress made by entomologist-plant breeder team efforts. At the same time, our ability to study the plant morphological characters affecting various insect behaviors has greatly progressed with the development, and now routine use, of both transmission and scanning electron microscopes.

These techniques have given researchers entirely new perspectives into the cellular and organelle bases of plant resistance to insects.

The commercial development and use of the semiconductor microchip in the mid 1960s has also had a direct impact on the study of plant resistance to insects. Digital balances now routinely measure insect and plant tissue weights in the microgram range. Microcomputers process the plant and insect data and textual information of plant resistance researchers at rates never thought possible 40 years ago.

Other relatively newer techniques have also facilitated the rapid development of techniques to identify crop plants resistant to insects. The electronic feeding monitor, a relatively inexpensive piece of equipment, is routinely used to determine the actual intracellular tissue site at which insects feed and to compare the feeding responses of insects on different genetic sources of resistance. The photovoltaic cell, a "space age" technology product, is a major component of the leaf area meter, a device also used regularly to measure insect defoliation on plant foliage evaluated for insect resistance. The invention and widely accepted use of a very simple device, the plastic larval plant inoculator, has greatly improved the accuracy of many crop plant insect-resistance breeding programs and tremendously accelerated the rate of progress of identifying sources of resistance in the major world food crops damaged by foliage-feeding Lepidoptera.

As plant breeding programs involving insect resistance have expanded, all of these techniques have become invaluable in detecting insect resistance in crop plants. Success in breeding for insect resistance depends, as always, on evaluating and identifying sources of insect-resistant plant materials. Unfortunately, published and unpublished records on these techniques are scattered and are often not easily available to scientists working in developing countries. The purpose of this book is to serve as a comprehensive overview of the entomological techniques developed during the past 40 years and to demonstrate how each has been utilized to evaluate insect resistance in crop plants and to understand the mechanisms and causes of this resistance.

We owe special thanks to several individuals who were instrumental in the development of this book. We are fortunate to have had the assistance of Evelyn Kennedy, who typed the entire manuscript. We also thank Debbie White for her assistance with typing certain chapters of the manuscript. Kimberly Schow carefully read several drafts of the manuscript for errors. Many of the figures were accurately drawn or redrawn by a gifted artist, Mr. Boying Rubio. In addition, several original photographs were provided by F. M. Davis, W. W. Hargrove, R. Nishida, N. Panda, H. D. Pierce, M. Sakuma, and B. R. Wiseman.

We are very indebted to Frank M. Davis for his long-time interest in developing and refining accurate techniques to measure plant resistance to insects. This interest served as a source of guidance and inspiration to us in the development and writing of this book.

Mike Smith and Zeyaur Khan, Manhattan, Kansas
Mano Pathak, Lucknow, India
May 1993

The Authors

Charles Michael Smith is professor and head of the Department of Entomology at Kansas State University. He graduated from Southwestern Oklahoma State University with a B.S. in Biology and received M.S. and Ph.D. degrees in entomology from Mississippi State University. He was a Postdoctoral Research Associate in the Department of Entomology at North Carolina State University from 1976 to 1978.

Dr. Smith was a member of the faculty in the Department of Entomology at Louisiana State University from 1978 to 1988 and Chairman of the Division of Entomology, in the Department of Plant, Soil and Entomological Sciences, at the University of Idaho and Louisiana State University and continues to present guest lectures on the subject. He is the author of *Plant Resistance to Insects—A Fundamental Approach* and author or coauthor of over 50 scientific articles and several book chapters.

Zeyaur R. Khan is Senior Research Scientist at the International Centre of Insect Physiology and Ecology (ICIPE), Nairobi, Kenya. From 1987 to 1991 he worked at the International Rice Research Institute (IRRI) in the Philippines as a team leader of ICIPE-IRRI Collaborative Program. His research interests center on the allelochemical basis of insect/plant interactions. Dr. Kahn is the author of *World Bibliography of Rice Stem Borers* and *Insect Pests of Rice*. He wrote this book during his sabbatical leave at the Kansas State University. Dr. Khan received his Ph.D. in 1980 from the Indian Agricultural Research Institute, New Delhi, India.

M. D. Pathak got his Ph.D. from Kansas State University in 1958 working with Dr. R. H. Painter. His research has been primarily on insect resistance in host plants in India, and at the International Rice Research Institute in the Philippines where he worked as Head of the Entomology Department for 14 years and as Director of Research and Training for 11 years. He was a visiting Professor at Cornell University for 1 year in 1969–70. For the last 4 years he has been working as Director General, U.P. Council of Agricultural Research, Lucknow, India.

Dr. Pathak's work has included screening nearly 80,000 samples of rice germplasm for resistance to common insect pests of rice and investigating various aspects of insect host-plant interactions, genes for resistance, and insect biotype development. He was the leader of the team which developed IR 20, the first high-yielding short-statured rice variety resistant to several insect pests, and also to a few diseases and soil problems. Dr. Pathak is the recipient of several national and international awards.

Contents

CHAPTER 1

Introduction

There is a constant need to develop accurate insect-plant bioassay methods and techniques, and to refine and improve existing techniques. Maxwell (1977) cited the need for a better understanding of insect behavior in order to design assays that accurately assess insect responses to plant chemicals. Kennedy (1977) reviewed many aspects of insect olfactory behavior and suggested ways to make insect olfactory bioassays more discriminating. Cook (1976) provided a similar review of the methodology of phagostimulants in insect-plant bioassays.

There is also a need to consider plant growth characteristics as determining factors in plant-insect bioassays, since resistance to insects is plant age specific (Chapman 1974). Researchers should attempt to develop bioassays that use plant growth and development stages similar to that when plants normally come under insect attack. This may be difficult, however, due to the differences in plant growth brought about by the conditions of artificial climate (light, temperature, humidity) under which plants are often grown (Tingey and Singh 1980).

Many challenges exist to develop continually refined techniques that measure plant resistance to insects, while giving careful consideration to the many sources of variation that arise from the insect species and plant material that are being evaluated, as well as the environment in which the bioassay is conducted. Each of these variables can have a pronounced effect on the expression of plant resistance and the ultimate success or failure of a bioassay. Examples of each of these types of stresses will be discussed in this introductory chapter. Additional information may also be obtained from reviews by Dahms (1972), Davis (1985), Heinrichs (1988), Fernandes (1990), Smith (1989), Tingey (1968), and Wolfson (1988).

1.1. PLANT VARIABLES

1.1.1. Plant Tissue Age

The expression of insect resistance in different plant tissues varies tremendously during the life of a plant. In some crops, plants are less resistant to insects in the early stages of development. Resistance to the southwestern corn borer, *Diatraea grandiosella* (Dyar), and European corn borer *Ostrinia nubilalis* (Hubner) in resistant corn hybrids is greater in whorl (vegetative) stages than in tassel (reproductive) stages (Klun and Robinson 1969, Videla et al. 1992). In sorghum, younger plants have greater resistance to the aphid, *Rhopalosiphum maidis* (Fitch), the planthopper, *Peregrinus maidis* (Ashm.) (Fisk 1978) and the migratory locust, *Locusta migratoria migratoroides* (R. and F.) (Woodhead and Bernays 1977) than older plants. Young leaves of chrysanthemum varieties resistant to the two-spotted spider mite, *Tetranychus urticae* (Koch), contain higher mono- and polyphenols than mature leaves (Kielkiewicz and van de Vrie 1990).

Conversely, resistance to the rice leaffolder, *Cnaphalocrocis medinalis* (Guenée), is more pronounced in older foliage of resistant rice varieties than in young foliage (Ramachandan and Khan 1991). Similar phenological relationships have been shown to exist in rice resistance to the brown planthopper, *Nilaparvata lugens* Stål, (Velusamy et al. 1986) and the green leafhopper, *Nephotettix virescens* (Distant) (Rapusas and Heinrichs 1987); the resistance of pasture grasses to the grass aphid, *Metopolophium festucae cerealium* (Dent and Wratten 1986); the resistance in tomato foliage to the Colorado potato beetle, *Leptinotarsa decemlineata* (Say) (Sinden et al. 1978); the resistance of sweet corn to the corn earworm, *Heliothis zea* (Boddie) (Wann and Hills 1966), and in barley resistant to bird cherry-oat aphids, *Rhopalosiphum padi* (L.) (Leather and Dixon 1981).

1.1.2. Plant Tissue Type

In the canopy of the soybean plant, younger, more succulent leaves near the plant apex are preferred for feeding by the bollworm, *Heliothis zea;* the two-spotted spider mite (McWilliams and Beland 1977, Rodriguez et al. 1983); and the soybean looper, *Pseudoplusia includens* (Walker) (Reynolds and Smith 1985) than lower leaves. Foliage of sweet clover resistant to feeding by the sweet clover weevil, *Sitona cylindricollis* (F.) (Beland et al. 1970), is also greater in older foliage. This is not a comprehensive trend however, as the upper leaves of the sweet pepper variety "CIND" are more resistant to feeding damage from the greenhouse whitefly, *Trialeurodes vaporariorum* (Westwood), than older, more mature leaves (Laska et al. 1986).

1.1.3. Induced Plant Responses

The expression of insect resistance in plant foliage is also affected by previous wounding by insect or mechanical means. This wounding often causes an induced resistance (Kogan and Paxton 1983) in many crop plants to insect damage. Induction can occur within hours after plant damage and may remain in effect for several days. In some tree species the induction may remain in effect for as long as 3 years. Feeding of the tobacco budworm, *H. virescens,* on cotton plants induces the production of increased levels of phenolic compounds that cause antibiotic symptoms in bollworm larvae (Guerra 1981). Feeding by the two-spotted spider mite on cotton and soybean plants induces resistance that effectively limits further increase of mite populations (Hildebrand et al. 1986, Karban and Carey 1984, Karban 1985, 1986). Resistance in cotton to the beet armyworm, *Spodoptera exigua* (Hubner), is also induced from feeding by the two-spotted spider mite (Karban 1988). Soybeans contain an induction mechanism triggered by the feeding of two-spotted spider mites (Brown et al. 1992), the Mexican bean beetle, *Epilachna varivestis* Mulsant (Lin et al. 1990), and the soybean looper (Reynolds and Smith 1985).

Induced resistance is expressed by *Solanum dulcamara* after feeding by the gall mite, *Aceria cladophthirus* (Nalepa) (Westphal et al. 1991). However, this induction caused by one family of phytophagous mites induces an increased susceptibility to another family of phytophagous mites (Westphal et al. 1992). Several species of trees, including apple (Ferree and Hall 1981), birch (Wallner and Walton 1979, Werner 1979), oak (Schultz and Baldwin 1982, West 1985), pine (Nebeker and Hodges 1983), and willow (Raupp and Denno 1984, Rhoades 1983) also become resistant to insect damage following defoliation or mechanical damage.

The induced response in plants damaged by insect feeding has also been shown to stimulate the production of higher levels of green-leaf volatiles in both attached and excised foliage. Both insect parasites and predators perceive these plant odors to more effectively locate their host insects (Dicke and Dijkman 1992, Turlings et al. 1991, Whitman and Eller 1990). Thus, the importance of knowing the preassay condition of plant foliage in evaluating tritrophic level interactions is also important.

Pronounced phytochemical changes occur in damaged plants and plant tissues after wounding and these changes accompany the induction response. Following insect feeding, the concentration of phenolics increases at the site of damaged tissues, a trend documented in insect-resistant *Solanum dulcamara* (Bronner et al. 1991, Westphal et al. 1981), wheat (Leszczynski et al. 1985), and several trees species, including birch, maple, pine, poplar, and spruce (Baldwin and Schultz 1983, Leather et al. 1987, Niemela et al. 1979, Rohfritsch 1981, Shain and Hillis 1972, Thielges 1968, Tjia and Houston 1975, Wratten et al. 1984). Mechanical wounding or feeding by the flea beetle, *Phyllotreta cruciferae,* causes the concentration of indole glucosinolates in leaves of the oilseed rape to increase by as much as threefold (Bodnaryk 1992). The above examples indicate how induced plant responses alter plant allelochemical content which may directly affect the outcome of plant-insect bioassays.

1.1.4. Response of Diseased Plant Tissues

The insect resistance of plants may also be enhanced by their immune response to invasion by infectious diseases. Ryegrass foliage highly infected by the endophytic fungus, *Acremonium loliae,* is more resistant than uninfected foliage to the Argentine stem weevil, *Listronotus bonariensis* (Kuschel) (Gaynor and Hunt 1983); the bluegrass billbug, *Sphenophorus parvulus* Gyllenhal (Ahmad et al. 1986); the fall armyworm, *Spodoptera frugiperda* (J. E. Smith) (Hardy et al. 1985); the southern armyworm, *Spodoptera eridania* Cramer (Ahmad et al. 1987); the Russian wheat aphid, *Diuraphis noxia* (Mordvilko) (Clement et al. 1992); and several species of sod webworms (Funk et al. 1983). Resistance is also increased in endophyte-infected tall fescue to bird cherry-oat aphids (Eichenseer et al. 1991). Researchers should have a good knowledge of the health of plant tissues prior to the conduct of plant resistance bioassays and attempt to use healthy tissues if at all possible.

1.1.5. Evaluation of Excised and Intact Plant Tissues

It is often helpful, in large-scale plant resistance screening programs, to remove plant tissues from whole plants for laboratory evaluations. Different researchers have attempted to determine if this practice has an effect on the expression of resistance. Sams et al. (1975) found no differences between damage to excised leaflets of various potato species in laboratory tests and population growth of the green peach aphid, *Myzus persicae* Sulzer, on plants of the same species in the field. Raina et al. (1980) reported similar results in evaluating the responses of feeding by the Mexican bean beetle on excised and intact leaves of green bean plants. Floating leaf discs and whole detached leaves of muskmelon varieties are actually superior to intact leaves placed in clip cages for determination of resistance to the two-spotted spider mite (East et al. 1992).

However, negative effects of foliage removal have also been observed. Thomas et al. (1966) noted higher survival of the spotted alfalfa aphid, *Therioaphis maculata* (Buckton), on excised alfalfa trifoliates than on intact trifoliates. Pronounced differences were noted in the level of feeding by chrysomelid beetles on whole leaves; excised leaves; and leaf discs of corn, bean, and squash plants (Risch 1985). In a related study, Jones and Coleman (1988) found that even the size of leaf disks influences the level of resistance to the chrysomelid beetle, *Plagiodera versicolora* (Laich.).

Removal of plant tissues also affects allelochemicals involved in the expression of insect resistance. Leaves of cassava are normally unpalatable to the locust, *Zonocercus variegatis* (L.). When leaves are excised from plants, however, locusts consume them within 1 hr, due to a sudden decrease in the cyanide content from the normally intact resistant leaves (Bernays et al. 1977).

Although the use of excised leaves in insect-resistance bioassays has become common, differences in the physiology of intact and excised plant tissues should not be taken for granted, especially when there is a known allelochemical basis to the resistance.

1.2. INSECT VARIABLES

Test insects themselves are also subject to a wide degree of variation, and the potential effects of this variation on the bioassay must also be determined. The effects of insect age, gender-based differences, and the type of preassay dietary conditioning should be well-defined before attempting to standardize plant-resistance bioassays.

1.2.1. Insect Age

The age of the test insect is directly proportional to the amount of plant biomass it will consume during the bioassay. For this reason, it is important to determine an insect age that shows the greatest difference between resistant and susceptible varieties. Procedures to determine the optimum age for foliage-feeding Coleoptera have been established by Smith et al. (1979) and Schalk and Stoner (1976).

1.2.2. Insect Gender

Gender-based differences in the behavior of test insects also affects the outcome of plant-resistance evaluations. Female plant-feeding insects often consume greater amounts of foliage than males because of the need for the high-protein requirements of egg production. This trend has been well documented in several Coleoptera, including the Colorado potato beetle (Schalk and Stoner 1976); the rice water weevil, *Lissorhoptrus oryzophilus* Kuschel (Cook et al. 1988); and the Mexican bean beetle (Smith et al. 1979). Gender-based responses to plant allelochemicals also exist, and female clover head weevils, *Hypera meles* (F.), and ponderosa pine cone beetles, *Conopthorus ponderosae* Hopkins, are more strongly attracted to host volatiles than males (Kinzer et al. 1972, Smith et al. 1976).

1.2.3. Preassay Conditioning

The pretest conditions in which insects are held prior to plant-resistance bioassays directly affect the outcome of these tests. Test insects used in short-term behavioral bioassays should be removed from food for several hours with access to water to increase their general level of sensory receptivity (Saxena 1967, Moorehouse 1971).

For insects used in longer-term feeding and developmental assays, they should be fed on the host plant they will be tested on for a period immediately prior to assay. Several studies have documented that susceptible varieties predispose insects to feed on them, a tendency which may mask the most accurate results of feeding bioassays involving other varieties on which they may be tested. This condition has been documented in the greenbug, *Schizaphis graminum* Rondani (Schweissing and Wilde 1979a); the corn earworm (Wiseman and McMillian 1980); the Russian wheat aphid (Schotzko and Smith 1991), the pear psylla, *Cacopsylla pyricola* Foerster (Horton and Krysan 1991); and the geometrid moth, *Epirrita autumnata* (Borkhausen) (Hanhimaki and Senn 1992). The need for preconditioning is especially important for antibiosis experiments, where longer term population development is often a criteria for evaluating resistance.

The same need for caution in pretest host selection exists in evaluations of tritrophic level interactions. The host plant on which the predatory mite, *Phytoseiulus persimilis* Athias-Henriot, consumes its prey, the two-spotted spider mite, may affect future responses to volatiles from either prey-infested or uninfested plants. Apparently, predatory mite host habitat searching behavior becomes conditioned to the olfactory cues emitted by prey host plants (Takabayashi and Dicke 1992).

The type of artificial diet on which an insect is reared may also induce changes in insect food preference and impact the definition or identification of insect-resistant plant material. Quisenberry and Whitford (1988) observed that the Lepidopteran artificial diet used to rear and precondition fall army-worm larvae influences the level of armyworm resistance expressed in resistant and susceptible bermuda grass varieties. Diet-induced changes in the food preference of tobacco hormworm, *Manduca sexta* (Johan.), larvae also result in an enhancement of larval preference for the dietary plant species that is stronger than that for acceptable non-host plants (de Boer 1992). Rose et al. (1988), noted that when the composition of the artificial diet used to rear the soybean looper was reduced to nutrient levels approximating those of the soybean plant, the soybean allelochemical coumestrol had a pronounced effect on looper growth, whereas standard (superoptimal) diets containing coumestrol had no effect on larval growth.

1.3. ENVIRONMENTAL VARIABLES

Changes in lighting, temperature, relative humidity, and soil nutrient conditions commonly affect the expression of insect resistance in crop plants. Examples cited below are presented to describe each of these types of environmental effect. For more exhaustive reviews, see Tingey and Singh (1980) and Smith (1989).

1.3.1. Light Quantity and Quality

Both the quantity and quality of light affect the expression of plant resistance to insects. Increased light quantity enhances resistance in tomato to the tobacco hornworm (Kennedy et al. 1981). Exposure of plants to continuous light, however, decreases resistance in soybeans to defoliation by the cabbage looper, *Trichoplusia ni* Hubner (Khan et al. 1986). The growth of plants produced under reduced light intensity decreases insect resistance in wheat (Roberts and Tyrell 1961), alfalfa (Shade et al. 1975), sorghum (Woodhead 1981), corn (Manuwoto and Scriber 1985a), sugarbeet (Lowe 1967), and soybean (Reynolds and Smith 1985).

1.3.2. Temperature

Plant resistance to insects may not be expressed at abnormally low or high temperatures. Higher than normal ambient temperatures diminish the expression of resistance in several small grain crops. This relationship has been established in wheat resistance to the Hessian fly, *Mayetiola destructor* Say (Sosa 1979, Tyler and Hatchett 1983, Ratanatham and Gallun 1986), greenbug resistance in barley (Salas and Corcuera 1991) and sorghum (Wood and Starks 1972), and Russian wheat aphid resistance in sorghum (Girma et al. 1990).

Reduced insect resistance occurs at temperatures 10° to 15° below normal in alfalfa (Karner and Manglitz 1985, Schalk et al. 1969). The amount of exudate from the glandular trichomes controlling two-spotted spider mite resistance in geranium also decreases with increased temperature (Walters et al. 1991).

Kindler and Staples (1970a) also noted that alfalfa resistance to the spotted alfalfa aphid was expressed more fully at fluctuating temperatures than at a constant temperature equivalent to the mean of the fluctuating regime. Similar results were recorded by van de Klashorst and Tingey (1979) when evaluating potato plant material for resistance to the potato leafhopper, *Empoasca fabae* (Harris). These examples indicate the need to grow plants at temperatures which are similar to field growing conditions during their evaluation for insect resistance.

1.3.3. Soil Fertility

The application of micronutrients to the medium in which plants are grown prior to evaluation also affects the expression of insect resistance. Increasing the level of nitrogen fertilizer decreases armyworm resistance in maize (Manuwoto and Scriber 1985b, Wiseman et al. 1973), grasses (Chang et al. 1985, Gaynor and Hunt 1983), peanuts (Leuck and Hammons 1974), alfalfa (Kindler and Staples 1970b), tomato (Barbour et al. 1991), and sorghum (Archer et al. 1990). Increasing the amount of potassium and phosphorous fertilizer, however, increases spotted alfalfa aphid resistance in alfalfa (Kindler and Staples 1970b), greenbug resistance in sorghum (Schweissing and Wilde 1979b), and fall armyworm resistance in pearl millet (Leuck 1972). The addition of silica or aluminum to the fertilization regime of rice plants increases resistance to the whitebacked planthopper, *Sogatella furcifera* (Horvath) (Salim and Saxena 1992). These results indicate the need for researchers to standardize micronutrient regimes used to grow plants for insect-resistance bioassays.

1.3.4. Relative Humidity

Relative humidity is especially important to the process of measuring plant resistance to insect pests of stored grain. The expression of resistance in stored grain sorghum to the lesser rice weevil, *Sitophilus oryzae* (L.) and the maize weevil, *Sitophilus zeamais* (Motschulsky) at differing humidities has been documented by Russell (1966) and Rogers and Mills (1974).

1.4. CONCLUSIONS

The preceding discussions demonstrate that both plant and insect factors directly affect the accuracy and outcome of plant-resistance bioassays. Researchers should bear in mind that plant resistance to insects is a relative phenomenon that is highly variable. As such, resistance is dependent on several interacting factors involving the test insect, the test plant, and test environment. These interactions must be defined as clearly as possible, in order to minimize the variations of these organisms in plant resistance assays.

Researchers should have a good knowledge of the condition of plant tissues prior to the conduct of plant resistance bioassays and attempt to use healthy, undamaged tissues if at all possible. Damaged tissues are likely to have undergone an induced response that alters their allelochemical content and which may directly affect the outcome of plant-insect bioassays. Differences between the response of insects to intact and excised resistant and susceptible plant tissues should also be determined before adopting a bioassay protocol for use in large-scale germplasm evaluations, especially when there is a known allelochemical basis to the insect resistance.

It is also important for researchers to determine the insect age and gender that show the greatest difference between resistant and susceptible varieties. In longer term bioassays, insects should be preconditioned prior to assay on the plant on which they will be tested, to allow the resistance, where present, to be expressed to the maximum levels possible.

Finally, researchers should also bear in mind that changes in lighting quantity and quality, temperature, relative humidity, and soil nutrient conditions commonly affect the expression of insect resistance in crop plants. These variables should be standardized in such a way that they mimic actual plant growing conditions as nearly as possible, so that results from experiments conducted at different times or locations can be compared.

REFERENCES

Ahmad, S., Johnson-Cicalese, J. M., Dickson, W. K. and Funk, C. R., Endophyte-enhanced resistance in perennial ryegrass to the bluegrass billbug, *Sphenophorus parvulus, Entomol. Exp. Appl.,* 41, 3, 1986.

Ahmad, S., Govindarajan, S., Johnson-Cicalese, J. M. and Funk, C. R., Association of a fungal endophyte in perennial ryegrass with antibiosis to larvae of the southern armyworm, *Spodoptera eridania, Entomol. Exp. Appl.,* 45,

Archer, T. L., Onken, A. B., Bynum, E. D., Jr. and Peterson, G. C., Banks grass mite (*Oligonychus pratensis*) abundance on sorghum cultivars with different levels of nitrogen use and metabolism efficiency, *Exp. Appl. Acarol.,* 9, 177, 1990.

Baldwin, I. T. and Schultz, J. C., Rapid changes in tree leaf chemistry induced by damage: evidence for communication between plants, *Science,* 221, 277, 1983.

Barbour, J. D., Farrar, R. R., Jr. and Kennedy, G. G., Interaction of fertilizer regime with host-plant resistance in tomato, *Entomol. Exp. Appl.,* 60, 289, 1991.

Beland, G. L., Akeson, W. R. and Manglitz, G. R., Influence of plant maturity and plant part on nitrate content of the sweet clover weevil-resistant species *Melilotus infesta, J. Econ. Entomol.,* 63, 1037, 1970.

Bernays, E. A., Chapman, R. F., Leather, E. M., McCaffery, A. R. and Modder, W. W. D., The relationship of *Zonocerus variegatus* (L.) (Acridoidea: Pyrgomorphidae) with cassava (*Manihot esculenta*), *Bull. Entomol. Res.,* 67, 391, 1977.

Bodnaryk, R. P., Effects of wounding on glucosinolates in the cotyledons of oilseed rape and mustard, *Phytochemistry,* 31, 2671, 1992.

de Boer, G., Diet-induced food preference by *Manduca sexta* larvae: acceptable non-host plants elicit a stronger induction than host plants, *Entomol. Exp. Appl.,* 63, 3, 1992.

Bronner, R., Westphal, E. and Dreger, F., Enhanced peroxidase activity associated with the hypersensitive response of *Solanum dulcamara* to the gall mite *Aceria cladopthirus* (Acari: Eriophyoidea), *Can. J. Bot.,* 69, 2192, 1991.

Brown, G. C., Nurdin, F., Rodriguez, J. G. and Hildebrand, D. F., Inducible resistance of soybean (var 'Williams') to twospotted spider mite (*Tetranychus urticae* Koch), *J. Kans. Entomol. Soc.,* 64, 388, 1992.

Chang, N. T., Wiseman, B. R., Lynch, R. E. and Habeck, D. H., Resistance to fall armyworm: influence of nitrogen fertilizer application on nonpreference and antibiosis in selected grasses, *J. Agric. Entomol.,* 2, 137, 1985.

Chapman, R. F., The chemical inhibition of feeding by phytophagous: a review, *Bull. Ent. Res.,* 64, 339, 1974.

Clement, S. L., Lester, D. G., Wilson, A. D. and Pike, K., Behavior and performance of *Diuraphis noxia* (Homoptera: Aphididae) on fungal endophyte-infected and uninfected perennial ryegrass, *J. Econ. Entomol.,* 85, 583, 1992.

Clement, S. L., Pike, K. S., Kaiser, W. J. and Wilson, A. D., Resistance of endophyte-infected plants of tall fescue and perennial ryegrass to the Russian wheat aphid (Homoptera: Aphididae), *J. Kans. Entomol. Soc.,* 63, 646, 1990.

Cook, A. G., A critical review of the methodology and interpretation of experiments designated to assay the phagostimulator activity of chemicals to phytophagous insects, in *The Host Plant in Relation to Insect Behavior,* T. Jermy, ed., Plenum Press, New York, 1976, 322.

Cook, C. A., Smith, C. M. and Robinson, J. F., Categories of Resistance in Rice to the Rice Water Weevil (Coleoptera: Curculionidae). M.S. Thesis, Louisiana State University, Baton Rouge, 1988.

Dahms, R. G., Techniques in the evaluation and development of host-plant resistance, *J. Environ. Quality,* 1, 254, 1972.

Davis, F. M., Entomological techniques and methodologies used in research programmes on plant resistance to insects, *Insect Sci. Applic.,* 6, 391, 1985.

Dent, D. R. and Wratten, S. D., The host-plant relationships of apterous virginoparae of the grass aphid *Metopolophium festucae cerealium, Ann. Appl. Biol.,* 108, 567, 1986.

Dicke, M. and Dijkman, H., Induced defense in detached uninfested plant leaves: effects on behavior of herbivores and their predators, *Oecologia,* 91, 554, 1992.

East, D. A., Edelson, J. V., Cox, E. L. and Harris, M. K., Evaluation of screening methods and search for resistance in muskmelon, *Cucumis melo* L., to the twospotted spider mite, *Tetranychus urticae* Koch, *Crop Prot.,* 11, 39, 1992.

Eichenseer, H., Dahlman, D. L. and Bush, L. P., Influence of endophyte infection, plant age and harvest interval on *Rhopalosiphum padi* survival and its relation to quantity of N-formyl and N-acetyl loline in tall fescue, *Entomol. Exp. Appl.,* 60, 29, 1991.

Fernandez, G. W., Hypersensitivity: a neglected plant resistance mechanism against insect herbivores, *Environ. Entomol.,* 19, 1173, 1990.

Ferree, D. C. and Hall, F. R., Influence of physical stress on photosynthesis and transpiration of apple leaves, *J. Am. Soc. Hort. Sci.,* 106, 348, 1981.

Fisk, J., Resistance of *Sorghum bicolor* to *Rhopalosiphum maidis* and *Peregrinus maidis* as affected by differences in the growth stage of the host, *Entomol. Expl. Appl.,* 23, 227, 1978.

Funk, C. R., Halisky, P. M., Johnson, M. C., Siegel, M. R., Stewart, A. V., Amhad, S., Hurley, R. H. and Harvey, I. C., An endophytic fungus and resistance to sod webworms: association in *Lolium perenne* L., *Bio/Techno.,* 1, 189, 1983.

Gaynor, D. L. and Hunt, W. F., The relationship between nitrogen supply, endophytic fungus, and argentine stem weevil resistance in ryegrasses, *Proc. N. Z. Grassland Assn.,* 44, 257, 1983.

Girma, M., Wilde, G. and Reese, J. C., Influence of temperature and plant growth stage on development, reproduction, life span, and intrinsic rate of increase of the Russian wheat aphid (Homoptera: Aphididae), *Environ. Entomol.,* 19, 1438, 1990.

Guerra, D. J., Natural and *Heliothis zea* (Boddie)-induced levels of specific phenolic compounds in *Gossypium hirsutum* (L.), M.S. Thesis, University of Arkansas, 1981, 85.

Hanimaki, S. and Senn, J., Sources of variation in rapidly inducible responses to leaf damage in the mountain birch-insect herbivore system, *Oecologia,* 91, 318, 1992.

Hardy, T. N., Clay, K. and Hammond Jr., A. M., Fall armyworm (Lepidoptera: Noctuidae): a laboratory bioassay and larval preference study for the fungal endophyte of perennial ryegrass, *J. Econ. Entomol.,* 78, 571, 1985.

Heinrichs, E. A., Ed., *Plant Stress Interactions,* John Wiley & Sons, New York, 1988, 600.

Hildebrand, D. F, Rodriquez, J. G., Brown, G. C., Luu, K. T. and Volden, C. S., Peroxidative responses of leaves in two soybean genotypes injured by twospotted spider mites (Acari: Tetranychidae), *J. Econ. Entomol.,* 79, 1459, 1986.

Horton, D. R. and Krysan, J. L., Host acceptance behavior of pear psylla (Homoptera: Psyllidae) affected by plant species, host deprivation, habituation, and eggload, *Ann. Entomol. Soc. Am.,* 84, 612, 1991.

Jones, C. G. and Coleman, J. S., Leaf disc size and insect feeding preference: implications for assays and studies on induction of plant defense, *Entomol. Exp. Appl.,* 47, 167, 1988.

Karban, R., Resistance against spider mites in cotton induced by mechanical abrasion, *Entomol. Exp. Appl.,* 37, 137, 1985.

Karban, R., Induced resistance against spider mites in cotton: field verification, *Entomol. Exp. Appl.,* 42, 239, 1986.

Karban, R., Resistance to beet armyworms (*Spodoptera exigua*) induced by exposure to spider mites (*Tetranychus turkestani*) in cotton, *Am. Midl. Nat.,* 119, 77, 1988.

Karban, R. and Carey, J. R., Induced resistance of cotton seedlings to mites, *Science,* 225, 53, 1984.

Karner, M. A. and Manglitz, G. R., Effects of temperature and alfalfa cultivar on pea aphid (Homoptera: Aphididae) fecundity and feeding activity of convergent lady beetle (Coleoptera: Coccinelidae), *J. Kan. Entomol. Soc.,* 58, 131, 1985.

Kennedy, J. S., Behaviorally discriminating assays of attractants and repellents, in *Chemical Control of Insect Behavior,* H. H. Shorey and J. J. McKelvey, eds., John Wiley and Sons, New York, 1977, 304.

Kennedy, G. G., Yamamoto, R. T., Dimock, M. B., Williams, W. G. and Bordner, J., Effect of daylength and light intensity on 2-tridecanone levels and resistance in *Lycopersicon hirsutumf. glabratum* to *Manduca sexta, J. Chem. Ecol.,* 7, 707, 1981.

Khan, Z. R., Norris, D. M., Chiang, H. S., Weiss, N. E. and Oosterwyk, S. S., Light-induced susceptibility in soybean to cabbage looper, *Trichoplusia ni* (Lepidoptera: Noctuidae), *Environ. Entomol.,* 15, 803, 1986.

Kielkiewicz, M. and van de Vrie, M., Within-leaf differences in nutritive value and defence mechanism in chrysanthemum to the two-spotted spider mite (*Tetranychus urticae*), *Exp. Appl. Acarol.,* 10, 33, 1990.

Kindler, S. D. and Staples, R., The influence of fluctuating and constant temperatures, photoperiod, and soil moisture on the resistance of alfalfa to the spotted alfalfa aphid, *J. Econ. Entomol.,* 63, 1198, 1970a.

Kindler, S. D. and Staples, R., Nutrients and the reaction of two alfalfa clones to the spotted alfalfa aphid, *J. Econ. Entomol.,* 63, 938, 1970b.

Kinzer, H. G., Ridgill, B. J. and Reeves, J. M., Response of walking *Conophthorus ponderosae* to volatile attractants, *J. Econ. Entomol.,* 65, 726, 1972.

van de Klashorst, G. and Tingey, W. M., Effect of seedling age, environmental temperature, and foliar total glycoalkaloids on resistance of five *Solanum* genotypes to the potato leafhopper, *Environ. Entomol.,* 8, 690, 1979.

Klun, J. A. and Robinson, J. F., Concentration of two 1,4-benzoxazinones in dent corn at various stages of development of the plant and its relations to resistance of the host plant to the European corn borer, *J. Econ. Entomol.,* 62, 214, 1969.

Kogan, M. and Paxton, J., Natural inducers of plant resistance to insects. In *Plant Resistance to Insects.* Am. Chem. Soc. Symp. Ser. 208, American Chemical Society, Washington, DC, 1983, 375.

Laska, P., Betlach, J. and Havrankova, M., Variable resistance in sweet pepper, *Capsicum annuum,* to glasshouse whitefly, *Trialeurodes vaporariorum* (Homoptera, Aleyrodidae), *Acta Entomol. Bohemoslov.,* 83, 347, 1986.

Leather, S. R. and Dixon, A. F. G., The effect of cereal growth stage and feeding site on the reproductive activity of the bird-cherry aphid, *Rhopalosiphum padi, Ann. Biol.,* 97, 135, 1981.

Leather, S. R., Watt, A. D. and Forrest, G. I., Insect-induced chemical changes in young lodgepole pine (*Pinus contorta*): the effect of previous defoliation on oviposition, growth and survival of the pine beautymoth, *Panolis flammea, Ecol. Entomol.,* 12, 275, 1987.

Leszczynski, B., Changes in phenols content and metabolism in leaves of susceptible and resistant winter wheat cultivars infested by *Rhopalosiphum padi* (L.) (Hom., Aphididae). *Z. Angew. Entomol.,* 100, 343, 1985.

Leuck, D. B., Induced fall armyworm resistance in pearl millet, *J. Econ. Entomol.,* 65, 1608, 1972.

Leuck, D. B. and Hammons, R. O., Nutrients and growth media: influence on expression of resistance to the fall armyworm in the peanut, *J. Econ. Entomol.,* 67, 564, 1974.

Lin, H., Kogan, M. and Fischer, D., Induced resistance in soybean to the Mexican bean beetle (Coleoptera: Coccinellidae): comparisons of inducing factors, *Environ. Entomol.,* 19, 1852, 1990.

Lowe, H. J. B., Interspecific differences in the biology of aphids (Homoptera: Aphididae) on leaves of *Vicia faba.* II. Growth and excretion, *Entomol. Exp. Appl.,* 10, 413, 1967.

Manuwoto, S. and Scriber, J. M., Neonate larval survival of European corn borers, *Ostrinia nubilalis,* on high and low DIMBOA genotypes of maize: effects of light intensity and degree of insect inbreeding, *Agric. Ecosyst. Environ.,* 14, 221, 1985a.

Manuwoto, S. and Scriber, J. M., Differential effects of nitrogen fertilization of three corn genotypes on biomass and nitrogen utilization by the southern armyworm, *Spodoptera eridania, Agric. Ecosyst. Environ.,* 14, 25, 1985b.

Maxwell, F. G., Host-plant resistance to insects-chemical relationships, in *Chemical Control of Insect Behavior,* H. H. Shorey and J. J. McKelvey, eds., John Wiley and Sons, New York, 1977, 304.

McWilliams, J. M. and Beland, G. L., Bollworm: Effect of soybean leaf age and pod maturity on development in the laboratory, *Ann. Entomol. Soc. Am.,* 70, 214, 1977.

Moorehouse, J. E., Experimental analysis of the locomotor behavior of *Schistocerca gregaria* induced by odor, *J. Insect Physiol.,* 17, 913, 1977.

Nebeker, T. E. and Hodges, J. D., Influence of forestry practices on host-susceptibility to bark beetles, *Z. Angew. Entomol.,* 96, 194, 1983.

Niemela, P., Aro, E. M. and Haukioja, E., Birch leaves as a resource for herbivores. Damaged-induced increase in leaf phenols with trypsin-inhibiting effects, *Rep. Kevo Subarctic Res. Stat.,* 15, 37, 1979.

Quisenberry, S. A. and Whitford, F. D., Evaluation of bermudagrass resistance to fall armyworm (Lepidoptera: Noctuidae): influence of host strain and dietary conditioning, *J. Econ. Entomol.,* 81, 1463, 1988.

Raina, A. K., Benepal, P. S. and Sheikh, A. Q., Effects of excised and intact leaf methods, leaf size, and plant age on Mexican bean beetle feeding, *Entomol. Exp. Appl.,* 27, 303, 1980.

Ramachandran, R. and Khan, Z. R., Feeding site selection of first-instar larvae of *Cnaphalocrocis medinalis* on susceptible and resistant rice plants, *Entomol. Exp. Appl.,* 60, 43, 1991.

Rapusas, H. R. and Heinrichs, E. A., Plant age effect on resistance of rice 'IR36' to the green leafhopper, *Nephotettix virescens* (Distant) and rice tungro virus, *Environ. Entomol.,* 16, 106, 1987.

Ratanatham, S. and Gallun, R. L., Resistance to Hessian fly (Diptera: Cecidomyiidae) in wheat as affected by temperature and larval density, *Environ. Entomol.,* 15, 305, 1986.

Raupp, M. J. and Denno, R. F., The suitability of damaged willow leaves as food for the leaf beetle, *Plagiodera versicolora, Ecol. Entomol.,* 9, 443, 1984.

Reynolds, G. W. and Smith, C. M., Effects of leaf position, leaf wounding, and plant age of two soybean genotypes on soybean looper (Lepidoptera: Noctuidae) growth, *Environ. Entomol.,* 14, 475, 1985.

Rhoades, D. F., Responses of alder and willow to attack by tent caterpillars and webworms: Evidence for pheromonal sensitivity of willows. *Plant Resistance to Insects,* P. A. Hedin, ed., Am. Chem. Soc. Symp. Ser. 208, American Chemical Society, Washington, DC, 56, 375, 1983.

Risch, S. J., Effects of induced chemical changes on interpretation of feeding preference tests, *Entomol. Exp. Appl.,* 39, 81, 1985.

Roberts, D. W. A. and Tyrrell, C., Sawfly resistance in wheat. IV. Some effects of light intensity on resistance, *Can. J. Plant Sci.,* 41, 457, 1961.

Rodriguez, J. G., Reicosky, D. A. and Patterson, C. G., Soybean and mite interactions: effects of cultivar and plant growth stage, *J. Kans. Entomol. Soc.,* 56, 320, 1983.

Rogers, R. R. and Mills, R. B., Reactions of sorghum varieties to maize weevil infestation under relative humidities, *J. Econ. Entomol.,* 67, 692, 1974.

Rohfritsch, O., A defense mechanism of *Picea excelsa* L. against the gall former *Chormes abietis* L. (Homoptera: Adelgidae), *Z. Angew. Entomol.,* 92, 18, 1981.

Rose, R. L., Sparks, T. C. and Smith, C. M., Insecticide toxicity to the soybean looper and the velvetbean caterpillar (Lepidoptera: Noctuidae) as influenced by feeding on resistant soybean (PI 227687) leaves and coumestrol, *J. Econ. Entomol.,* 81, 1288, 1988.

Russell, M. P., Effects of four sorghum varieties on the longevity of the lesser rice weevil, *Sitophilus oryzae* (L.), *J. Stored Prod. Res.,* 2, 75, 1966.

Salas, M. L. and Corcuera, L. J., Effect of environment on gramine content in barley leaves and susceptibility to the aphid *Schizaphis graminum, Phytochemistry,* 30, 3237, 1991.

Salim, M. and Saxena, R. C., Iron, silica, and aluminum stresses and varietal resistance in rice: effects on whitebacked planthopper, *Crop Sci.,* 32, 212, 1992.

Sams, D. W., Lauer, F. I. and Radcliffe, E. B., Excised leaflet test for evaluating resistance to the green peach aphid in tuber-bearing *Solanum* plant material, *J. Econ. Entomol.,* 68, 607, 1975.

Saxena, K. N., Some factors governing olfactory and gustatory responses in insects, in *Olfaction and Taste II,* T. Hayashi, ed., Pergamon Press, Oxford, 1967, 835.

Schalk, J. M., Kindler, S. D. and Manglitz, G. D., Temperature and preference of the spotted alfalfa aphid for resistant and susceptible alfalfa plants, *J. Econ. Entomol.,* 62, 1000, 1969.

Schalk, J. M. and Stoner, A. K., A bioassay differentiates resistance to the Colorado potato beetle on tomatoes, *J. Am. Soc. Hort. Sci.,* 101, 74, 1976.

Schotzko, D. J. and Smith, C. M., Effects of preconditioning host plants on population development of Russian wheat aphids (Homoptera: Aphididae), *J. Econ. Entomol.,* 84, 1083, 1991.

Schultz, J. C. and Baldwin, I. T., Oak leaf quality declines in response to defoliation by Gypsy moth larvae, *Science,* 221, 149, 1982.

Schwessing, F. C. and Wilde, G., Predisposition and nonpreference of greenbug for certain host cultivars, *Environ. Entomol.,* 8, 1070, 1979a.

Schwessing, F. C. and Wilde, G., Temperature and plant nutrient effects on resistance of seedling sorghum to the greenbug, *J. Econ. Entomol.,* 72, 20, 1979b.

Shade, R. E., Thompson, T. E. and Campbell, W. R., An alfalfa weevil larval resistance mechanism detected in *Medicago, J. Econ. Entomol.,* 68, 399, 1975.

Shain, L. and Hillis, W. E., Ethylene production in *Pinus radiata* in response to *Sirex amylostereum* attack, *Phytopathology,* 62, 1407, 1972.

Sinden, S. L., Schalk, J. M. and Stoner, A. K., Effects of day length and maturity of tomato plants on tomatine content and resistance to the Colorado potato beetle, *J. Am. Soc. Hort. Sci.,* 103, 596, 1978.

Smith, C. M., *Plant Resistance to Insects: A Fundamental Approach.* John Wiley & Sons, New York, 1989, 286.

Smith, C. M., Frazier, J. L. and Knight, W. E., Attraction of clover head weevil, *Hypera meles,* to flower bud volatiles of several species of *Trifolium, J. Insect Physiol.,* 22, 1517, 1976.

Smith, C. M., Wilson, R. F. and Brim, C. A., Feeding behavior of Mexican bean beetle on leaf extracts of resistant and susceptible soybean genotypes, *J. Econ. Entomol.,* 72, 374, 1979.

Sosa, O., Jr., Hessian fly: resistance of wheat as affected by temperature and duration of exposure, *Environ. Entomol.,* 8, 280, 1979.

Takabayashi, J. and Dicke, M., Response of predatory mites with different rearing histories to volatiles of uninfested plants, *Entomol. Exp. Appl.,* 64, 187, 1992.

Thielges, B. A., Altered polyphenol metabolism in the foliage of *Pinus sylvestris* associated with European pine sawfly attack, *Can. J. Bot.,* 46, 724, 1968.

Thomas, J. G., Sorenson, E. L. and Painter, R. H., Attached vs. excised trifoliate for evaluation of resistance in alfalfa to the spotted alfalfa aphid, *J. Econ. Entomol.,* 59, 444, 1966.

Tingey, W. M., Techniques for evaluating plant resistance to insects, in *Insect Plant Interactions,* J. R. Miller and T. A. Miller, eds., Springer-Verlag, New York, 1968, 342.

Tingey, W. M. and Singh, S. R., Environmental factors influencing the magnitude and expression of resistance, in *Breeding Plants Resistant to Insects,* F. G. Maxwell and P. R. Jennings, eds., John Wiley & Sons, New York, 1980, 683.

Tjia, B. and Houston, D. B., Phenolic constituents of Norway spruce resistant or susceptible to the eastern spruce gall aphid, *For. Sci.,* 211, 180, 1975.

Turlings, T. C. J., Tumlinson, J. H., Eller, F. J. and Lewis, W. J., Larval-damaged plants: source of volatile synomones that guide the parasitoid *Cotesia marginiventris* to the micro-habitat of its hosts, *Entomol. Exp. Appl.,* 58, 75, 1991.

Tyler, J. M. and Hatchett, J. H., Temperature influence on expression of resistance to Hessian fly (Diptera: Cecidomyiidae) in wheat derived from *Triticum tauschii, J. Econ. Entomol.,* 76, 323, 1983.

Velusamy, R., Heinrichs, E. A. and Medrano, F. G., Greenhouse techniques to identify field resistance to the brown planthopper, *Nilaparvata lugens* (Stål) (Homoptera: Delphacidae), in rice cultivars, *Crop Pro.,* 5, 328, 1986.

Videla, G. W., Davis, F. M., Williams, W. P. and Ng, S. S., Fall armyworm (Lepidoptera: Noctuidae) larval growth and survivorship on susceptible and resistant corn at different vegetative growth stages, *J. Econ. Entomol.,* 85, 2486, 1992.

Wallner, W. E. and Walton, G. S. Host defoliation: A possible determination of Gypsy moth population quality, *Ann. Entomol. Soc. Am.,* 72, 62, 1979.

Walters, D. S., Harman, J., Craig, R. and Mumma, R. O., Effect of temperature on glandular trichome exudate composition and pest resistance in geraniums, *Entomol. Exp. Appl.,* 60, 61, 1991.

Wann, E. V. and Hills, W. A., Earworm resistance in sweet corn at two stages of ear development, *Proc. Am. Soc. Hortic. Sci.,* 89, 491, 1966.

Werner, R. A., Influence of host foliage on development, survival, fecundity, and oviposition of the spear-marked black moth, *Rheumaptera hastata* (Lepidoptera: Geometridae), *Can. Entomol.,* 111, 317, 1979.

West, C., Factors underlying the late seasonal appearance of the lepidopterous leaf-mining guild on oak, *Ecol. Entomol.,* 10, 111, 1985.

Westphal, E., Bronner, R. and LeRet, M., Changes in leaves of susceptible and resistant *Solanum dulcamara* infested by the gall mite *Eriophyes cladophthirus* (Acarina: Eriophyoidea), *Can. J. Bot.,* 59, 875, 1981.

Westphal, E., Dreger, F. and Bonner, R., Induced resistance in *Solanum dulcamara* triggered by the gall mite *Aceria cladophthirus* (Acari: Eriophyoidea), *Exp. Appl. Acarol.,* 12, 111, 1991.

Westphal, E., Perrot-Minnot, M. J., Kreiter, S. and Gutierrez, J., Hypersensitive reaction of *Solanum dulcamara* to the gall mite *Aceria cladophthirus* causes an increased susceptibility to *Tetranychus urticae, Exp. Appl. Acarol.,* 15, 15, 1992.

Whitman, D. W. and Eller, F. J., Parasitic wasps orient to green leaf volatiles, *Chemoecology,* 1, 69, 1990.

Wiseman, B. R., Leuck, D. B. and McMillian, W. W., Effects of fertilizers on resistance of Antigua corn to fall armyworm and corn earworm, *Fla. Entomol.,* 56, 1, 1973.

Wiseman, B. R. and McMillian, W. W., Feeding preferences of *Heliothis zea* larvae preconditioned to several host crops, *J. Ga. Entomol. Soc.,* 15, 449, 1980.

Wolfson, J. L., Bioassay Techniques: An Ecological Perspective, *J. Chem. Ecol.,* 14, 1951, 1988.

Wood, E. A., Jr. and Starks, K. J., Effect of temperature and host plant interaction on the biology of three biotypes of the greenbug, *Environ. Entomol.,* 1, 230, 1972.

Woodhead, S., Environmental and biotic factors affecting the phenolic content of different cultivars of *Sorghum bicolor, J. Chem. Ecol.,* 7, 1035, 1981.

Woodhead, S. and Bernays, E. A., Changes in release rates of cyanide in relation to palatability of sorghum to insects, *Nature,* 270, 235, 1977.

Wratten, S. D., Edwards, P. J. and Dunn, I., Wound-induced changes in the palatability of *Betula pubescens* and *B. pendula, Oecologia,* 61, 372, 1984.

CHAPTER **2**

Evaluation of Plants for Insect Resistance

Accurate, efficient techniques that identify plants with insect resistance are essential to all insect-resistance, plant-breeding programs. Plant resistance to insects is normally measured through the exposure of a plant or plant parts to a pest insect and is generally evaluated as the percentage of damage to the plant foliage or fruiting parts, reduction in stand or yield, and general vigor of the plant. The relative susceptibility or resistance of plants to insects can also be determined by factors that influence insect establishment on plants (Saxena 1969), depending on the interaction between insect responses and plant stimuli.

2.1. EVALUATION OF INSECT RESISTANCE BASED ON PLANT DAMAGE

Evaluation techniques based on the measurement of insect damage to plants vary with crop plant, pest insect, and site (laboratory, greenhouse, and field) of experiments. Selection of an efficient, simple, and accurate screening method is extremely important in a varietal evaluation program. The screening method selected should give distinctly different reactions for plants of susceptible, moderately resistant, and resistant cultivars. Plant reaction to insect attack may also depend on the number of insects per plant, plant vigor, plant age, and environmental factors. When insect populations are too high, cultivars with low and moderate levels of resistance may appear susceptible, whereas too few insects may prevent the separation of resistant and susceptible cultivars (Heinrichs et al. 1985, Davis 1985). Plants that lack vigor because of soil or water deficiencies or plants that are very young may also be rated susceptible although under normal conditions they would be resistant. Therefore, to establish an efficient screening method, researchers should understand the biology of the insect and its potential plant damage, the number of insects necessary to infest each plant, the number of insect releases required, the site where insects should be released, the growth stage of the plant, and the most appropriate time interval between infestation and evaluation (Davis 1985). Techniques for screening large numbers of varieties for insect resistance differ for each insect and each crop. However, there are several standard procedures to consider when a screening program is initiated. This chapter deals with important entomological techniques developed and employed for screening for insect resistance during the last 40 years. The chapter uses selected examples as appropriate illustrations and does not list techniques developed for each insect and each crop.

2.1.1. Field Screening

Field screening is usually conducted initially by selecting an area with a predictable high infestation level and by planting the germplasm to be tested so that it coincides with when the pest is abundant at such locations.

Uniform infestations are critical to a successful screening program (Davis 1985). Researchers should make every effort to insure that each plant is infested or has the same opportunity of being infested as any other plant in the test. The level of insect pressure applied to each test plant is also critical. The appropriate insect infestation level is the minimum number of insects required to place the susceptible check cultivar consistently in the susceptible category (Davis 1985).

2.1.1.1. Using Field Populations

Field populations of pest insects are normally used by researchers to evaluate plant materials in early stages of a plant-resistance program (Painter and Grandfield 1935, Jones and Sullivan 1979, Lyman and Cardona 1982, Dickson et al. 1990, Pfannenstiel and Meagher, Jr. 1991). However, the final proving ground for plant material found to be resistant is in replicated field trials. Field evaluations of plant material have some inherent problems that may directly affect the search for resistance (Smith 1989). Unmanaged insect populations may be either too low or too high or unevenly distributed in space or time to inflict a consistent level of damage. Year-to-year variation in population levels of the target pest insect may also make interpreting the results of field evaluations difficult. Finally, unmanaged field populations may be contaminated with nontarget pest insects that cause feeding damage similar to the target insect.

There are a number of ways in which the effectiveness of natural populations can be manipulated by researchers to increase the probability of having uniform and sufficiently high pest population levels at the desired plant growth stage for screening. A trap crop consisting of a mixture of susceptible cultivar may be planted as border rows with the plot (Starks 1970, Doggett et al. 1970, Jackai 1982, Sharma et al. 1988a, Anderson and Brewer 1991). If large-scale populations of pest insects fail to build up in the test plots, trap crop rows can be mechanically cut to force insects onto test plants (Laster and Meredith 1974).

Insecticides with selective properties have been used to eliminate the insect pest's predators and parasites, thereby aiding the buildup of the pest insect population (Sams et al. 1975, Chelliah and Heinrichs 1980, Prakasha Rao 1985, Kushwaha and Singh 1986, Armstrong 1991). The effectiveness of this technique largely depends on the availability of an insecticide relatively inactive against the target pest, but toxic to nontarget pests and natural enemies (Tingey 1986).

Pest insect populations can also be augmented by using a higher dose of nitrogenous fertilizer (Panda 1979, Prakasha Rao 1985, Gahukar 1990), and placing light traps, pheromone traps, or kairomone traps to draw insects into experimental plots (Smith 1989). Mass collection of indigenous pest insect populations collected from surrounding areas and releasing them on test plants may also increase infestation levels (Branson et al. 1981, Ellsbury and Davis 1982, Davis 1983).

2.1.1.2. Using Artificial Infestation

Mass rearing of test insects on natural host plants or on artificial diets offers the most dependable and preferred method of obtaining large and continuous supplies of insects for evaluating plant material. Significant progress has been made in the last 40 years in the mass rearing of insects (Davis 1985). Whether it is more feasible to rear an insect on a natural host or an artificial diet depends upon the insect, the crop, and the facilities available to the researcher. Many phytophagous hemipterans, homopterans, and dipterans can be successfully mass reared on their natural hosts (Starks and Burton 1977, Heinrichs et al. 1985). Mass rearing lepidopterous insects on artificial diets has increased significantly in recent years due to the availability of improved and practical artificial diets (Guthrie et al. 1965, Chatterji et al. 1968, Burton 1969, Dang et al. 1970, Davis 1976, Fery et al. 1979, Mihm 1983a,b, 1989, Khan 1987). For complete discussions on artificial diets, readers are referred to Smith (1966), Singh (1977), Singh and Moore (1985), and Anderson and Leppla (1992). However, continued production of test insects on artificial diets often decreases their genetic diversity (Guthrie et al. 1971, Berenbaum 1986). To avoid these problems, quality control measures must be made a part of the rearing program to ensure that the behavior and metabolism of the laboratory reared insect is similar to that of wild individuals. An effective means of avoiding the development of these problems is to infuse wild individuals into the laboratory colony and to ensure that the artificial diet closely resembles the nutritional and allelochemical composition of the host plant (Smith 1989).

Mechanical techniques have been developed to greatly reduce the amount of time required to mix, dispense, and inoculate artificial diets (Davis 1980a,b), remove pupae from diets (Davis 1982), and harvest insect eggs (Davis 1982, Davis et al. 1985). Techniques to infest plants have also been developed and refined. Early methods made use of agar-based suspensions containing corn earworm, *Helicoverpa zea* (Boddie), eggs that were injected into maize silk masses (Widstrom and Burton 1970). Similar techniques have been used to apply bollworm, *Helicoverpa virescens* (F.), eggs to the fruiting structures of cotton plants (Dilday 1983). A recent refinement in plant inoculation technology is a plastic manual larval plant inoculator (Figure 2.1) that dispenses a suspension of immature larvae and fine-mesh sterilized carrier onto plant tissue (Wiseman et al. 1980). This inoculation method allows the rapid, accurate placement of insect larvae onto plants (Wiseman and Widstrom 1980, 1984, Davis and Williams 1980, Sutter and Branson 1980, Nwanze and Reddy 1991). The larval inocular, first termed a "bazooka" by Mihm et al. (1978), has been used to successfully infest test plants with several species of insects (Davis 1980b, Mihm 1983a,b,c, Pantoja et al. 1986a, Nwanze and Reddy 1991, Diawara et al. 1992).

2.1.2. Field Cage Screening

In spite of all of the efforts outlined above, caging insects on test plants may be necessary for insect-resistance screening (Chalfant and Mitchell 1970, Jones, Jr. and Sullivan 1979, Schalk and Jones 1982, Birch 1989). Cages limit emigration of the test insect from plants being evaluated as well as protecting these insects from predation and parasitism. Cages can be relatively large and enclose several plants in a field or small and enclose a plant part only. Whole plants can be placed in cages constructed of wood, Plexiglass®, or metal frames supporting screened aluminum panels of nylon or saran (Figure 2.2). Cage size and shape are determined by the type, age, and number of test plants that must be evaluated. Dimensions vary from small field cages (Figure 2.3) to large field cages that are placed over galvanized metal frames to cover entire

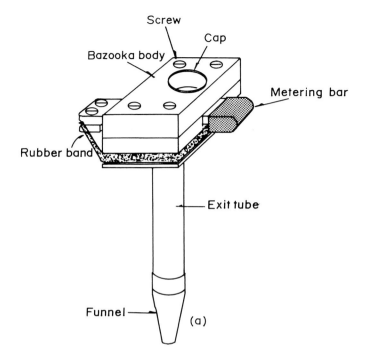

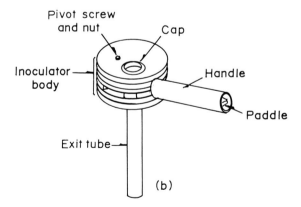

FIGURE 2.1. Manual dispensers for inoculating plant tissues with immature lepidopterous larvae. (a) Modified Bazooka inoculator. (From Mihm, J. A. et al.,
1978, courtesy CIMMYT); (b) Davis larval inoculator. (From Davis, F. M.
and Oswalt, T. G., 1979, Courtesy the U.S. Department of Agriculture.)

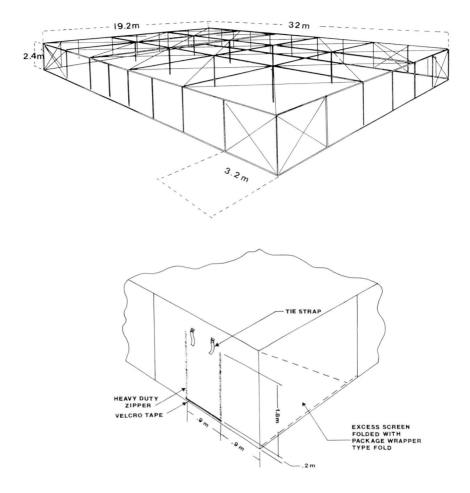

FIGURE 2.2. Galvanized permanent metal cage frame supporting saran screen cover used to confine insect populations on field plantings of soybean. (From Smith, C. M., *Plant Resistance to Insects: A Fundamental Approach,* John Wiley & Sons, New York, Copyright © 1989. With permission. Modified from Lambert, 1984.)

experimental plots (Figure 2.4) (Lambert 1984). Plant parts can be enclosed in sleeve cages made of dialysis tubing (Dimock et al. 1986) or polyester organdy or nylon cloth (Hollay et al. 1987, Sharma et al. 1992) (Figures. 2.5 through 2.7).

Despite their advantages, cages also have some inherent disadvantages that should be anticipated and that require compensation. Some cages may cause abnormal environmental conditions and can alter plant growth, insect

FIGURE 2.3. Nylon cloth field cages used to confine insect populations on field plantings of rice. (Photograph provided by Z. R. Khan.)

FIGURE 2.4. Large screen cages placed over rows of small grain plants to confine cereal leaf beetles. (Photo courtesy of J. A. Webster.)

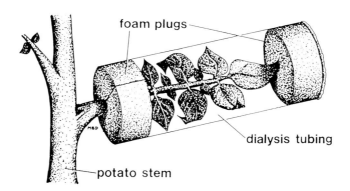

FIGURE 2.5. Dialysis tubing sleeve cage with polyurethane foam end plugs used to confine *Leptinotarsa decemlineata* larvae on leaves of *Solanum* spp. (From Dimock, M. B. et al., *J. Econ. Entomol.,* 79, 1269, 1986. With permission of the Entomological Society of America.)

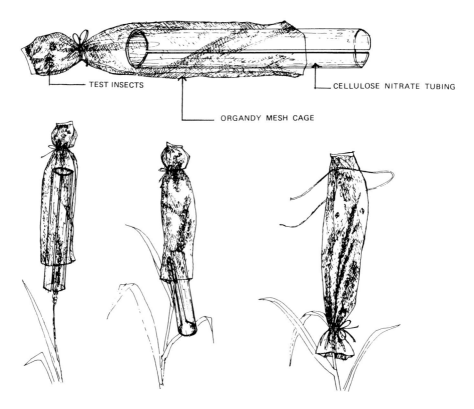

FIGURE 2.6. Organdy mesh sleeve cage used to confine *Oebalus pugnax* adults on rice plants. (From Smith, C. M., *Plant Resistance to Insects: A Fundamental Approach,* John Wiley & Sons, New York, Copyright © 1989. With permission.)

FIGURE 2.7. Nylon cloth cage used to confine insect population on sorghum panicles. (From Sharma, H. C. et al., 1992; photo courtesy of H. C. Sharma and ICRISAT.)

behavior, or can cause foliar disease outbreaks. Obviously, not all plants are affected similarly and cage effects, if any, must be determined on a case-to-case basis.

2.1.3. Greenhouse Screening

Greenhouse screening permits greater control in selecting resistant plants but restricts the amount of material that can be screened over a given period of time. Greenhouse mass screening of seedlings also offers a time saving tool for those crops that have a long growing cycle. However, in all greenhouse tests, it is important to control insect infestation rates so that population pressure does not destroy or obscure potential resistance sources.

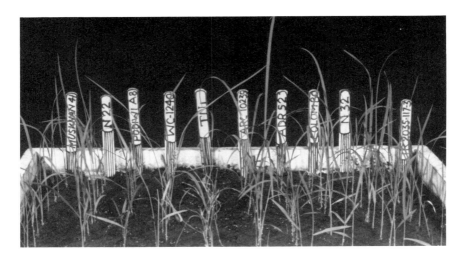

FIGURE 2.8. A wooden seedbox used to screen rice varieties in greenhouses. (From Khan, Z. R. and Saxena, R. C., *Crop Prot.*, 5, 15, 1986. With permission of Butterworth-Heinemann Ltd.)

2.1.3.1. Standard Seedbox Screening Test

A seedbox screening test is the most commonly used technique to screen plants for insect resistance in greenhouses (Wood, Jr. 1961, Pathak 1969b, Starks and Mirkes 1979, Pathak and Saxena 1980, Wiseman and Gourley 1982, Lambert et al. 1982, Saxena and Khan 1984, Hansen et al. 1985, Hansen 1986, Velusamy et al. 1986, Webster 1990). Seedbox screening tests can be undertaken following the method described by Starks and Burton (1977) for the greenbug, *Schizaphis graminum* (Rondani), and by Pathak and Saxena (1980) for rice leafhoppers and planthoppers.

The test cultivars are sown in wooden or metal flats filled with soil (Figure 2.8). A susceptible check and a resistant variety are also sown in random rows in each seed box. Test insects, in numbers sufficient to kill the susceptible check cultivars, are uniformly distributed onto the seedlings. The damage is graded upon the death of the susceptible check cultivar. This screening procedure is replicated four or five times.

2.1.3.2. Modified Seedbox Screening Test

Because the seedbox screening test is mostly qualitative, cultivars with moderate levels of resistance are usually rated as susceptible. For this reason, the seedbox test has been modified in some cases to detect cultivars with moderate levels of resistance to the brown planthopper, *Nilaparvata lugens*

(Stål) (Heinrichs et al. 1985, Velusamy et al. 1986, Zhang et al. 1986). In the modified test, plants are older at the time of infestation and the infestation rate is lowered. In addition, the progeny, rather than the initial source of infestation, are the insects that cause the plant damage. With these exceptions, all other methods used are similar to the conventional test. The modified seedbox screening test has primarily been used for evaluation of cultivars resistant to *N. lugens,* but with modification the method can also be used for other hopper species.

Velusamy et al. (1986) compared the seedbox screening and modified seedbox screening methods to determine the levels of resistance of rice cultivars to three *N. lugens* biotypes. The cultivars ASD11, Wagwag, Utri Rajapan, IR46, and Kencana were susceptible in the standard seedbox screening test but moderately resistant or resistant in the modified seedbox screening test, indicating an increase in the level of resistance with plant age. Similar differences between the seedbox screening and the modified seedbox screening tests were repeated by Zhang et al. (1986). Velusamy et al. (1986) concluded that the modified seedbox screening test provides a rapid method for identifying field-resistant cultivars in the greenhouse.

2.1.3.3. No-Choice Screening Test

Choice (preference) screening tests are carried out routinely in the evaluation of multiple or paired-plant genotypes. The method is relatively simple, inexpensive, and is extremely useful in the preliminary evaluation of resistance. However, insect behavioral responses sometimes differ between choice and no-choice assays. A plant genotype classified as "resistant" in a choice test may be susceptible in a no-choice test (Tingey 1986).

While screening rice cultivars for resistance to the whitebacked planthopper, *Sogatella furcifera* (Horváth), using a choice test, Saxena and Khan (1984) reported that significantly more hoppers moved to the susceptible check cultivar soon after infestation. This uneven distribution of insects caused an unbalanced infestation among the test cultivars and resulted in the escape of some cultivars from insect damage. To avoid this shortcoming, Saxena and Khan (1984) conducted a no-choice screening test in which each row of test cultivar seedlings was excluded from the other by an intervening, vertical, Mylar® plastic partition, and exposed to equal insect infestation. The no-choice test ensures an even distribution of test insects on all the cultivars, irrespective of their susceptibility or resistance, from the beginning to the termination of the experiment and excludes any possibility of escape of the test cultivars from insect damage. From a comparison of 10 genetically diverse cultivars from free-choice and no-choice tests, Saxena and Khan (1984) reported that Podini A8 and N22, rated as moderately resistant and resistant in the choice test, proved to be susceptible and moderately resistant, respectively, in the no-choice test.

The no-choice technique has been widely used to complement free-choice procedures to both identify and confirm the presence or absence of insect resistance in a wide variety of plants (Wiseman et al. 1961, Overman and McCarter 1972, Goonewardene et al. 1979, Pantoja et al. 1986a, Sharma et al. 1988b, Meehan and Wilde 1989, Echendu and Akingbohungbe 1990, Rezaul Karim and Saxena 1990). To maximize the identification and measurement of insect resistance, the use of both free-choice and no-choice screening methods is suggested to provide reliable results.

2.1.4. Laboratory Screening

Techniques for evaluating resistance under more controlled laboratory conditions are often necessary, since field and greenhouse tests are affected by a number of environmental factors that cannot always be controlled. Laboratory screening methods should only be viewed as a reliable and rapid method for confirming insect resistance before or after field or greenhouse testing.

It is generally impractical to use whole plants for laboratory evaluations of insect resistance; therefore, excised leaves and leaf disks are often utilized. Various investigators have conducted research to determine if removing tissues from plants for laboratory evaluation has an effect on the expression of resistance. Sams et al. (1975) compared the results of evaluations of *Solanum* sp. plant material for resistance to the green peach aphid, *Myzus persicae* (Sulzer), using excised leaflet bioassays in the laboratory and aphid population counts on plants in the field. Their results suggested the use of excised leaf assays as a means of rapid assessment of *Solanum* sp. for aphid resistance. Similarly, Raina et al. (1980) found no differences in the amount of feeding of the Mexican bean beetle, *Epilachna varivestis* Mulsant, on excised and intact leaves of green bean plants. However, during the last 10 years it has become increasingly clear that many plants which are physically damaged or diseased undergo significant changes and that these changes can decrease or increase plant palatability and the fitness of insects consuming the plants (Kogan and Paxton 1983, Reynolds and Smith 1985, Risch 1985). Also, the size of leaf disks used in insect feeding tests was shown to affect the behavior of the chrysomelid beetle, *Plagiodera versicolora* (Laicharting) (Jones and Coleman 1988).

Leaf disks or plant tissues are commonly used in insect feeding bioassays of chewing insects (Barnes and Ratcliffe 1967, Barnes et al. 1969, Kogan and Goeden 1970, Kogan 1972a,b, Schalk and Stoner 1976, Khan et al. 1986, 1989, Jackai 1991). Although choice tests are generally used to evaluate insect resistance (Figure 2.9), no-choice forced feeding tests have also been conducted (Barnes and Ratcliffe 1967). Plant damage due to insect feeding has been measured based on area fed (Kogan and Goeden 1970, Kogan 1972a,b,

FIGURE 2.9. Plastic petri dish with divided base for evaluating soybean leaf disks for resistance to *Heliothis zea.* (From Smith, C. M., *Plant Resistance to Insects: A Fundamental Approach,* John Wiley & Sons, New York, Copyright © 1989. With permission.)

Schalk and Stoner 1976, Khan et al. 1986, 1989), dry weight of control and damaged tissue (Barnes and Ratcliffe 1967, Barnes et al. 1969), and estimating the damage by visually scoring the amount of uneaten material on a damage scale (Barnes and Ratcliffe 1967, Jackai 1991).

Resistance against sucking insects can be evaluated following the method developed by Thomas (1970) and Roof et al. (1976) for evaluating resistance of alfalfa clones against potato leafhopper, *Empoasca fabae* (Harris), and the spotted alfalfa aphid, *Therioaphis maculata* (Buckton). Petri dishes (1.5 cm diameter, 2.5 cm deep) are used as test cages. The bottom of the petri dish has 10 holes large enough to insert stems of cuttings, drilled 1.5 cm from the border of the dish, and 3.5 cm apart from each other (Figure 2.10). The bottom is coated with flat black paint. The cuttings are furnished with tap water

Top view with cutting in place

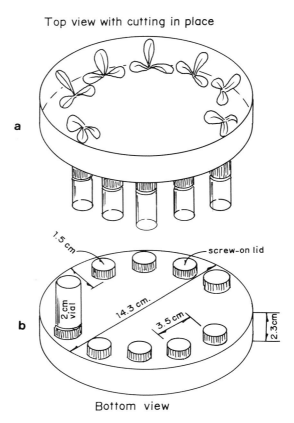

Bottom view

FIGURE 2.10. (a) Round clear plastic dish served as *Empoasca fabae* nymphal preference cage; and (b) plant cuttings placed in tap water in glass vials with
screw-on lids cemented to the bottom cages. (From Roof, M. E. et al.,
Environ. Entomol., 5, 295, 1976. With permission of the Entomological

in 1.7×6 cm vials with screw-on lids cemented to the bottom of the cage. A
2-mm hole drilled in the center of the lid of the vials correspond in the position to the hole in the bottom of the dish. Plant cuttings in petri-dish cages are
infested with first-instar nymphs of the test insect. Visual damage rating is
recorded daily until the susceptible check is dead.

2.1.5. Damage Rating

Host reaction scales are generally developed in germplasm screening programs to accurately describe insect damage levels (Bellotti and Kawano 1980).
These scales should be capable of differentiating among small differences in

FIGURE 2.11. A visual damage rating scale for sorghum midge incidence. (From Sharma, H. C. et al., 1992; photo courtesy of H. C. Sharma and ICRISAT.)

plant damage and of clearly defining resistant, intermediate, and susceptible plants. The rating techniques associated with the scales should be fast and easy to execute because thousands of plants must normally be evaluated during the process of screening and developing resistant germplasm (Davis et al. 1992). Descriptions of damage rating scales should be concise and clear, and if possible, supplemented with illustrations (Figure 2.11) since researchers located in different areas may use the same scale to evaluate germplasm. A uniform scale permits accurate and rapid comparison of native varieties with standard check varieties. Several such scales are being used to evaluate damage by important insect pests of maize (Guthrie et al. 1960, 1978, Wiseman and Davis 1979, Mihm 1983a,b, Davis et al. 1992), rice (Heinrichs et al. 1985, IRRI 1988), sorghum (Johnson and Teetes 1979, Wiseman and Gourley 1982, Nwanze and Reddy 1991, Sharma et al. 1992), etc. (Tables 2.1, 2.2, 2.3, 2.4).

Basically, two types of insect damage rating scales are used. The first is usually based on a 0 to 5 (Linduska and Harrison 1977, Bellotti and Kawano 1980), 1 to 5 (Jackai 1982, Kugler and Ratcliffe 1983, Dickson et al. 1990), or 1 to 6 (Starks and Mirkes 1979) scale. A rating of 0 to 2 or 1 to 2 suggests some resistance and need for further testing; a rating of 4 to 5 or 5 to 6 represents highly susceptible cultivars. Here, the researcher judges whether a cultivar merits future testing or should be discarded. A second

Table 2.1. Rating Scale Used to Evaluate Sorghum for Resistance to Common Insect Pests[a]

Score	Chilo partellus[a]		Sorghum midge (% damage)[b]	Greenbug damage[b]
	Leaf area eaten (mm^2)	% deadheart		
0	No damage	No damage	None	—
1	< 150	< 10	1 – 10	No red spots on leaves
2	150 – 300	11 – 20	11 – 20	Red spots on leaves
3	300 – 450	21 – 30	21 – 30	Part of one leaf dead
4	450 – 600	31 – 40	31 – 40	One leaf dead
5	600 – 750	41 – 50	41 – 50	Two leaves dead
6	750 – 900	51 – 60	51 – 60	Four leaves dead
7	900 – 1050	61 – 70	61 – 70	Six leaves dead
8	1050 – 1200	71 – 80	71 – 80	Eight leaves dead
9	> 1200	> 80	81 – 90	Entire plant dead
10	—	—	91 – 100	—

[a]From Sharma et al. (1992).
[b]From Johnson and Teetes (1979).

Table 2.2. Rating Scale Used to Evaluate Rice for Resistance to Common Insect Pests[a]

Score	Stem borers		Gall midge (% silver shoots)	Leafhoppers/ planthoppers damage
	% deadheart	% whitehead		
0	No damage	No damage	None	None
1	1 – 10	1 – 5	< 5	Slight
3	11 – 20	6 – 10	6 – 10	Leaves 1 and 2 yellow
5	21 – 30	11 – 15	11 – 20	Plants stunted; > 1/2 leaves yellow
7	31 – 60	16 – 25	21 – 50	> 1/2 plants dead; 1/2 severely stunted
9	61 – 100	> 26	> 50	All plants dead

[a]From IRRI (1988).

scale is generally used for cultivars identified for further evaluation. In this case, a rating scale is needed with more classes, for example, 0 to 8 (Lambert et al. 1982), 0 to 9 (Wiseman and Davis 1979, Wiseman and Gourley 1982, Wiseman and Widstrom 1984, Ng et al. 1985, Diawara et al. 1991, Davis et al. 1992), 1 to 9 (Starks and Doggett 1970, Roof et al. 1976, Wilde and Apostol 1983, Bing et al. 1990, Webster 1990, Nwanze and Reddy 1991), or 0 to 10 (Hansen et al. 1985). The differences between the classes, in terms of damage level, are smaller than in 0 to 5 or 1 to 5 scale, allowing a more accurate definition of varietal reaction. These differences are also very important when dealing with low levels of resistance or when trying to increase resistance by combining low or intermediate levels through crossing (Bellotti and Kawano 1980). Several workers have used only 5 or 6 (0, 1, 3, 5, 7, and 9) digit scales used in varietal screening programs (Heinrichs et al. 1985, IRRI 1988). Before scoring for damage,

Table 2.3. Rating Scale Used to Evaluate Cassava for Resistance to Common Insect Pests[a]

Score	Thrips (damage)	Whiteflies (damage)
0	No symptoms	No damage
1	Yellow irregular leaf spots only	Slight speckling of lower leaves
2	Leaf spots, light leaf deformation, part of leaf lobes missing, brown wound tissue in spots on stem and petiole	Heavy speckling of lower leaves
3	Severe leaf deformation and decoloration, poorly expanded leaves, internode stunted and covered with brown wound tissue	Mosaic like symptoms on leaves but little wrinkling, sooty mold on lower and lateral leaves
4	As above, but with growing points dead, sprouting of lateral buds	Wrinkling and yellowish mottling of lower and apical leaves, some leaf necrosis, sooty mold
5	Lateral buds also killed, plants stunted	Severe wrinkling of apical leaves, leaf necrosis, death of plant

[a]From Bellotti and Kawano (1980).

Table 2.4. Rating Scale Used to Measure Leaf Feeding Resistance in Maize Against *Heliothis, Ostrinia, Chilo, Spodoptera,* and *Diatraea*[a]

Score	Description
1	No damage or few pinholes
2	Few short holes on several leaves
3	Short holes on several leaves
4	Several leaves with short holes and a few long lesions
5	Several holes with long lesions
6	Several leaves with lesions < 2.5 cm
7	Long lesions common on one half of the leaves
8	Long lesions common on one half to two thirds of leaves
9	Most leaves with long lesions

[a]From Guthrie et al. (1960, 1978); Mihm (1983a,b).

comparative reactions of resistant and susceptible check varieties should be combined in making a final judgement on varietal reaction to insect damage. If insect pressure is extremely low on susceptible checks, the trial should not be scored for test cultivars.

When developing an insect damage rating scale, the researcher should determine the most appropriate time interval between insect infestation and damage evaluation (Davis 1985). For screening fall armyworm, *Spodoptera frugiperda* (J. E. Smith), resistance in maize, Davis et al. (1992) compared ratings taken 7 days and 14 days after infestation. The scales for both ratings are based on the types of feeding lesions and the numbers of these lesions present on whorl leaves 7 and 14 days after infesting each plant with 30 neonate larvae (Table 2.5, Figure 2.12). Davis et al. (1992) reported that 7- and 14-day ratings were highly correlated in their ability to separate resistant from susceptible genotypes.

Table 2.5. Comparision of 7-Day and 14-Day Rating Scales for Measuring Leaf Feeding Resistance in Maize Against *Spodoptera frugiperda* Feeding[a]

Score	7-day damage rating description	14-day damage rating description
0	No visible damage	No visible damage
1	Only pinhole damage present on whorl leaves	Only pinhole damage present on whorl leaves
2	Pinholes and small circular lesions present on whorl leaves	Pinholes and small circular lesions present on whorl leaves
3	Pinholes, small circular lesions, and a few small elongated lesions up to 1.3 cm in length present on whorl and furl leaves	Pinholes, small circular lesions, and a few small elongated lesions up to 1.3 cm in length present on whorl and furl leaves
4	Small elongated lesions present on whorl leaves and a few midsized elongated lesions of 1.3 to 2.5 cm in length present on whorl and/or furl leaves	Several small to midsized 1.3 to 2.5 cm in length elongated lesions present on few whorl and furl leaves
5	Small elongated lesions and several midsized elongated lesions present on whorl and furl leaves	Several large lesions greater than 2.5 cm in length present on whorl and furl leaves and/or a few small to midsized uniform- to irregular-shaped holes eaten from the whorl and/or furl leaves
6	Small and midsized elongated lesions of greater than 2.5 cm in length present on whorl and/or furl leaves	Several large elongated lesions present on several whorl and furl leaves and/or several large uniform- to irregular-shaped holes eaten from furl and whorl leaves
7	Many small and midsized elongated lesions present on whorl leaves and several large elongated lesions present on furl leaves	Many elongated lesions of all sizes present on several whorl and furl leaves, and several large uniform- to irregular-shaped feeding holes present on whorl and furl leaves
8	Many small and midsized elongated lesions present on whorl leaves and many large elongated lesions present on furl leaves	Many elongated lesions of all sizes present on several whorl and furl leaves and many mid- to large-sized uniform to irregular feeding holes present on whorl and furl leaves
9	Many elongated lesions of all sizes on whorl and furl leaves plus a few uniform- to irregular-shaped holes eaten from the base of the whorl and/or furl leaves	Whorl and furl leaves almost totally destroyed

[a]From Davis et al. (1992).

Other methods of measuring insect damage to plants have also been used, including yield reduction (Peterson et al. 1989, Macfarlane 1990), soft X-ray photographs (George et al. 1983), and root growth (Villani and Gould 1986). Tissue damage in plants can also be determined by measuring the loss in photosynthetic leaf area due to chlorosis or necrosis (Smith, Jr. et al. 1985). The severity of virus-related stunting, yellowing, or curling can indicate resistance to the insect virus vector or resistance to the virus itself (Smith 1989).

FIGURE 2.12. Comparison of visual damage rating scales for screening maize plants
for resistance to *Spodoptera frugiperda* 7 days after infestation (DAI)
and 14 DAI: (a) rating score of 4, 7 DAI; (b) rating score of 9, 7 DAI; (c)
rating score of 5, 14 DAI; and (d) rating score of 9, 14 DAI. (From Davis,
F. M. et al., 1992; photo courtesy of F. M. Davis and the Mississippi
Agricultural and Forestry Experiment Station.)

Insect feeding injury can be simulated by mechanical defoliation (Bautista et al. 1984, Smith, Jr. et al. 1985, Lye and Smith 1988). However, plants respond somewhat differently to artificial defoliation than to actual insect tissue removal (Smith 1989). Therefore, the relationship between the results of artificial and natural defoliation should be determined before accepting results based on artificial defoliation exclusively.

Chang et al. (1973) used artificial defoliation in maize to explore the usefulness of aerial photography in detecting crop-insect infestation. Aero infrared film exposed through a Wratten 89B filter gave the best results, and morning flights at a scale of 1:15,840 were recommended. Defoliation on top of the plant was easily detected, while that on the base was less easily seen (Chang et al. 1973).

Feeding injury can also be simulated by injection or application of phytotoxic insect secretions in plant tissues. This technique holds great promise as a means of evaluating plant material against sucking insect pests. The application of crude extract of greenbug, *S. graminum,* to sorghum foliage causes plant reactions similar to those resulting from actual greenbug feeding (J. C. Reese, unpublished data). Similar reactions are caused on wheat foliage after applying extracts of the Russian wheat aphid, *Diuraphis noxia* (Mordvilko), to callus tissues (Zemetra et al. 1993).

2.2. EVALUATION OF INSECT RESISTANCE BASED ON INSECT RESPONSES TO PLANTS

Six main categories of insect behavioral and physiological responses are normally considered when developing plant-resistance bioassays based on insect responses. These include orientation, contact, settling, feeding, metabolism of ingested food, growth, adult longevity, and fecundity and oviposition (Saxena 1969, Saxena et al. 1974, Visser 1983). Egg hatchability is also an important factor for the establishment of insects, particularly those that deposit their eggs inside the plant tissue (Saxena and Pathak 1977). The degree to which an insect responds positively in each of these response categories determines the degree of its establishment on different plants. It appears that interruption of any one or more of these insect responses due to unfavorable plant characters renders some degree of plant resistance (Visser 1983). Thus, for a realistic determination of the resistance of a plant to an insect, it is desirable to understand various behavioral and physiological responses of the insects during different phases of establishment on the plant. The interaction of the sum of all these responses determines the overall susceptibility or resistance of a test plant to an insect species.

From the view point of Painter's (1951) classification of the components of plant resistance (see Chapter 3), "nonpreference" of insects for different plants or the "antixenosis" (Kogan and Ortman 1978) of resistant plants may be observed in respect to insect orientation, contact, settling, feeding, and ovipositional responses. Similarly, the term "antibiosis" (Painter 1951) refers to those plant characteristics that inhibit insect survival, growth, fecundity, and egg hatchability. In this chapter, we will discuss various techniques to identify resistance in plants based on the behavioral and physiological responses of insects to plants.

2.2.1. Orientation and Settling

An insect must first locate and remain on a plant before it can feed, oviposit, and become established. The insect's orientation involves various visual or volatile chemical stimuli which emanate from the plant and which may be perceived by the insect at varying distances prior to its contact with a plant (Saxena 1969). A positive orientation response of the insect to a plant results in its arrival at and possible settling on the plant, whereas a negative orientational insect response to a plant results in its repulsion from it.

Orientation and settling responses of an insect to a plant are generally measured in choice tests by observing the numbers of individuals which initially orient toward a plant (orientation), and then remain settled for some time for feeding or oviposition (settling). While the orientation response should be measured within a few minutes to an hour of the release of the test insect (depending upon its ability to find a suitable host), settling responses are generally measured for longer time intervals. The following methods have been used to measure orientation and settling responses of insects.

2.2.1.1. Choice Tests

Tests insects are provided a choice of plants or plant parts that have been randomly but equidistantly placed from the center of a circular test arena. Insects are released in the center of the test arena and the number of individuals orienting to and settling on different cultivars is recorded at several hours or days after their infestation on plants (Kennedy and Schaefers 1974, Wiseman et al. 1982, Webster and Inayatullah 1988, Khan et al. 1989, Inayatullah et al. 1990, Dixon et al. 1990b, Smith et al. 1992). Others have also provided information on the orientational behavior of the insects soon after their release in test arenas (Kennedy and Kishaba 1977, Khan and Saxena 1985b).

In experiments with the whitebacked planthopper, *S. furcifera,* Khan and Saxena (1985b) reported that the visual attraction of *S. furcifera* females was identical on resistant and susceptible rice cultivars, but after 8 hr, significantly more individuals settled on susceptible plants than on resistant ones. Similarly, when the melon aphid, *Aphis gossypii* Glover, was given a choice between a susceptible and a resistant muskmelon cultivar, more aphids settled on the susceptible cultivar (Kennedy and Kishaba 1977). This preference was not apparent after 4 hr but was significant after 24 hr and remained so after 49 hr.

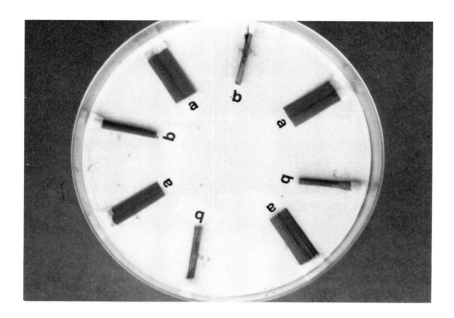

FIGURE 2.13. Settling response of *Cnaphalocrocis medinalis* third-instar larvae on leaf-cuts of susceptible IR36 (b) vs. resistant TKM6 (a). Larvae show a greater preference for IR36 leafcuts. (Photograph provided by Z. R. Khan.)

In a petri dish arena (Figure 2.13) larvae of rice leaffolder, *Cnaphalo-crocis medinalis* (Guenée), showed a strong preference for settling on suscep-tible IR36 rice leaves as compared to leaves of resistant wild rice, *Oryza aus-traliensis* Domin, and *Oryza nivara* Sharma et Shastry (Khan et al. 1989). In a greenhouse choice test, however, Smith et al. (1992) found no difference in the settling response of Russian wheat aphid, *Diuraphis noxia* (Mordvilko), to susceptible and resistant wheat cultivars.

Webster and Inayatullah (1988) reported high variability in commonly conducted choice tests with the greenbug, *S. graminum,* and compared this procedure with a completely randomized design (CRD) and a completely ran-domized design with a central composite arrangement (CRD-CCA).

Individual tests arranged in the CRD and CRD-CCA were conducted in flats in a greenhouse. Four test cultivars were arranged randomly as shown in Figure 2.14. There were 24 experimental plants in the CRD, 6 of each culti-var, with a plant-to-plant distance of 5 cm (Figure 2.14a). On the borders, ad-jacent to experimental plants, 24 nonexperimental plants were randomly se-lected from 1 of the 4 test cultivars and planted. Known numbers of insects were released on the soil in the center of each set of 4 plants and the number of insects on each plant was recorded 48 hr after infestation.

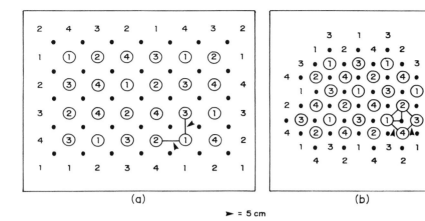

> = 5 cm

FIGURE 2.14. Randomization plans for settling preference tests. (a) A completely randomized design, and (b) a completely randomized design with central composite arrangement. Numbers represent different test varieties; encircled numbers represent experimental plants while numbers without circles represent nonexperimental plants. Black dots represent insect release sites. (From J. A. Webster and C. Inayatullah, *J. Econ. Entomol.*, 81, 1246, 1988. With permission of the Entomological Society of America.)

In the CRD-CCA, each of the 4 test cultivars were sown in a diamond-shaped pattern (Figure 2.14b), with a plant-to-plant distance of 5 cm. The test consisted of 24 experimental plants surrounded by 27 nonexperimental plants. Known numbers of test insects were released on the soil in the center of each set of 4 plants and the number of insects settling on each plant was recorded 48 hr after infestation.

When the experimental error, coefficient of variation (CV), and variance of treatment means were compared among the three test methods, the experimental error and CV for the CRD and the CRD-CCA tests were half of those in the standard choice test. However, considering the impracticality of conducting the CRD-CCA test with more than five or six test cultivars, the CRD was recommended as the best of the three designs.

2.2.1.2. Attraction Tests

Methods to measure insect orientation to plant odors or colors have been studied in both the laboratory and the field. Saxena et al. (1974) studied the orientation of the cotton leafhopper, *Amrasca biguttula biguttula* (Ishida), to leaves of various plants in a horizontal clear plastic chamber (7 cm long, 12 cm diameter), whose open ends were covered with detachable clear nylon net wall. Leaves of a test plant are placed across the end of the chamber 1 to 2 mm outside the nylon net walls. The other end is left blank. Test insects are introduced into the middle of the chamber, observed for 60 min, and the chamber is turned 90° every 5 min. The number of times the insect arrives on each end wall is recorded. The orientation response of *A. biguttula biguttula* is expressed as the percentage of arrivals at the leaf-facing end wall and is calculated as 100(A/A+B), where A is the number of arrivals on the leaf-facing end wall and B is the number of arrivals on the blank. Values above 50% indicate an insect's greater attraction to the leaves than to the blank end wall. Saxena et al. (1974) reported that *A. biguttula biguttula* orientation to host plants was in response to certain stimuli from the leaves perceivable at a distance, prior to contact.

Saxena (1990) determined the attraction of larvae of the stem borer, *Chilo partellus* (Swinhoe), to various susceptible and resistant sorghum cultivars in a field test. Plants of each test cultivar are grown in 3.0 m × 2.5 m plots in five rows parallel to the wind direction. A rectangular tray (40-cm-long × 25-cm-wide), with the two longer sides continuing upward as 10-cm-high vertical walls, is lined with filter paper. The tray is placed 20 cm from the downwind end of each plot with its long axis parallel to and in line with the central row of plants so that distance-perceivable stimuli from the plants can reach the tray. Twenty first-instar *C. partellus* larvae are released across the middle of the tray and the number of larvae that move to the two ends of the tray in 30 min are recorded. The percentage of larvae reaching the end nearest the plants reflects the larval attraction to the plants. A higher percentage of *C. partellus* larvae moved toward the susceptible sorghum cultivar IS 18363 than toward the highly resistant IS1044 sorghum (Saxena 1990).

Okech and Saxena (1990) compared the attraction of bean pod borer larvae, *Maruca testulalis* (Geyer), to susceptible and resistant cowpea cultivars under no-choice and free-choice situations in the laboratory and the field. In the no-choice experiment, five plants of each cultivar in the desired stage are randomly arranged 60 cm apart, and 50 first-instar *M. testulalis* larvae are released uniformly around each plant 15 cm from the stem. In the free-choice experiment, 1 plant of each of the 3 test cultivars is arranged in a triangle equidistant from each other and 15 cm from its center, and 50 first-instar larvae are released in the center of the triangle. In both experiments, plants are dissected after 24 hr and the percentage of larvae arriving and settling is recorded.

2.2.1.3. Arrest and Dispersal Tests

The settling response of lepidopterous larvae to different cultivars can be compared with respect to their arrest and dispersal on plants or plant parts. Robinson et al. (1978) placed a sticky trap around corn plants in the laboratory and field to measure arrestment or dispersal behavior of the European corn borer, *Ostrinia nubilalis* (Hübner). Thirty first-instar *O. nubilalis* larvae are placed in the whorl of each plant. The number of larvae that move off the plant is recorded daily for 4 days, then each plant is dissected and the remaining

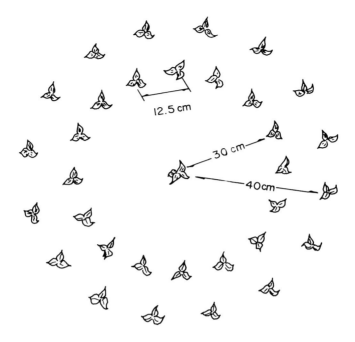

12.5 cm

30 cm

40cm

FIGURE 2.15. Arrangement of a test plant (center) infested with lepidopterous larvae and surrounding uninfested border plants as a means of indicating insect resistance based on larval movement off the test plant. (From B. R. Wiseman et al., *Prot. Ecol.,* 5, 135, 1983. With permission of Elsevier Science Publishers.)

larvae are counted. Robinson et al. (1978) reported that more larvae consistently settled on the susceptible, inbred WF9 than on the highly resistant inbred CI31A.

Wiseman et al. (1983) designed a field cage experiment to evaluate fall armyworm, *S. frugiperda,* larval movement off of both resistant and susceptible corn plants. A test plant is surrounded by susceptible plants spaced alternately at 30 and 40 cm from the central test plant (Figure 2.15) and test plants are spaced about 12.5 cm apart. Test plants are infested with a known number of neonate larvae and the number of larvae present on the surrounding plants 4, 6, 8, and 10 days after infestation serve as an indicator of larval movement off of the test plants. Wiseman et al. (1983) reported that many larvae of *S. frugiperda* crawled off the resistant Antigua 2D-118, MP SWCB-4, and Mp496XMpSWCB-1 plants to surrounding uninfested susceptible plants, whereas larvae remained settled on the susceptible Cacahuacintle plants.

Ampofo (1986) studied arrestment and dispersal of *Chilo partellus* (Swinhoe) larvae on susceptible and resistant corn plants in field plots. Each test cultivar is planted in 5-m-long rows and the infested plants are surrounded by plants of the same cultivar or by different cultivars. Dispersal of first-instar larvae increased twofold when the infested resistant cultivar, IC22-CM, was surrounded by a susceptible cultivar, and decreased when the infested susceptible cultivar, Inbred A, was surrounded by IC22-CM (Ampofo 1986).

Saxena (1990) studied arrestment and dispersal behavior of *C. partellus* on susceptible and resistant sorghum plants in a no-choice situation. Each test plant is infested with 20 neonate or fourth-instar *C. partellus* larvae. The plants are then covered with wire-mesh cages (250 × 80 × 80 cm). The infested plants are dissected after 72 hr for first-instar larvae and after 24 hr for fourth-instar larvae, and the percentages of the larvae recovered from the plants are recorded, reflecting "arrestment." Saxena (1990) reported that *C. partellus* settling was equal on susceptible and moderately resistant sorghum cultivars and low on the resistant cultivar IS1044. Using the same method, Okech and Saxena (1990) reported that under field conditions, the percentage of *M. testulalis* larvae settling on the susceptible cowpea cultivar, VITA 1, was significantly higher than on the moderately resistant VITA 5 and resistant TVU946 cultivars.

Ramachandran and Khan (1991) recorded arrestment and dispersal of the rice leaffolder, *C. medinalis,* on cut leaves of susceptible and resistant rice plants. A 6-cm-long piece of leaf from a susceptible or a resistant plant is placed individually inside a petri dish (9 cm diameter) lined with a moist filter paper. Ten first-instar larvae are released on each leaf piece. The petri dishes are kept under uniform light and at 10-min intervals for 1 hr, the number of larvae remaining on the leafcut are counted. Subsequently, similar observations are taken at hourly intervals for up to 5 hr.

Using this method, Ramachandran and Khan (1991) reported that 10 min after release, significantly fewer larvae remained settled on resistant wild rice, *Oryza brachyantha* A. Chev. et Roehr, and the F$_1$ hybrid of *O. brachyantha* and the susceptible IR31917-45-3-2 rice, compared with larvae on moderately resistant TKM6 and susceptible IR31917-45-3-2. At the end of the experiment only 1 and 15% of the test larvae remained settled on *O. brachyantha* and the F$_1$ hybrid, respectively.

2.2.1.4. Olfactometers

The orientational responses of insects to the odor of plants can be studied using the various kinds of olfactometers described in Chapter 4.3.1.1. Chang et al. (1985) used a Y-shaped olfactometer to study the orientational response of fall armyworm, *S. frugiperda,* to grass food sources. The olfactometer is constructed from two plastic rearing cups (4 cm diameter, 4.5 cm high) connected with a 10-cm, Y-shaped tube. Grass clippings (4 g/sample) are placed in one or both of the rearing cups and the cups are capped. Twenty newly emerged larvae are placed in the entrance of the Y-tube after which the tube is sealed with a cork. The olfactometer is kept in darkness and the number of larvae

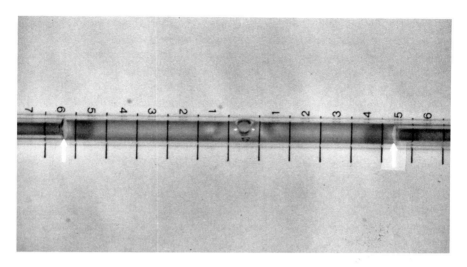

FIGURE 2.16. A simple tube olfactometer with two plungers (see white arrows), one holding crushed leaf and the other a control. (Photo courtesy of R. Ramachandran.)

reaching each rearing cup is recorded after 3 hr. Larvae showed a significant orientation to all grasses but showed no preference in paired comparisons.

Ramachandran and Khan (1991) studied the orientational response of first-instar *C. medinalis* larvae to crushed leaves of susceptible and resistant rice plants in a simple tube olfactometer consisting of a glass tube (1 cm diameter × 30 cm long) with a median hole (90.3 cm diameter) (Figure 2.16). Two plungers (glass tubes of 0.7 cm diameter × 15 cm long), 1 holding 0.5 g of crushed leaf and the other holding a rolled paper towel of equal dimension, are pushed into either end of the tube and placed such that their ends are 5 cm away from the median hole. The experiment is conducted in a dark room under red light. Ten newly emerged *C. medinalis* larvae are released into the tube and air is pushed through the plungers at the rate of 1 ml/min with a peristaltic pump. After 3 min of odor release, the distribution of the larvae on either side of the center is recorded. A significantly higher proportion of *C. medinalis* larvae oriented toward crushed leaves of susceptible IR31917-45-3-2 than to the control side. However, no difference was observed in the number of larvae orienting to the resistant *O. brachyantha* wild rice and those orienting to the control.

2.2.1.5. Use of Systemic Insecticides

The settling response of sucking insects on plants can also be determined using systemic insecticides, following the method described by van Emden et al. (1991) for the bird cherry-oat aphid *Rhopalosiphum padi* (L.). This method identifies aphids which reach the phloem of the host plant and feed there long enough to be killed. The settling response of the insect is measured by pairing plants of any 2 cultivars, with 1 plant of each pair treated with a soil drench of the systemic insecticide pirimicarb (40 ml of 0.5 g/l per plant). Equal areas of leaf, still attached to a plant, are crossed and held in a 2-cm-diameter plastic clip cage together with 20 test insects. After 48 hr the cages are opened and aphid mortality is assessed.

Van Emden et al. (1991) compared the results obtained by this technique to those obtained without insecticide. By either technique, the Mogham2 and Ommid wheat cultivars were less preferred for aphid settling, compared with Timmo. This technique helps to distinguish insects which have accepted the plant from those which have not settled but happen to be on the plant at the time of assessment.

2.2.2. Feeding

Techniques that record subtle changes in insect feeding behavior on susceptible and resistant plants can be useful in identification of germplasm with insect resistance. Such changes in insect feeding behavior on susceptible and resistant plants can be determined either through the measurement of damaged plant parts, or in terms of insect feeding response.

Insects damage plants by excessive feeding on plant tissues. Insect-resistant plants are less damaged than susceptible plants because less insect feeding occurs on them. Insects with chewing mouthparts cause damage by directly feeding on various plant parts while insects with piercing and sucking mouthparts damage plants by either removing vascular sap or by transmitting disease organisms during feeding.

2.2.2.1. Piercing and Sucking Insects

Damage by insects with piercing and sucking mouthparts (aphids, leafhoppers, planthoppers, whiteflies) to susceptible plants results from an excessive loss of sap from the vascular tissue, particularly phloem, and from the transmission of associated viruses into the phloem tissue. Resistant cultivars are less damaged than susceptible cultivars due to low intake of phloem sap or because of being restricted to feeding on xylem or mesophyll tissue. The following techniques that measure insect feeding responses have been used by various researchers to differentiate between susceptible and resistant plants.

2.2.2.1.1. Parafilm® Sachet Technique

Saxena and Pathak (1977) and Pathak et al. (1982) described a Parafilm® sachet technique for collection and quantitative determination of honeydew secreted by the brown planthopper, *N. lugens,* feeding on rice plants. The quantity of honeydew per insect in 24 hr is used as a parameter in comparing hopper feeding activity on susceptible and resistant plants. Parafilm® is cut into a piece 10×5 cm and folded in the middle (Figure 2.17a). It is placed on

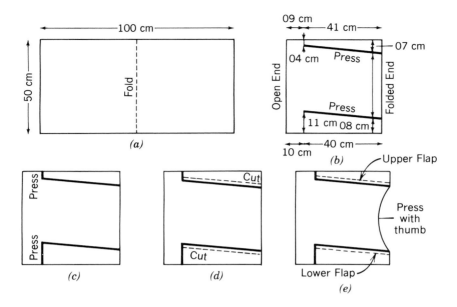

FIGURE 2.17. Steps to prepare a parafilm sachet for collecting honeydew excreted by *Nilaparvata lugens* while feeding on rice plants. (From Pathak, P. K. et al., *J. Econ. Entomol.*, 75, 194, 1982. With permission of the Entomological Society of America.)

a smooth surface and sealed by pressing with the edge of a microslide to form grooves at specified distances (Figure 2.17b,c). Upper and lower flaps are then cut along and parallel to the oblique grooves with a pair of scissors (Figure 2.17d). The sachet is dilated by inserting the blunt end of a pencil through the open end and pressing the folded end with the thumb (Figure 2.17e). This increases the capacity of the sachet and permits the test insect to move freely. The edges of the open end are placed on each side of the stem, and the lower portion is pressed together. The lower flap is stretched and wrapped around the stem. The insect is then released into the sachet with a blowing tube, and the sachet is sealed by pressing its upper edges together and by winging the upper flap around the stem (Figure 2.18a,b).

The excreted honeydew collects within the sachet and slowly builds up in the outer lower corner. When the feeding period is completed, the insect is removed by introducing an aspirator tube through a vertical incision made in the sachet parallel to the stem. The volume of honeydew is measured by calibrated micropipettes of various sizes (1 to 100 µl), and its weight is determined by differential weighing of the micropipettes before and after the collection of honeydew on a 0.001-mg sensitive weighing balance. From this,

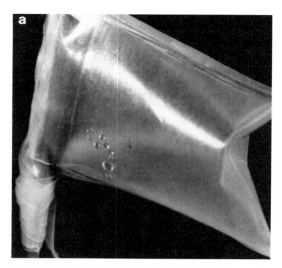

FIGURE 2.18. (a) A parafilm sachet attached to rice plant showing honeydew excreted by *Sogatella furcifera* in 24 hr on a susceptible rice variety. (From Smith, C. M. et al., in *Rice Insects: Management Strategies,* Springer-Verlag, New York, Copyright © 1991. With permission.) (b) arrangement of several parafilm sachets on rice plants grown in a pot. (Photograph provided by Z. R. Khan.)

the specific gravity of honeydew is determined by comparing its weight and volume with that of distilled water. The weight of a known volume of honeydew compared with that of water gives a specific gravity for honeydew estimated at 1.009 ± 0.001. A known volume of honeydew is then multiplied by its determined specific gravity to estimate its weight. Another technique to determine the weight of excreted honeydew is to first weigh the sachet with honeydew and then blot out the honeydew with a filter paper, and to reweigh the sachet.

The Parafilm® sachet method proved useful for determining the levels of resistance of rice cultivars to three biotypes of brown planthopper (Pathak et al. 1982). The authors reported significant differences in the amount of honeydew excreted by *N. lugens* biotypes 1, 2, and 3 on susceptible Taichung Native 1 and resistant Mudgo and ASD7 rice cultivars. On resistant cultivars, significantly less honeydew was excreted than on susceptible cultivars, suggesting significantly reduced feeding.

This technique has also been used successfully in collecting honeydew from the green leafhopper, *Nephotettix virescens* (Distant) (Khan and Saxena 1985b), and the white-backed planthopper, *S. furcifera* (Khan and Saxena 1984b, 1985c), on rice and is applicable to other sucking insects that feed on other crop plants.

2.2.2.1.2. Filter Paper Technique

The relative amounts of feeding by insects on susceptible and resistant plants can be assessed by measuring the area of honeydew excreted by them on a filter paper disk. This technique was first demonstrated by Auclair (1958, 1959) to assess resistance and susceptibility of green peas to the pea aphid, *Acyrthosiphon pisum* (Harris).

Pea plants are encircled at the level of the 4th internode with a circular piece of cardboard 45 cm in diameter, slit from the periphery to the center and covered with 2 halves of filter paper disks (Whatman No. 1, 15 cm diameter). Filter paper is replaced every 12 hr. The number of dried honeydew droplets deposited is revealed by dipping the filter papers in a solution of 0.2% ninhydrin in n-butanol, or in a benzidine-trichloracidic acid mixture prepared in acetone as described by Harris and MacWilliam (1954). Development with ninhydrin is standardized by carrying out the reaction in a CO_2 atmosphere as described by Auclair et al. (1957). Development with the benzidine mixture is carried out in an oven at 60°C for 10 min. Treatment of the dried droplets with ninhydrin produces purple spots because of the presence of amino acids. Treatment with benzidine mixture produces brownish spots due to the presence of sugars such as glucose, fructose, and sucrose. The optical densities of the spots are determined under the white light of a photoelectric densitometer. The densitometer is set at a 1-mm aperture on the transmission unit and a galvanometer reading of 0.24 as a blank for the paper when reading the amino acid spots, and 0.46 for the sugar spots. This procedure provides rapid information on the relative concentrations of amino acids

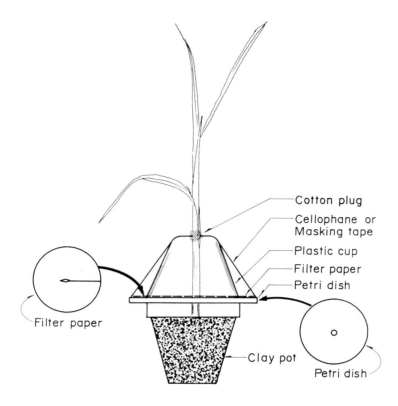

FIGURE 2.19. A feeding chamber for *Nilaparvata lugens*. (From Paguia, P. et al., *J. Econ. Entomol.*, 73, 35, 1980. With permission of the Entomological Society of America.)

and sugars in individual droplets. Measurements of the diameters of the spots also gives information on the approximate ages of the aphids excreting the honeydew.

A technique to quantify the feeding activity of *N. lugens* on rice cultivars based on honeydew excretion was described by Paguia et al. (1980). The honeydew of *N. lugens,* excreted on a filter paper disk, is stained with ninhydrin and quantified by measuring the area (mm^2) of the honeydew spot. Honeydew excreted by *N. lugens* is collected in a feeding chamber consisting of an inverted transparent plastic cup placed over a filter paper disk resting on a plastic petri dish (Sogawa and Pathak 1970) (Figure 2.19). Five 2-day-old adult females, previously starved for about 5 hr, are placed into the chamber through a hole at the top of the cup. A cotton plug in the hole prevents the escape of the insects. The insects are allowed to feed overnight and then the filter paper disks are collected and treated with 0.001% ninhydrin in acetone solution. After being oven-dried for 5 min at 100° C, the honeydew stains appear as violet or purple due to their amino acid content. The spots are traced on a tracing paper and the area is measured over a millimeter square graph paper.

The ninhydrin method of assessing *N. lugens* feeding was modified by Pathak and Heinrichs (1982) to determine the feeding activity of leafhoppers and planthoppers. A Whatman No. 1 filter paper disk is dipped in a 0.02% bromocresol green solution in ethanol, allowed to dry for 1 hr and dipped again. Bromocresol green is a pH indicator that is yellow at pH 3.8 and turns blue-green at pH 5.4 (Anonymous 1976). The yellow filter paper is then placed on an inverted petri dish at the base of the plant. The cage is placed over the filter paper and a known number of leafhoppers or planthoppers are placed in the cage and allowed to feed for 24 hr. Immediately upon contact with honeydew, blue spots appear on the treated filter paper. As the concentration of the honeydew increases, the spots turn white in the center with blue edges where the honeydew concentration is less. The area of honeydew spot is then measured and the spots from susceptible and resistant rice plants are compared.

Pathak and Heinrichs (1982) compared the bromocresol green method to the ninhydrin method of Paguia et al. (1980) and the honeydew weight methods of Saxena and Pathak (1977) and Pathak et al. (1982). The three methods all indicated that feeding on the susceptible rice cultivar TN1 was significantly greater than feeding on resistant cultivars.

The advantages of bromocresol as an indicator over ninhydrin are the ability to estimate feeding activity during the course of the test and the permanency of the obtained spots, which permits measurements to be made *ad libitum*. However, both the ninhydrin and bromocresol techniques are solely based on measurements of fecal spots rather than fecal weights. Such results are sometimes questionable because of the possibility of overlapping deposits of honeydew.

The area of stained honeydew spots on filter paper disks can also be expressed gravimetrically (Paguia et al. 1980). The stained spots are cut out, weighed, and results expressed as the weight (mg) of filter paper containing honeydew deposits.

2.2.2.1.3. Feeding Site Techniques

Several species of cercopids, cicadids, and cicadellids can feed both on xylem and phloem vessels (Wiegert 1964, Carlé and Moutous 1965, Hagley and Blackman 1966, Cheung and Marshall 1973, Chang 1978, Auclair et al. 1982, Khan and Saxena 1984a). *N. virescens* is primarily a phloem feeder but switches to xylem drinking if the offered host plant is intrinsically unsuitable (Auclair et al. 1982, Khan and Saxena 1984a, Saxena and Khan 1985a). This type of switching behavior prohibited identification of a marked quantitative difference in the uptake of sap by the green leafhopper fed susceptible and resistant rice plants (Auclair et al. 1982). Since then, techniques have been developed to determine the feeding sites of such insects, in order to evaluate plants for resistance to pests and pest-transmitted diseases.

Khan and Saxena (1984a) developed a simple technique for monitoring the feeding behavior of *N. virescens* on susceptible and resistant rice cultivars using safranine, a dye that is highly selective to lignin and is translocated in xylem vessels. The susceptible cultivar "TN1" and *N. virescens* resistant cultivar "ASD7" are grown separately in clay pots in a greenhouse. When seedlings are 10 days old, they are removed without damaging their roots and washed thoroughly to remove soil particles. They are then immersed in an aqueous solution of 0.2% safranine for about 6 hr, and the translocated dye colors the xylem vessels red throughout the entire length of the seedlings. The treated seedlings are then removed and the excess dye is washed off. Two to three seedlings of each variety are placed in separate, 250-ml glass beakers which contain enough water to immerse their roots. Each beaker is then covered with a medially perforated, 12 cm diameter, plastic petri dish through which seedling shoots emerge (Figure 2.20). A 9-cm-diameter Whatman filter paper disk is placed on each petri dish around the base of the seedlings and the seedlings and dish are then enclosed in a cylindrical Mylar® cage (15 cm high, 9 cm diameter). Meanwhile, newly emerged females of *N. virescens* are collected from stock cultures, kept starved but water satiated for 4 hr and released into the Mylar® cages on either resistant or susceptible seedlings (10 leafhoppers per cage). A set of untreated seedlings are infested similarly and serve as controls. The honeydew excreted by the hoppers drops onto the filter paper disks and is readily absorbed. Red honeydew spots on the filter paper disks indicate xylem feeding by the hopper females on the safranine-treated seedlings (Figure 2.21). Bluish amino acid spots, however, indicate phloem feeding by the insects when filter paper disks from the control seedlings are treated with a 0.1% ninhydrin-acetone solution.

Using this technique, Khan and Saxena (1984a) reported that *N. virescens* is primarily a phloem feeder on susceptible rice cultivars, although it occasionally also fed on xylem. On a resistant cultivar, the hopper switched to xylem feeding but occasionally fed on phloem.

Auclair et al. (1982) analyzed the honeydew of *N. virescens* using paper chromatography-densiometry and pH determinations to describe the *N. virescens* feeding sites on susceptible and resistant rice plants. The technique is described as follows:

N. virescens adults are enclosed in Parafilm® sachets (as described by Pathak et al. 1982) attached to the middle portion of the two terminal leaves of susceptible (TN1) or resistant (A5D7 or P203) plants in order to collect honeydew. In each experiment, 5 or 10 adults are confined per sachet, 10 to 15 sachets being mounted individually on plants of each of the varieties used. Honeydew excreted by the feeding leafhoppers appears as minute droplets on the Parafilm® membrane about 0.5 to 1 hr after the insects have been confined in the sachets. These freshly excreted droplets are collected by making

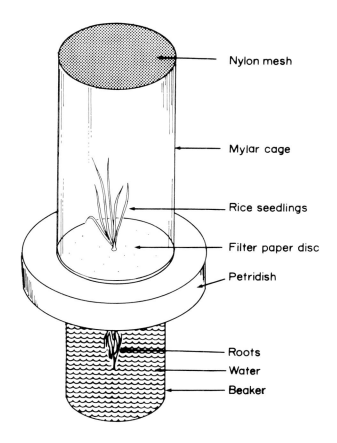

FIGURE 2.20. Assembly for collecting honeydew of leafhoppers and planthoppers. (From Khan, Z. R. and Saxena, R. C., *J. Econ. Entomol.,* 77, 550, 1984. With permission of the Entomological Society of America.)

a 0.5- to 1-cm slit in the wall of the sachet with a razor blade and inserting the tip of a disposable micropipette (Drummond Scientific Co., Broomall, PA) until it comes into contact with a droplet. The pH of the honeydew is evaluated by pouring a small fraction onto one of a series of pH special indicator sticks of pH range 4.0 to 7.0 and 6.5 to 10.0, in 0.2 to 0.4 increments (Color-phast, E. Merck, Darmstadt, West Germany). The volume of honeydew remaining in the pipette is estimated by measuring its length relative to the whole pipette length, and the samples are finally transferred onto filter paper chromatograms for amino acid or sugar separations and quantitative estimations.

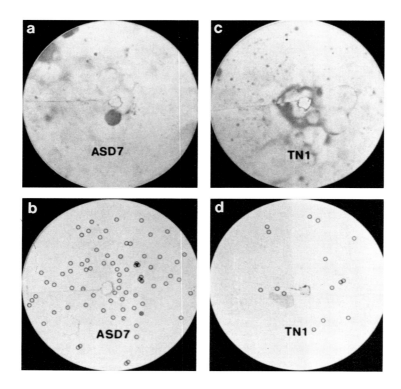

FIGURE 2.21. Filter paper disks on which honeydew of *Nephotettix virescens* was col
lected when the leafhoppers fed on seedlings of susceptible TN1 anc
the resistant ASD7 rice varieties. Bluish amino acid spots on ninhydrin
treated filter paper disks indicate phloem feeding (a, c). Encircled spots
(b, d) on filter paper disks indicate phloem feeding. (From Khan, Z. R.
and Saxena, R. C., *J. Econ. Entomol.,* 77, 550, 1984. With permission of
the Entomological Society of America.)

Honeydew samples from all experiments are spotted on one- and two-
dimensional paper chromatograms. The chromatograms are irrigated by capillary ascent in 80% neutral phenol in water in the long dimension for 15 hr,
air dried, and n-butanol-glacial acetic acid-water (15:3:7 parts, respectively)
in the second dimension for 9 hr and air dried. Development of amino compounds is accomplished by dipping the chromatograms in 0.25% ninhydrin in
acetone, or in a benzidine mixture (Harris and MacWilliam 1954) for sugars.

Ninhydrin development is carried out for about 18 hr at room temperature;
whereas, chromatograms treated with benzidine are air dried in a fume hood
for 5 min, and then placed in an oven at 100° C for 5 min. Quantitative estimations of amino acids and sugars on developed chromatograms are made by
total scanning of spots with a Fujiox densitometer, model FL-A IV, using a

550-μm filter. The integrated readings of the optical densities are compared with those obtained from known amounts of pure amino acids or sugars separated on control chromatograms, and the amount of compounds occurring in the honeydew samples are estimated.

The results revealed higher free amino acid and sugar concentrations in honeydew excreted on the susceptible cultivar than on the resistant cultivars. Honeydew excreted on the resistant cultivars is acidic, whereas honeydew from the susceptible cultivar is predominantly basic. In herbaceous plants, phloem sap is usually alkaline and contains higher concentrations of sugars and free amino acid than xylem sap, which is acidic. The biochemical evidence from this study strongly suggests that *N. virescens* is primarily a xylem feeder on the resistant plants tested, but feeds predominantly on phloem sap in susceptible cultivars.

2.2.2.1.4. Honeydew Clock

Padgham and Woodhead (1988) devised and used a honeydew clock to examine the periodicity of honeydew production in *N. lugens* feeding on different susceptible and resistant rice cultivars. The clock is constructed of a lower container holding rice plant roots and an upper chamber containing the plant stem and the insect (Figure 2.22). A filter paper strip (Whatman no. 1) slides between the lower and upper chambers to collect honeydew produced during the experimental period. The paper is notched so that it fits around one side of the rice plant stem and is moved (manually) by sliding it lengthways one division and re-engaging the next notch. Honeydew drops collected on the filter paper strip are sprayed with 0.1% ninhydrin in acetone and the area of blue amino acid spots are measured.

Padgham and Woodhead (1988), using a honeydew clock, reported that on a susceptible (IR22) rice cultivar, 65% of the *N. lugens* individuals tested produced honeydew regularly after 4 hr. A similar number of insects produced honeydew regularly on a moderately resistant cultivar (IR46) after 8 to 10 hr. On IR62, a highly resistant cultivar, only 30% of the test insects produced honeydew after 13 to 17 hr.

2.2.2.1.5. Electronic Recording of Insect Feeding

An electronic feeding monitor, developed by McLean and Kinsey (1964, 1965) and more fully described by McLean and Weight (1969), greatly facilitates the study of the feeding behavior of piercing and sucking insects. The electronic feeding monitor is a device that detects the changing electrical impedance in the insect and the substrate on which the insect is probing or feeding. To accomplish this, a small alternating-current (AC) or direct-current (DC) voltage is applied across the insect and the substrate, which results in a small flow of charge. During feeding, the insect interrupts the flow of current, producing a complex of electrical signals consisting of different patterns which presumably reflect different elements of feeding behavior. By converting the

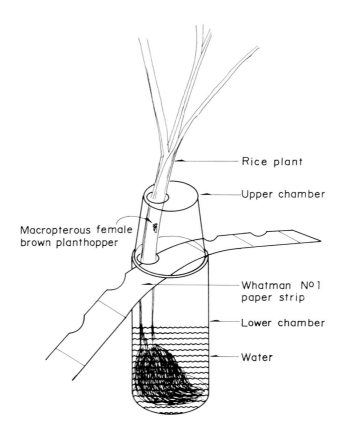

Rice plant

Upper chamber

Macropterous female
brown planthopper

Whatman No1
paper strip

Lower chamber

Water

FIGURE 2.22. Honeydew clock apparatus used for quantifying patters of honeydew production. (From Padgham, D. E. and Woodhead, S., *Bull. Entomol. Res.*, 78, 339, 1988. With permission of International Institute of Entomology.)

current fluctuations to voltage fluctuations and amplifying the voltage levels, the impedance changes ultimately become DC signals, which may be recorded on a chart recorder, oscilloscope, or stored in microcomputer memory. Thus, the electronic monitor provides information on the occurrence, duration, and sequence of insect feeding behaviors, and on the location of insect penetration of different plant tissues.

Both AC systems (McLean and Kinsey 1964, McLean and Weight 1969, Brown and Holbrook 1976, Kawabe 1985, Velusamy and Heinrichs 1986, Kabrick and Backus 1990, Dixon et al. 1990a), and DC systems (Kashin and Wakeley 1965, Schaefers 1966, Smith and Friend 1971, Tjallingii 1978, 1988, Khan and Saxena 1984a, 1985a, 1988) have been used to study the probing and feeding behavior of insects.

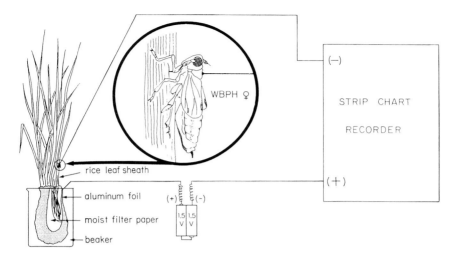

FIGURE 2.23. A schematic diagram of circuit and equipment for electronically recording feeding behavior of leafhoppers and planthoppers on rice plants. (From Khan, Z. R. and Saxena, R. C., *J. Econ. Entomol.*, 77, 1479, 1984. With permission of the Entomological Society of America.)

Electronic monitoring systems have been used to investigate plant resistance to insect pests (Nielson and Don 1974, Shanks, Jr. and Chase 1976, Kennedy et al. 1978, Campbell et al. 1982, Khan and Saxena 1984b, 1985a, Velusamy and Heinrichs 1986, Dixon et al. 1990a) to confirm the occurrence of insect biotypes (Nielson and Don 1974, Khan and Saxena 1988), and to study the transmission of viral diseases by piercing-sucking insects (Haniotakis and Lange 1974, Chang and Ota 1978, Scheller and Shukle 1986, Dahal et al. 1990).

Khan and Saxena (1984b, 1985b, 1988) used a DC electronic monitoring system to record the feeding behaviors of leafhoppers and planthoppers on susceptible and resistant rice plants. A 10-cm long, 18-μm gold wire is attached to the dorsum of 1- or 2-day-old females with silver paint. The insect is starved but water satiated for 2 hr and then placed on an intact leaf blade or leaf sheath of a 30- to 45-day-old plant. The gold wire is connected directly to the negative input terminal of a transistorized, automatic, and null-balancing DC chart recorder with a 250-mm recording width and input resistance of 1 MΩ (Figure 2.23). The voltage source consists of 2 1.5V DC batteries connected in series. The positive battery terminal is connected to the plant roots through moistened filter paper and aluminum foil. The negative battery terminal is connected directly to the positive input terminal of the chart recorder. The recorder pen is adjusted to the chart baseline and insect feeding is monitored for 180 min. A chart speed of 1.5 cm/min at 500-mV amplifier power is generally adequate for distinguishing various voltage

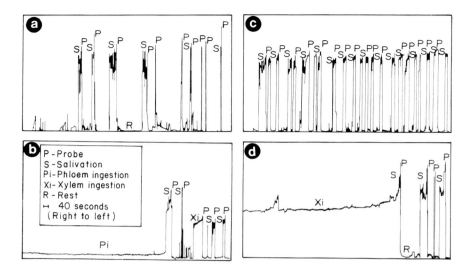

FIGURE 2.24. Waveforms recorded during *Nephotettix virescens* feeding on susceptible (a, b) and resistant (c, d) rice plants, using an electronic monitoring
device. (From Khan, Z. R. and Saxena, R. C., *J. Econ. Entomol.,* 78,
583, 1985. With permission of the Entomological Society of America.)

reversals during insect feeding. Each cultivar is tested ten times using ten
new plants and ten new insects. All recordings are generally performed at 27
± 2°C and 65 to 70% RH.

Waveforms recorded during insect feeding are interpreted as probing, salivation, phloem feeding, and xylem drinking (Figure 2.24). Voltage signals
produced during xylem drinking are generally stronger than those produced
during phloem feeding. Khan and Saxena (1984b) reported that *S. furcifera*
was a phloem feeder on both susceptible and resistant rice plants but probed
readily and fed longer on the susceptible variety. In contrast, *S. furcifera*
made brief and repeated probes on the resistant cultivar, consequently reducing its effective ingestion period (Figure 2.25). Khan and Saxena (1985a) reported that the *N. virescens* also made more repeated and brief probes on resistant rice cultivars than on susceptible ones. The waveforms indicated that
the insect fed primarily from the phloem of susceptible plants but switched to
xylem drinking on resistant plants (Figure 2.24).

An AC electronic monitoring device has also been used to study the feeding behavior of *N. lugens* biotype 3 (Velusamy and Heinrichs 1986). Using
insects starved but water satiated for 2 hr, a gold wire (20 μm diameter) is attached to the dorsum of 1-day-old, brachypterous females with an electrocon-
ductive paint. The opposite end of the gold wire is attached to a larger wire
leading to the input of a current detection amplifier. The insect is then placed

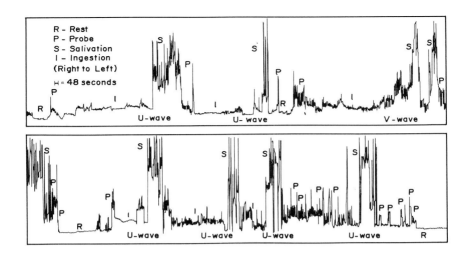

FIGURE 2.25. Waveforms recorded during *Sogatella furcifera* feeding on susceptible TN1 (top) and resistant IR2035-117-3 (bottom) rice varieties, using an electronic monitoring device. (From Khan, Z. R. and Saxena, R. C., *J. Econ. Entomol.*, 77, 1479, 1984. With permission of the Entomological Society of America.)

on the leaf sheath of a potted plant that is electrified by a 500-Hz, 0.5-V alternating current from the oscillator into the potting soil. The final amplifiers are adjusted to 500-mV, full-scale output and the speed of the strip-chart recorder is 2 cm/min. Velusamy and Heinrichs (1986) reported distinct differences in waveforms produced by *N. lugens* during probing, salivation, and feeding on susceptible and resistant rice plants. *N. lugens* probed repeatedly, salivated for a long time, and ingested for very short periods on resistant cultivars as compared to their feeding on susceptible plants.

To study changes in the feeding behavior of *Aphis fabae* Scopoli tethered on gold wire, Hardie et al. (1992) made simultaneous electronic and close-up video recordings. Behavioral differences were recorded between tethered (as required for electronic monitoring) and freely moving insects. The tether decreased the number of penetrations, delayed the first penetration, but promoted longer penetration periods.

2.2.2.1.6. Radiolabeled Plant Tissues

The use of radioactivity to determine the feeding rate of *N. lugens* was utilized by Lee et al. (1981a,b), Chelliah and Heinrichs (1980), and Hopkins (1990). The roots of 25- to 30-day-old susceptible and resistant rice plants are immersed in flasks containing 100 ml of distilled water with 0.5 µc/ml ^{32}P for 24 or 48 hr, and each plant is then transferred to a 100-ml flask of distilled water. Each flask is then covered with a medially perforated, 12-cm-diameter plastic petri dish. A 9-cm-diameter Whatman filter paper disk is placed on each petri dish around the base of the plant which is then enclosed in a cylindrical Mylar® cage.

Five 4-day-old female hoppers, previously starved for 4 hr, are introduced into each cage and allowed to feed for 24 hr. The honeydew excreted by the hoppers drops on the filter paper disks and is readily absorbed. After 24 hr, hoppers and filter paper disks are removed. Hoppers are anesthetized and the counts of samples consisting of five hoppers and one filter paper disk are recorded in a Geiger counter (5 s/sample). The total sap ingestion by hoppers in 24 hr is indirectly assessed by summing the radioactivity in the hoppers and filter papers.

Using the radioactive tracer technique, Lee et al. (1981b) reported that *N. lugens* fed radiolabeled susceptible rice plants contained about 100 times more ^{32}P than that absorbed by *N. lugens* fed resistant plants. Hopkins (1990) recorded similar results using the electronic monitoring system of Tjallingii (1978).

2.2.2.1.7. Honeydew Production

An instant bioassay for evaluating the green peach aphid, *M. persicae,* resistance in lettuce based on honeydew droplet counts was reported by Eenink et al. (1984). A feeding cage (Figure 2.26) is placed in the center of the lower side of the 5th or 6th leaf of 3-week-old lettuce plants with 6 to 10 leaves. Three to 5 aphids are released in each cage and after the aphids settle on to leaves, a Parafilm® "M" disk (5 cm diameter) is placed on the bottom of each cage to collect honeydew droplets excreted by aphids. Disks are removed after 3 hr and the number of honeydew droplets are counted under a binocular microscope at 6× magnification. A similar feeding cage has been used by van Helden et al. (1993) to study resistance in lettuce to an aphid, *Nasonovia ribisnigri* (Mosley). Eenink et al. (1984) demonstrated that the honeydew production of *M. persicae* in 3 hr on 5 lettuce plant leaves allowed effective selection for aphid resistance. However, selection was limited to genotypes with high or low levels of resistance, since there was not a strong correlation between aphid infestation and honeydew production for genotypes with intermediate resistance.

2.2.2.1.8. Stained Plant Tissues

Traynier and Hines (1987) reported a method of staining plant tissue for callose, a polysaccharide (ß-1,3-glucan), to reveal aphid feeding probes. Callose is known for its rapid deposition within the cell walls of plants as a local reaction to wounds. Probes by piercing-sucking insects are revealed by this technique of detecting damage to the plant. To reveal the probes, leaves are excised from the plant without further damaging their surfaces and are mounted in 0.5% W/V aqueous aniline blue in 0.067-M phosphate buffer at pH 8.5 with a trace of wetting agent. The preparations are examined immediately with an epifluorescence optics microscope under incident UV illumination.

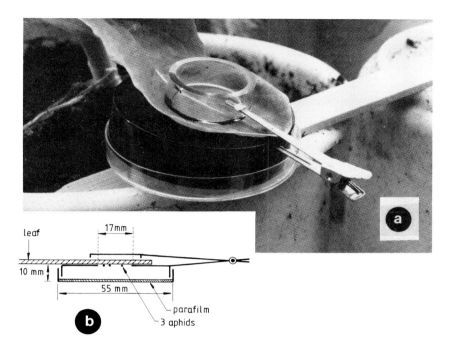

FIGURE 2.26. (a) A cage for measuring honeydew production on plants; (b) cross section of the cage. (From Eenink, A. H. et al., *Euphytica,* 33, 825, 1984. With permission of Kluwer Academic Publishers.)

Wounds made by the mouthparts of aphids in plant cell walls appear on fluorescent spots. This method is useful when extensive information on feeding behavior is not required, and can be used to identify plants susceptible to insect feeding as indicated by a greater density of probes.

Backus et al. (1988) stained whole salivary sheaths in situ without sectioning the plant. Broad bean leaves, on which leafhoppers have fed, are transversely cut into thirds and placed into a bottle and covered with McBride's (1936) stain (0.2% acid fuchsin in 95% ethanol and glacial acetic acid [1:1 vol/vol]) for 20 to 24 hr. The leaf pieces are removed from the stain and placed in a beaker with a clearing agent (glycerine, lactic acid, water [1:1:1, vol/vol/vol]) for several hours. Salivary sheaths can be observed under a microscope. The salivary sheath is a lipoproteinaceous substance deposited during feeding which surrounds the stylets as they penetrate the plant.

2.2.2.1.9. Collection of Plant Sap via Stylectomy

Several leafhoppers switch from phloem feeding to xylem drinking on resistant plants. Stylectomy and collection of the vascular sap from the stylets of piercing-sucking insects can reveal the site of feeding within a plant.

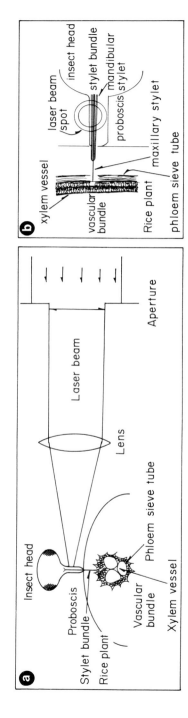

FIGURE 2.27. (a) Schematic diagram of a cross section of the rice plant, insect head, and YAG laser beam focused on insect proboscis for stylectomy; (b) diagrammatic representation of a longitudinal section of rice plant, insect mouth parts, and the laser beam spot. (Courtesy S. Kawabe and IRRI.)

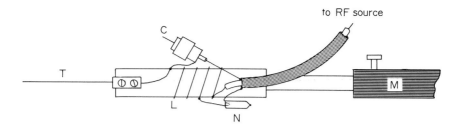

FIGURE 2.28. A radio frequency probe used to cut mouth parts of feeding aphids. T, tungsten needle; C, 3-30 pF concentric capacitator; L, copper inductor coil; N, neon tuning indicator; M, micromanipulator. (From N. Downing and D. M. Urwin, *Physiol. Entomol.,* 2, 275, 1977. With permission of Blackwell Scientific Publications Ltd.)

Stylectomy can be performed using a Yttrium-Aluminum Garnet (YAG) laser (Kawabe et al. 1980) or a radio frequency probe (Downing and Unwin 1977).

Kawabe et al. (1980) used YAG laser to sever the stylets of *N. cincticeps* and *N. lugens* (Figure 2.27). A rice plant with an insect feeding on its culm is set in front of the condensing lens of the YAG laser system. The horizontal beam is focused at a right angle on the proboscis of the feeding insect on the culm by positioning the plant. A video camera attached to a microscope may also be used to record and display the insect sucking the plant sap on a video monitor while the laser beam is focused. Observing insects in this manner allows the researcher to sever the stylet at the proper point on the proboscis. The stylet bundle is severed by a pulse of a 0.13-mm-diameter focused beam on the insect proboscis. The energy of the focused beam is 0.1 joule, the pulse width is 0.2 msec, and the wavelength is 1.064 μm. The diameter of the focused beam is controlled by the size of the replaceable apertise placed in front of a 6-mm-diameter laser rod. The rate of sap exudation after stylet severance varies according to the size of the insect. Kawabe et al. (1980) reported that exudation reaching 0.2 μl/hr was maintained over 3 hr from large female planthoppers.

Downing and Unwin (1977) severed the mouthparts of feeding aphids with a radio frequency probe consisting of a parallel-tuned resonant circuit fed with power from a VHF radio telephone (Figure 2.28). The inductor (L) is wound on a 1-cm-diameter perspex rod and is tuned to resonate at source frequency by a 3 to 30 pF concentric capacitor (C). A miniature neon lamp (N) is used to indicate resonance. The capacitor is adjusted according to the length of the needle (T). The outer sheath of the coaxial cable which feeds the probe is earthed to the micromanipulator (M). The radio frequency source is the transmitter of a high band VHF radio telephone at 145.0 MHz. The interchangeable needles are 2 to 3 cm in length and are prepared from 0.5-mm-diameter tungsten wire. The final centimeter of the wire is electrolytically sharpened. The sharpened portion tapers from about 0.15 mm in diameter to a tip of about 1 μm.

The feeding insect is approached from the front with the probe mounted on a micromanipulator. In order to sever the stylet, the probe is cautiously advanced until the tip of the needle touches the proboscis. At the moment of contact, a radiofrequency pulse ranging from 0.25 to 1.05 pF, depending on the size of the proboscis and the position of the probe, is activated. Plant exudes are collected using a micropipette held in a micromanipulator.

2.2.2.2. Chewing Insects

The assessment of foliar losses due to insect feeding is particularly necessary when evaluating plants for resistance to chewing insects. Measurements of direct insect feeding injury are often more useful than measurements of insect growth or population development on plants because reduced plant damage and the resulting increases in plant yield or quality are the ultimate goals of most crop improvement programs.

Insect feeding in a choice assay (Figure 2.29) involves the determination of insect feeding preference on multiple plant genotypes. Choice experiments are very useful in the preliminary evaluation of plants. However, no-choice experiments are necessary to maximize the identification and measurement of resistance. Insect feeding can be measured either on excised or on intact plants. Although several workers have reported differences in the feeding preference of insects on excised vs. intact plants (Thomas et al. 1966, Bernays et al. 1977), others have found no difference between insect feeding on intact and excised foliage (Sams et al. 1975, Raina et al. 1980). Barnes and Ratcliffe (1967) reported that alfalfa weevil, *Hypera postica* (Gyllenhal), feeding on alfalfa leaves was influenced by light and by moisture levels. Weevils ate more in light than in darkness, but feeding was more erratic under light. Barnes and Ratcliffe (1967) recommended constant darkness and uniform plant moisture level for alfalfa weevil feeding bioassays. Insect feeding can be measured using the following methods.

2.2.2.2.1. Photometric Devices

Several photometric devices have been used to measure leaf surface area (Frar 1935, Mitchell 1936, Kramer 1937, Orchard 1958, Donovan et al. 1958, Viosey and Mason 1963, Pedigo et al. 1970, Jensen et al. 1977). These devices function on a relationship between leaf size, reduction in light by leaf coverage and the reception of reduced light with a photovoltaic or a photoconductive cell.

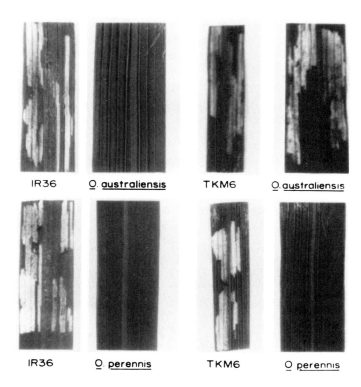

IR36   O. australiensis   TKM6   O. australiensis

IR36   O. perennis   TKM6   O perennis

FIGURE 2.29. Leaf cuts of rice plants obtained from two-choice feeding bioassays where *Cnaphalocrocis medinalis* was used as a test insect. Feeding tests were conducted using leafcuts of susceptible IR36 or resistant TKM6 rice varieties versus wild rices, *Oryza australiensis* or *O. perennis*. (From Khan, Z. R. and Joshi, R. C., *Crop Prot.*, 9, 243, 1990. With permission of Butterworth-Heinemann Ltd.)

A low-intensity light controlled by a constant voltage transformer and a rheostat is passed through damaged leaves laid on a specimen stage and received by a photoconductive cell. Photocell response is measured by a microammeter and converted to leaf surface area. Greater leaf damage results in higher microammeter readings. Leaves are measured before and after the experiment and the area consumed by insects is determined as the difference in the area before and after insect feeding.

2.2.2.2.2. Video Digitizer

Hargrove and Crossley, Jr. (1988) used a microcomputer equipped with a video-digitizer card interfaced with a standard video camera to measure the percentage of leaf area lost to herbivorous insects. A standard 16-mm c-mount video lens is adequate for larger leaf types, but a macro lens can be used to increase resolution and accuracy of small leaves and leaflets. The system digitizes a video image of a reconstructed leaf and displays the area of the leaf consumed by insect feeding. The system simultaneously measures area removed and area remaining without tracing these areas by the operator. Before digitizing, leaves to be measured are photocopied and any interrupted margins are manually completed with a dark line.

An edge-seeking algorithm (Morgan 1984) fills the background with red color around the perimeter of the leaf. The fill color serves to differentiate background from eaten leaf portions and paints around any closed irregular areas fed upon. Once the background has been filled, yellow pixels are counted in uneaten areas and blue pixels in insect-consumed areas. The pixel counts and percent feeding data appear on the screen and are also saved in a computer memory. Escoubas (1993) also used a video camera interfaced with a computer. The scanned image is stored and the eaten areas are measured using video-image analysis software.

2.2.2.2.3. Time Lapse Photography

Hargrove (1988) described a field photographic technique for following herbivory through time on living plants. This technique has merit, since most automated and video-integration photography methods require removal of leaves from plants, which may induce chemical changes in plants and affect subsequent consumption by insects.

The technique uses photographic printing paper or studio proof paper. Photographic developer is incorporated directly into the emulsion of the paper, causing a visible darkening with exposure to light. Thus, prints do not require any wet chemical development process for viewing. The leaf is gently pressed between two thin Plexiglass® sheets and the photographic paper is slipped in behind it. The Plexiglass® and paper "sandwich" are clipped together with small binder clips along the edges (Figure 2.30). Complete exposure takes 2 min in full direct sunlight and 5 min on overcast days. The finished print is a uniformly dark magenta duplicate image of the leaf. Insect-fed areas exposed to light are easily visible in the print. This method can be used serially through time to record insect feeding on individual leaves.

2.2.2.2.4. Microcomputer-Assisted Video Imagery

A system that integrates a video digitizer, light pen, and a microcomputer were designed by Nolting and Edwards (1985) to measure damage for those insects which do not produce a definite hole in a soybean leaf but leave behind a thin membrane that prevents rapid measurement of feeding by leaf area meters. To assess feeding damage, soybean leaves are back lit on a fluorescent light table (35×50 cm) to differentiate defoliated, partially defoliated, and undefoliated leaves.

FIGURE 2.30. Time-lapse photography for measuring insect damage. (a) Leaf pressed between two thin plexiglass sheets and a photographic paper slipped in behind it; (b) a full-size negative after complete exposure. (Photo courtesy of W. M. Hargrove.) (From Hargrove, W. M., *Ecol. Entomol.*, 13, 359, 1988. With permission of Blackwell Scientific Publications Ltd.)

A black plastic mat limiting light passing through to an area 15 cm × 15 cm is located at the center of the light table. The area of the light table digitized is delimited by a hole (11.6 × 13.2 cm) cut in a piece of clear acetate placed between the white acrylic top of the light table and the black mat. Leaf objects to be measured are held flat by a piece of single strength glass slightly larger than the digitized area. A video camera placed over the light table records the leaf image and a computer program integrates the functions of the digitizer and the light-pen system. A computer language program analyzes the digitized image based on the number of white and black pixels in the object. This system has been used to measure the damage caused by several insect pests on several plants.

2.2.2.2.5. Electronic Recording of Feeding Behavior

As previously described, the electronic monitoring system has been extensively used for recording the feeding behavior of numerous piercing-sucking insects. In addition, however, it has also been used to record the feeding of

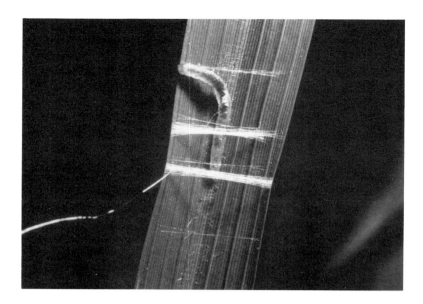

FIGURE 2.31. A fourth-instar *Cnaphalocrocis medinalis* larva on a rice leaf, attached to a gold wire for electronically measuring its feeding activity. (Photograph provided by Z. R. Khan.)

lepidopterous larvae. Saxena and Khan (1991) compared the larval feeding behavior of *C. medinalis* on susceptible and resistant rice plants using a DC electronic monitoring system. Because *C. medinalis* larval feeding takes place within a rolled-up leaf (Fraenkel et al. 1981), direct observation of feeding is not feasible. Also, damage produced by larval feeding is not easily detected with photometric devices because the insect consumes tissue only part of the way through the leaf.

Third- or fourth-instar larvae (starved but water satiated for 2 hr) are slightly anesthetized with CO_2 and a 10-μm gold wire (about 10 cm long) is attached by a drop of silver paint to the dorsal metathoracic segment of each anesthetized larva. The tethered larva is then placed on the leaf of a susceptible or resistant plant (Figure 2.31). The gold wire is connected directly to the negative input terminal of a transistorized, automatic, null-balancing DC recorder with a 250-mm recording width and input resistance of 1 mΩ. The voltage source consists of a 1.5V battery. The positive battery terminal is connected directly to the plant. The negative battery terminal is connected directly to the positive input terminal of the chart recorder. Feeding is recorded for at least 60 min at a chart speed of 5 cm/min.

In the feeding monitoring assembly, the rice plant and the insect each form part of a closed electrical circuit. When the insect feeds on the plant, slow changes in electrical activity are recorded which characterize feeding activity

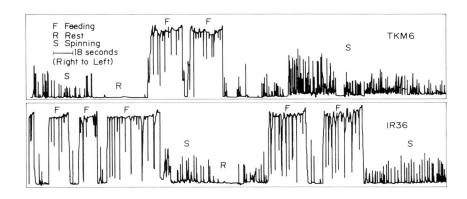

FIGURE 2.32. Electronically recorded waveforms produced by *Cnaphalocrocis medinalis* larva when spinning silk (S), feeding (F), and resting (R) on leaves of resistant TKM6 and susceptible IR36 rice plants. (From Saxena, R. C. and Khan, Z. R., *Ann. Entomol. Soc. Am.,* 84, 316, 1991. With permission of the Entomological Society of America.)

and allow counting of each bite. Saxena and Khan (1991) distinguished waveforms for *C. medinalis* larval spinning and feeding. During larval feeding bouts, a distinct undulate waveform pattern, each wave representing a bite, was recorded on both susceptible and resistant plants (Figure 2.32). On susceptible plants, larvae readily folded leaf blades and fed for sustained periods, but on resistant plants, excessive movement of larvae reduced the effective feeding period.

2.2.2.2.6. Automated Cafeteria

Bowdan (1984, 1988) developed an apparatus, known as an automated cafeteria, for continuously monitoring the feeding of lepidopterous larvae in choice or no-choice tests. In this device, each activity of the caterpillar generates a characteristic slow change in electric charge which is recorded. The apparatus (Figure 2.33) consists of a round, 9-cm-diameter paraffin platform resting on a shelf (S1) on a wooden base. Six pin plugs, simple electrical connectors with an insulated base and a single, printed prong, are arranged at equal intervals around the periphery of the platform. Color-coded wires lead from the pin plugs, below the platform, to one input of a chart recorder. Routing the wires under the platform keeps them away from caterpillars. Disks of the leaves of test plants being offered are impaled on the pins of the pin plugs. Good contact between the leaf and the pin plugs is ensured by covering both the base of the pin and leaf with 1% agar in 10 mM Nacl. The pin of the pin plug is then insulated with a small piece of polyethylene tubing. If the caterpillar climbs on the pin, the polyethylene tubing ensures that no electrical connection is made.

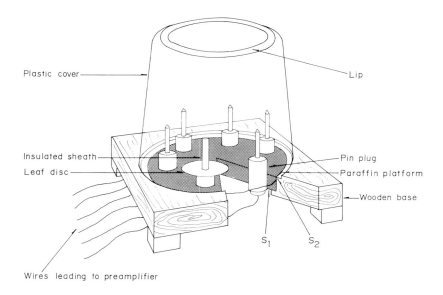

Plastic cover

Lip

Insulated sheath

Leaf disc

Pin plug

Paraffin platform

Wooden base

S_1 S_2

Wires leading to preamplifier

FIGURE 2.33. An automated cafeteria apparatus for continuous monitoring of feeding by caterpillars in choice or no-choice tests. S1 and S2 are two shelves. (From Bowdan, E., *Entomol. Exp. Appl.*, 36, 13, 1984. With permission of Kluwer Academic Publishers.)

Another part of the circuit consists of a 40-µm-diameter wire, insulated except at the ends, attached to the caterpillar using a conductive silver paint. The electrode is held in place using low-melting-point paraffin and must be long enough to allow freedom of movement to caterpillars. The other end of the electrode is held in the second input of the chart recorder. A transparent plastic cylinder prevents the caterpillar from escaping. If the aim of an experiment is to record feeding behavior on a single type of food, the wires from all pin plugs are connected together and fed into the input of a single chart recorder. If, on the other hand, a choice test is being run, then two different types of disks may be arranged on alternating pins around the arena. The wires from the pins holding one type of leaf are then connected together and led into one input of one chart recorder and the wires from the pin plugs holding the other type of leaf are connected together and led into one input of another chart recorder. In the choice test, the electrode from the insect leads into the second input of both chart recorders. Hence, the device can also be used for comparing insect feeding on resistant and susceptible plants. The chart recorder amplifier uses AC current, a band width of 0.1 Hz to 30 kHz, and amplification of 50 with a 60 cycle filter. An open circuit generates much 60 cycle noise, whereas a closed electrical circuit does not.

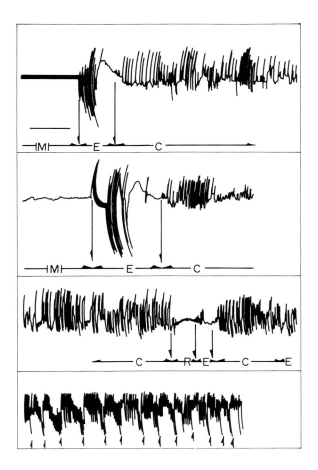

FIGURE 2.34. Recording of *Manduca sexta* feeding on tomato leaves using an automated cafeteria. E, exploration; c, chewing; IMI, intermeal interval; R, resting; (*) single bites; arrows, end of sweeping in a single bite. (From Bowdan, E., *Entomol. Exp. Appl.,* 47, 127, 1988. With permission of Kluwer Academic Publishers.)

Bowdan (1988) recorded feeding activity of the tobacco hornworm, *Manduca sexta* Johansson, on tomato leaves. Distinct waveforms were recorded for biting, exploring, and resting (Figure 2.34). Another electronic feeding monitor, devised by Blust and Hopkins (1990) to record the feeding activities of grasshoppers, also uses the varying resistance of the insect-plant contact to detect biting (Figure 2.35). The grasshopper is implanted with a fine, insulated, wire connected to one of the inputs of an operational amplifier (OP-AMP). A second wire connected to 0.5V DC is inserted into either moist

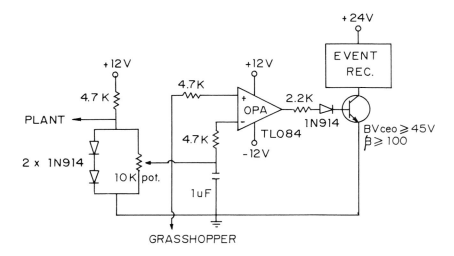

FIGURE 2.35. A schematic one-channel circuit diagram of an electronic feeding moni-
tor used to record grasshopper feeding behavior on an event recorder.
(From Blust, M. H. and Hopkins, T. L., *Physiol. Entomol.*, 15, 261, 1990.
With permission of Blackwell Scientific Publications Ltd.)

soil or water surrounding the test plant. A reference signal is provided to the
other input of the OP-AMP through a 10K ohm multiturn potentiometer and
serves as a sensitivity adjustment. When the signal from the grasshopper ex-
ceeds the reference voltage, the OP-AMP switches from its negative voltage
supply (−12 VDC) to its positive voltage supply of +12VDC. This output
triggers a transistor to conduct the +24VDC used by an event recorder. The
event recorder traces a straight line while periods of feeding appear as
"blocks" of activity that can be read from the chart. This technique can also
be used to monitor the feeding behavior of insects on susceptible and resistant plants.

2.2.2.2.7. Colorimetric Method

Krishna and Saxena (1962) devised a colorimetric method to estimate food
consumption by *Tribolium castaneum* (Herbst) or *Trogoderma granarium*
Everts because the feces of these insects and their flour diet could not be
completely separated. They described the method based on their assertion
that ethanolic extracts of larval feces do not contain sugars which react with

anthrone. They estimated the weight of uneaten flour in an unseparable mixture of flour and feces by the following formula:

$$\text{Weight of uneaten flour} = \frac{\begin{array}{c}\text{total sugar in ethanolic}\\ \text{extract of a sample of flour}\end{array}}{\begin{array}{c}\text{total sugar in ethanolic}\\ \text{extract of a sample of the}\\ \text{mixture feces and uneaten flour}\end{array}} \times \text{weight of flour introduced}$$

This method can be used to determine food intake in cases where it is difficult to separate feces from uneaten food. However, Bhattacharya and Waldbauer (1969b) found that the feces of *Tribolium confusum* Jacquelin du Val and several other stored-grain pests react strongly with anthrone, casting doubt as to the validity of the original method of Krishna and Saxena (1962).

Daum et al. (1969) reported a Calco Oil Red (COR) dye technique for measuring ingestion in adult cotton boll weevils, *Anthonomus grandis grandis* Boheman. COR is incorporated into an artificial diet medium at a rate of 1 g/l. Test insects are allowed to feed on the diet, removed, washed, and killed in acetone to avoid external contamination by COR. The amount of COR in the medium is measured from an aliquot of each replication.

The COR concentrated in test insects, feces, and the diet are extracted in tissue grinders with acetone. The solutions are filtered and concentrations are measured with a spectrophotometer at 510 μm against an acetone blank-color, intensities are converted to micrograms COR using standard curves, and the results are used to express the micrograms of dye found in the insect as micrograms of diet ingested.

2.2.3. Metabolism

Consumption of plant tissue by an insect must be followed by proper metabolism of the ingested food before the food can support insect growth, survival, and fecundity. Therefore, analyses of the utilization of food has proven useful in investigations of the mechanisms of plant resistance to insects and in the classification of levels of resistance (Kogan and Parra 1979). In many studies, attempts have been made to quantify the efficiency with which insects exploit their food plants. Metabolism of the food ingested by insects involves digestion of nonabsorbable food constituents, absorption of absorbable food constituents, and metabolism of food, resulting in assimilation or conversion of constituents into insect body tissues (Saxena 1969). Assimilation, which results in increased insect weight gain, depends upon the degree to which absorbed food constituents can satisfy the requirements of intermediary metabolism and the absence of any toxins in the absorbed food which might inhibit metabolism. Therefore, the overall nutritive value of a plant as a food is determined by its rate of absorption and assimilation by an insect. In addition, differences in the efficiencies of digestion and conversion of food by insects also contributes to the nutritive value of a plant to an insect.

In the majority of cases, the measurement of food utilization by insects has been determined using the gravimetric methods described by Waldbauer (1964, 1968). Other methods, such as colorimetric, radioisotopic, calorimetric, enzymatic, and respirometric, rely upon indirect measurements of metabolic utilization. The indices used in the analysis of food utilization generally require the measurement of initial and final insect weights, initial and final food weights, and the amount of feces accumulated during the experiment (Kogan and Parra 1979). Numerous methods have been used to quantify food utilization by insects using various piercing, sucking and chewing modes of feeding.

2.2.3.1. Chewing Insects

2.2.3.1.1. Gravimetric Method

The plant tissue consumed, excreta produced, and the weight gained by individual larvae over the entire larval instar can be determined on a dry-weight basis according to procedures described for artificial diets by Waldbauer (1964, 1968). The method involves cutting the leaves to be fed to the larvae into two symmetrical portions along the midrib. One half is weighed wet and fed to the larvae. The other half is weighed wet and then dried to a constant weight to determine the dry weight percentage. This permits calculation of the dry weight of the introduced food. The portions of leaves used to determine the dry weight are saved and analyzed for organic and ammoniacal nitrogen by a macro-Kjeldahl method. The amount of food consumed during the larval instar is equal to the weight of food provided minus the weight of the uneaten food. The difference between the amount of food consumed and of feces produced during the instar is the weight of digested food.

The fresh weight of each larva is determined at the beginning of each experiment. The dry weight of each larva is estimated by using the mean percentage dry matter of an aliquot of like larvae which have been killed by freezing and dried at 100° C to a constant weight. The difference between the final dry weight of the larva when sacrificed after the moult and the estimated initial dry weight is the net weight increase. The amount of food consumed or digested in relation to the increase in body weight is a measure of the efficiency of utilization.

The indices developed and used by Waldbauer (1968) and adapted for use by other researchers are described below:

The approximate digestibility (A.D.) of food by an insect is calculated as:

$$A.D. = \frac{\text{Dry weight of food ingeste} - \text{dry weight of excreta}}{\text{Dry weight of food ingested}} \times 100.$$

The efficiency of conversion of ingested food to body matter (E.C.I.), a measure of the overall ability of an insect to grow on a given food, is calculated as:

$$E.C.I. = \frac{\text{Dry weight gained by insect}}{\text{Dry weight of food ingested}} \times 100.$$

The efficiency of conversion of digested food to insect body matter (E.C.D.) is calculated as:

$$E.C.D. = \frac{\text{Dry weight gained by insect}}{\text{Dry weight of food ingested} - \text{dry weight of excreta}} \times 100.$$

Since larvae on susceptible and resistant leaves eat and grow at different rates over variable periods, it is necessary to express all measurements in a form which make possible the comparison of intake and efficiency of utilization. The three components to be considered are the amount of food consumed (F), the mean weight of larvae during the feeding period (A), and the duration of the feeding period in days (T). The consumption index is then calculated as C.I. = F/TA, with F and A expressed as dry weight.

The mean weight (A) of a larva is calculated either by summing the initial, final, and intermediate weights determined every 24 hr and dividing by the number of weighings (Soo Hoo and Fraenkel 1966) or by calculating mean weight from weighted averages of daily weights (Waldbauer 1964, 1968).

Waldbauer (1968) indicated a source of experimental error in nutritional indices due to the metabolism of excised leaves. Small discrepancies between the actual percentage dry matter of the insect's food source and the percentage dry matter determined from a sample aliquot resulted in errors in the estimated dry weight of the food at the beginning of the experiment. Waldbauer (1968) suggested that this factor can be minimized by frequent changes of food and the immediate drying of uneaten food. The potential error from this source can be further minimized by estimating the dry weight of eaten food from the mean of the percent dry matter of the aliquot and the percent dry matter of the uneaten food.

Nutritional indices are widely used to detect leaf-feeding resistance to insects in various crop plants, e.g., maize against southern armyworm, *Spodoptera eridania* (Cramer) (Manuwoto and Scriber 1982), cotton against cotton leafworm, *Alabama argillacea* (Hübner) (Montandon et al. 1987), on tobacco budworm, *H. virescens* (Mulrooney et al. 1985, Montandon et al. 1987), potato against Colorado potato beetle, *Leptinotarsa decemlineata* (Say) (Cantelo et al. 1987), and soybean against soybean looper, *Pseudoplusia includens* (Walker) (Reynolds et al. 1984).

A number of researchers have also used nutritional indices to study the assimilation and utilization of artificial diets by lepidopterous larvae (Chou et al. 1973, Dahlman 1977, Reese and Beck 1976a,b,c, Schmidt and Reese 1986, Parra and Kogan 1981, Mansour et al. 1990). Bhattacharya and Waldbauer (1972) introduced a correction factor for fecal urine in dietary studies with *T. confusum*. Because *T. confusum* fecal pellets contain metabolic wastes in addition to undigested food, the urine content of the end products should be subtracted from the total weight of the feces to provide a better estimate of food digested. Schmidt and Reese (1986) used an electronic spreadsheet to demonstrate the algebraic magnification of introduced errors made in calculating initial percentage dry matter of the food source, the amount of food eaten, and the weight of feces produced. These authors demonstrated that small errors in these measurements substantially alter the calculated values of food utilization efficiencies. Van Loon (1991) suggested that to minimize errors in the measurement of food utilization, several additional parameters should also be recorded. These include percentage of food consumed relative to the amount offered, duration of feeding periods, total duration of the experiment, respiration rates of plant tissue, homogeneity and dynamics of contents of plant dry matter, and nutritive elements.

2.2.3.1.2. Colorimetric Method

McGinnis and Kasting (1964a) developed a sensitive colorimetric analysis for chromic oxide which makes it possible to detect this compound in individual insects. Chromic oxide is incorporated into the diet and the percent food utilization is calculated from the concentrations of the index compound in the food and the excreta. The method is based on wet oxidation of Cr_2O_3 to $Cr_2O_7^{-2}$, followed by colorimetric determination of the dichromate ion with diphenylcarbazine.

Parra and Kogan (1981) adapted the chromic acid colorometric method to study food intake and utilization of the soybean looper, *P. includens*. They incorporated chromic oxide in an artificial diet at a concentration of 40%. The Cr_2O_3 was premixed with KOH of the standard artificial media to facilitate incorporation. Feces from larvae fed on the artificial media containing Cr_2O_3 is bulked and dried at 120°C for 2 hr. The amount of chromium in the medium, larvae and feces is then determined by atomic absorption. Ten- to 25-mg samples are weighed and placed in a 100-ml Kjeldahl flask. Ten ml of the digestion mixture [10g $Na_2M_0O_4$, $2H_2O$, 150 ml H_2SO_4 (conc.), and 200 ml 70% $HClO_4$ in 150 ml distilled water] (McGinnis and Kasting 1964a) are added to each sample and samples are treated for 30 min on a rotary Kjeldahl digestion rack. A NaOH gas trap is connected to the Kjeldahl rack and the digestion is carried out in an explosion-proof fume hood. After 30 min, the digests change color from green to orange, and the flasks are cooled to room temperature. Water (500 ml) is added to the diet and feces digests and 100 ml is added to larval digests. Analysis of chromium is done on an automatic absorption emission spectrophotometer following the method of Parra and Kogan (1981).

Computation of food consumed and utilized is calculated following the technique of (McGinnis and Kasting 1964b):

$$\text{MG}[Cr_2O_3]\text{*Insect} \atop \text{Dry matter (D.M.) consumed} = \frac{[Cr_2O_3]\text{*Excreta} \times \text{D.M. (mg) Excreta}+}{[Cr_2O_3]\text{*Food}}$$

$$\% \text{ Utilization} = (\frac{[Cr_2O_3]\text{*}}{[Cr_2O_3]\text{*Excreta}}) \times 100$$

where $[Cr_2O_3]\text{*}$ is Mg Cr_2O_3/mg dry matter.

McGinnis and Kasting (1964a) compared gravimetric and chromic oxide methods for measuring the percentage consumption and utilization of food for the two-striped grasshopper, *Melanoplus bilituratus* (Walker), and larvae of the pale western cutworm, *Agrotis orthogonia* Morrison. They concluded that the chromic oxide procedure gives more reliable results than the gravimetric method. According to Waldbauer (1968), however, this method is probably usable only with artificial diets which can be ground and mixed to assure even distribution of the marker. The utilization of whole leaves can be measured if it is possible to distribute the marker evenly on the leaf surface. However, the amount of chromic oxide ingested will depend upon the area of the leaf rather than on the weight of leaf eaten.

Heinrichs and Pruess (1966) used chromogens (naturally occurring plant pigments) as markers to measure the digestibility of dried ground plant material by grasshoppers. The authors reported a general correlation between the results of chromogen-ratio and gravimetric methods, although the results differed widely in 4 of 21 tests with different plants.

The use of fecal uric acid as an indicator has also been shown to yield accurate estimates of uneaten food and feces in a mixture left by stored grain pests. Bhattacharya and Waldbauer (1969a, 1970) developed a colorimetric method of estimating the proportions of uneaten food and feces in a mixture left by *T. confusum* which can be used to determine the weight of digested food. The method depends upon the fact that the feces contain uric acid while food does not. Thus, the weight of feces present can be calculated by dividing the total weight of uric acid in the mixture by the weight of uric acid per unit weight of feces. Uric acid is extracted from dried and preweighed samples with 3.0 to 3.5 ml of a 0.6% aqueous lithium carbonate solution as described by Bhattacharya and Waldbauer (1969b, 1970).

The weight of feces in the mixture is calculated as follows (Bhattacharya and Waldbauer 1969a,b):

$$\text{mg feces in mixture} = \frac{\text{mg uric acid in mixture}}{\text{mg uric acid per mg feces}} + \text{weight of feces sample.}$$

After the dry weight of uneaten food is determined, the weight of food eaten is calculated by subtracting the weight of the uneaten food from the weight of the food originally offered. The approximate weight of food digested can be calculated by subtracting the weight of the feces from the weight of food eaten.

2.2.3.1.3. Radioisotope Method

Radioisotopes have been used to some extent to measure food consumption and utilization in insects (Kasting and McGinnis 1965a,b, Buscarlet 1974). Sucrose-U-^{14}C or cellulose-U-^{14}C are added in an artificial diet with an activity of 2.1×10^6 cpm/ml as markers to estimate consumption and utilization of food (Kasting and McGinnis 1965a,b). At the end of the feeding period, the insects, their feces, and the CO_2 they expire are measured for radioactivity on a liquid scintillation counter (Cammen 1977). The quantity of food consumed is calculated following the method of Kasting and McGinnis (1965a):

$$\text{Dry matter (D.M.) consumed (mg)} =$$

$$\frac{\text{Total radio activity in insect, excreta and } CO_2 \text{ (c.p.m.)}}{\text{Specific activity of food (c.p.m.)/mg D.M.}} .$$

CO_2 is collected from cups with insects feeding on diets kept under vacuum bell jars (Parra and Kogan 1981). A hole is made in the lid of each cup and a glass tube is inserted through the hole. Air passing through a water bath is drawn into the cups by means of a single-stage vacuum pump set on a timer activated for 2 min every 15 min. The air removed from the rearing cups passes through two flasks of $CaCl_2$ pellets to remove most of the water. The CO_2 expired from the cups is collected in gas traps containing Carbosorb®. The Carbo-sorb® is changed every 48 hr and radioactivity is measured by adding 0.5 ml of Carbo-sorb® directly to a scintillation vial containing 10 ml Permafluor®. Radioactivity is measured for 10 min on a liquid scintillation spectrometer. Readings are converted to dpm's and results expressed as percentages of labeled materials. Computations of ECI, ECD, and AD are made following the methods of Waldbauer (1968).

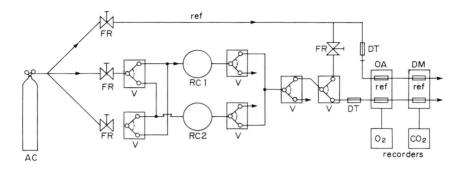

FIGURE 2.36. A schematic diagram of a flow-through gas analysis system used to monitor continuously insect respiratory metabolism. AC, air cylinder; FR, flow regulator; V, solenoid valve (two way); RC, respiration chamber; OA, paramagnetic oxygen analyzer; DM, carbon dioxide analyzer; DT, air drying tube. (From van Loon, J. J. A., *Entomol. Exp. Appl.*, 49, 265, 1988. With permission of Kluwer Academic Publishers.)

2.2.3.1.4. Respirometry

Respirometry is a form of indirect calorimetry which can be used to measure consumption and utilization of food in insects (Van Loon 1988, 1991). The method has been used for lepidopterous larvae and locusts in studies on assimilation efficiency of different nutrients (Simpson 1982, Simpson and Simpson 1989). Both closed-vessel respirometers (Gilson 1963, Keister and Buck 1974) and flow-through respirometers (Armstrong and Mordue 1985, Sell et al. 1985, Van Loon 1988, 1991) have been used. Closed-vessel respirometers require a closed system in which the oxygen concentration gradually decreases and the carbon-dioxide concentration is kept at zero by its absorption through an alkaline solution. The drawback of this system is that it does not permit measurement for longer than 1 to 2 hr because the carbon-dioxide-absorbing alkaline solution is strongly hygroscopic and the respirometer atmosphere becomes extremely low in humidity. This may result the closing of insect spiracles to prevent desiccation and thereby to the limitation of gaseous exchange. Flow-through respirometers operate with a flowing gas stream. The flow-through respirometer described by Van Loon (1988) is based on the use of two differential gas analyzers in series: a paramagnetic oxygen analyzer and a diaferometer (Figure 2.36). The flow-through respirometer makes it possible to continuously measure the respiration rate of undisturbed, feeding insects for longer periods of time.

To calculate consumption and utilization of food by insects, the units of gas exchange are converted to caloric units (Joules). Accurate data on oxygen consumption and carbon dioxide release are essential to calculate heat loss as well as respiratory quotient. The different chemical forms and the quantity of the products of nitrogen metabolism (that are excreted as parts of the feces or stored in the body) should also be taken into account, because nitrogen metabolism is largely determined by the quality of food protein.

The sensitivity of a flow-through respirometer prevents its use with smaller herbivores such as aphids (Van Loon 1991), although the measurement of respiration of groups of such insects is generally feasible. Van Loon (1988) measured the respiration rate of an actively feeding caterpillar of the cabbage butterfly, *Pieris brassicae* (L.), on an artificial diet and on leaves of its host plant, *Brassica oleracea* L. Caterpillar respiration (μ 10_2/individual/hr) during the course of final instar was lower on the artificial diet than on leaf material.

2.2.3.2. Piercing and Sucking Insects

Food utilization by sucking insects has been measured using the gravimetric methods described by Auclair (1958, 1959), Saxena (1969), Saxena et al. (1974), Saxena and Pathak (1977), Khan and Saxena (1985b), and Smith et al. (1991). In addition to the nutritional indices developed for chewing insects, Saxena (1969) also developed additional nutritional indices for piercing and sucking insects. These include an index of absorbability of food and an index of food assimilation. The ratio of ingested food [equivalent to the efficiency of conversion of ingested food of Waldbauer (1968)] and assimilated food indicates the overall "nutritive value" of the plant.

In a typical experiment of this type, newly emerged adults (females) or nymphs are starved but water satiated for 4 to 8 hr (depending upon the insect species tested) to clear their guts. After recording initial weights (W1) insects are enclosed singly in air-tight Parafilm® sachets (5 cm × 5 cm) sealed around the plant tissue (Saxena and Pathak 1977, Pathak et al. 1982, Khan and Saxena 1985b, Smith et al. 1991). In some cases, insects are kept individually in preweighed beakers with a watered plant leaf or stem placed across the mouth of the beaker and covered with nylon net (Saxena 1969, Saxena et al. 1974). Insects feed on the plant tissue and excrete in the Parafilm® sachet or in the beaker. To assess the loss in insect body weight due to catabolism, a control is similarly established in which the insect has access to a moist cotton swab to prevent desiccation. The initial weights of the control insects (C_1) are recorded. After 24 or 48 hr (depending on the experiment), the final weights of each plant-fed (W_2) and water-satiated but starved insect (C_2) is recorded. The starved as well as the plant-fed insects lose weight from the wear and tear on their tissues, but plant-fed insects also gain weight as a result of assimilation of ingested food. The weight lost per unit of initial weight of the starved insects is calculated as $(C_1 - C_2)/C_1$. Similarly, the weight lost by plant-fed insects is calculated as $W1[(C_1 - C_2)/C_1]$ and is added to the difference ($W_2 - W_1$) between final and initial weights to obtain the weight of the ingested food assimilated by its conversion into tissue:

$$\text{Weight of assimilated food} = W1\left(\frac{C_1 - C_2}{C_1}\right) + (W_2 - W_1).$$

The weight of excreta from each insect is recorded and added to the weight of assimilated food to obtain the weight of ingested food. The weight of absorbed food is calculated as:

$$\text{Weight of Absorbed Food} = \frac{\text{Weight ingested food} - \text{Weight excreta}}{\text{Weight ingested food}} \times 100.$$

Using these factors, Saxena et al. (1974) calculated the following indices of metabolic utilization of food:

$$\text{Absorbability of Food} = \frac{\text{Weight ingested food}}{\text{Weight absorbed food}}$$

$$\text{Assimilability of Food} = \frac{\text{Weight absorbed food}}{\text{Weight assimilated food}}$$

$$\text{Nutritive Value of Food} = \frac{\text{Weight ingested food}}{\text{Weight assimilated food}}.$$

Saxena (1969) reported that the nutritive value of the food ingested by the red cotton bug, *Dysdercus koenigii,* was high because of high absorbability as well as assimilability. The nutritive value of wheat seeds was quite low because of a low absorbability of wheat seed constituents.

Khan and Saxena (1985b) reported that values for ingestion and assimilation of food by *S. furcifera* were significantly higher on susceptible TN1 rice plants than on resistant rice plants.

2.2.4. Growth

Insect feeding and metabolism of ingested food are integral components of the antibiosis category of plant resistance. Therefore, the growth of an insect on susceptible or resistant plants is determined by measuring the weight gain of larvae or nymphs and the development of larvae or nymphs into adults or pupae. The latter is determined as (a) the percentage of immature individuals transforming into adults or pupae, and (b) the average time period required to do so. The ratio of a to b represents an insect "growth index" (Saxena et al. 1974). Since the survival of an insect on a plant depends on both development and weight gain, the growth index is a more suitable measurement of insect growth on susceptible and resistant plants. Generally, insects fed on the more susceptible host plants have higher growth index scores. The growth of foliage, stem, and root feeding insects has been measured in several different ways. The following descriptions demonstrate the diversity of both feeding methods and feeding measurements.

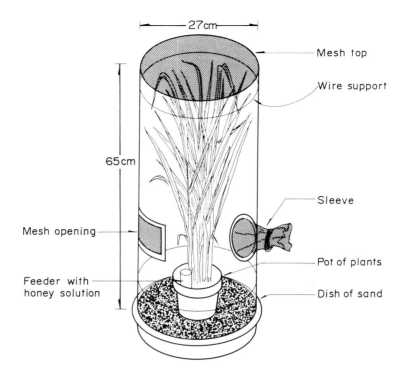

FIGURE 2.37. A Mylar film cage to monitor insect growth and to study ovipositional response on rice plants. (From Heinrichs, E. A. et al., *Genetic Evaluation for Insect Resistance in Rice,* International Rice Research Institute, 1985. Courtesy, IRRI.)

2.2.4.1. Stem and Foliage Feeders

Newly hatched first-instar larvae or nymphs are commonly caged on whole plants or on plant parts to limit emigration of test insects from plants and to protect test insects from predation and parasitism. Whole plants can be enclosed in cages constructed of Plexiglass® (Heinrichs et al. 1985) (Figure 2.37). Cage size and shapes are determined by the type and number of test plants that must be evaluated. In greenhouse and field tests, small insects may be housed in small clip-on cages placed over a section of intact plant leaf (Eenink et al. 1984) (Figure 2.26). Larger tissues of living plants can be enclosed in sleeve cages made of dialysis tubing (Dimock et al. 1986, Benedict et al. 1981a) or polyester, organdy or nylon cloth (Hollay et al. 1987) (Figure 2.6). Full or partial growth index measurements are then recorded, based on the developmental period of the test insect.

Insect growth can also be measured on freshly excised leaves in laboratory assays of resistant and susceptible plants (Wiseman et al. 1981, Rufener et al. 1987, Beach et al. 1985, Barney and Rock 1975, Reynolds et al. 1984, Ng et al. 1985, Cantelo et al. 1987). Freshly excised leaves are placed in plastic or

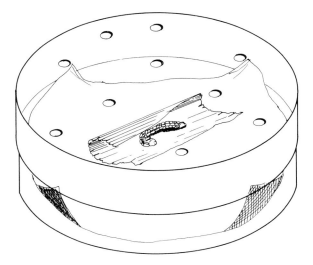

FIGURE 2.38. *Spodoptera frugiperda* larva on moistened paper towel in a growth chamber. (From Wiseman, B. R. et al., *J. Econ. Entomol.,* 74, 622, 1981. With permission of the Entomological Society of America.)

paper containers (commonly in petri dishes) lined with filter paper kept moist for the duration of the experiment (Figure 2.38). Known numbers of neonate larvae are placed in each container, the containers are covered, and then placed in an environmental growth chamber and incubated at a predetermined temperature, relative humidity, and photoperiod. Often, tests are conducted in a 12 hr light:12 hr dark cycle at 25° C temperature, with 60 to 70% relative humidity. Larvae dying within 24 hr are generally replaced with new neonate larvae because mortality at this point is often high due to mechanical injury suffered during transfer and not an antibiotic effect of the plant. Fresh foliage from the original plant is added to the test container every 3 to 5 days.

Depending upon the test insect species, observations are made 10 or 15 days after infestation to determine larval survival, larval weight, number of larval instars, length of larval life, percent larvae pupating, and pupal weight.

Rufener et al. (1987) assigned a point value to Mexican bean beetle, *E. varivestis,* larvae reared on susceptible and resistant soybean varieties. The point value was based on the condition or instar of larvae after 10 days of feeding, where dead larvae = 0; first- and second-instar larvae = 2; and third- and fourth-instar larvae = 3. The cummulative point value of each test container was then calculated. Plants with higher scores were found to be more suitable for insect growth and survival.

Growth rate of stem borers on resistant and susceptible varieties of rice and maize is frequently measured by removing infested plants after intervals from 7 to 35 days of infestation. Each plant is carefully dissected and the number of surviving larvae and their respective growth stages and weights are recorded (Pathak 1969a, Oliver and Gifford 1975, Robinson et al. 1978, Heinrichs et al. 1985, Davis and Williams 1986, Ampofo et al. 1986, Ampofo and Kidiavai 1987).

The larval growth of the corn earworm, *H. zea,* on intact soybean leaves has been shown to be positively correlated to growth on excised leaves (Hart et al. 1988). However, the mean weight of larvae fed on intact leaves differentiates resistance better than weights of larvae fed excised leaves.

Lukefahr et al. (1966) reported that growth of *H. zea* and *H. virescens* larvae was related to gossypol content in the pigment glands of cotton plants. Larval growth was significantly greater on susceptible glandless than on resistant glanded cotton strains. Saxena (1969) in comparing the growth of the red cotton bug, *D. koenigii,* on different host and nonhost plants, found that the high growth index of *D. koenigii* on cotton seeds was due to their high nutritive value as well as high *D. koenigii* food intake. Lower growth indices of *D. koenigii* on okra and peanut seeds was due to lower food intake, whereas failure of growth of the insect on wheat seeds was due to poor food intake as well as the deficient nutritive value of the seeds. Pathak (1969b) compared growth of *N. lugens* and the green leafhopper, *Nephotettix cincticeps* (Distant), on susceptible and resistant rice varieties. When a uniform number of newly emerged first-instar nymphs were caged on susceptible and resistant rice varieties, those infesting resistant plants had lower survival and grew more slowly than those caged on susceptible plants. Saxena et al. (1974) compared the growth of the cotton leafhoppers, *Empoasca devastans* Distant and *Empoasca kerri motti* Pruthi, on several species of cotton plants. The authors reported that one host plant may be better than another for leafhopper weight gain but not for moulting and metamorphosis. A plant good for larval weight gain but poor for moulting and metamorphosis will produce fewer adults than plants that support enhanced moulting and metamorphosis but poor weight gain. Therefore, the authors concluded that the percentage of larvae or nymphs completing development over an identical period, calculated as a percentage of the total developmental period, was a suitable parameter for comparison of insect growth.

Differences in larval weight gain on susceptible and resistant plants have been reported as the sole criterion for evaluating insect resistance by several authors. These include the rice stalk borer, *Chilo plejadellus* Zinken, and sugarcane borer, *Diatraea saccharalis* (F.) (Oliver and Gifford 1975) on rice; *S. frugiperda* (Wiseman et al. 1981), European corn borer, *O. nubilalis* (Robinson et al. 1978), and Southwestern corn borer, *Diatraea grandiosella* Dyar (Davis and Williams 1986) on maize; and *L. decemlineata* on potato (Cantelo et al. 1987).

Larval survival, percent larval metamorphosis, and larval developmental period have been used as measures of growth of *N. lugens* (Saxena and Pathak 1977), *N. virescens* (Distant) (Khan and Saxena 1985a), and *S. furcifera* (Khan and Saxena 1985b) in soybean against *P. includens* (Beach et al. 1985, Reynolds et al. 1984), *E. varivestis* (Rufener et al. 1986), *Helicoverpa punctigera* (Wallengren) and *Chrysodeixis argentifera* (Guenée) (Brier and Rogers 1991), in maize against *S. frugiperda* (Ng et al. 1985), *D. grandiosella* (Davis and Williams 1986), and *C. partellus* (Ampofo and Kidiavai 1987), and in sorghum against the chinch bug, *Blissus leucopterus leucopterus* (Say) (Mize and Wilde 1986).

2.2.4.2. Root Feeders

Ortman and Branson (1976) developed a hydroponic growth pouch technique for studying maize resistance to larvae of the western corn rootworm, *Diabrotica virgifera virgifera* LeConte, and southern corn rootworm, *Diabrotica undecimpunctata howardi* (Barber). This method permits visual observation of both host and larvae during the test period and has the potential for use with many other root-feeding insects.

The growth pouches are made of a durable, clear plastic and contain an absorbent paper wick which is folded at the top to form a trough with a perforated bottom. Seeds of test plants are treated with fungicides and then germinated on a filter paper in a petri dish. When the radical has emerged about 1 cm, a single seed is transferred to each growth pouch. The radical of the seed is placed in such a way that it projects down through the perforation. The wick is moistened with a nutrient solution without adding excess moisture in the growth pouch. When the maize roots grow about 10 cm in length, they are infested with 8 newly hatched rootworm larvae. The pouches are closed at the top with masking tape or by heat sealing (Figure 2.39), and held at 24° C, 50 to 60% RH, and 12:12 L:D for the entire experiment. Observations are recorded on larval survival and growth.

Byers and Kendall (1982) and Powell et al. (1983) used the hydrophonic slant board technique of Kendall and Leath (1974) for evaluation of clover genotypes for resistance to root-feeding clover root curculio, *Sitona hispidulus* (F.). In this technique, plastic cafeteria trays with a carrying surface of about 35×45 cm and 2 cm deep are used. Eight plants of each genotype were selected from potted plants for uniformity in plant size. Two sheets of polyester sheath lining (20×30 cm) are placed side by side on each cafeteria tray. One plant is placed on top of each cloth with the plant crown at the top of the tray. Five to eight neonate larvae or surface-sterilized eggs are placed on the roots of each plant. Plants are covered with another cloth of the same size. A sterile cloth bag (20×30 cm) containing vermiculite is placed on top of the second cloth of each plant. A strip of 46-cm-wide, heavy-duty aluminum foil, folded in half, is fastened tightly across the tray with rubber bands and binder clips in such a way that it can hold bags, cloths, and plants in place on the tray. The lower 5 cm of the tray is left uncovered to allow aeration and drainage of the system. Each plant and bag is saturated with nutrient solution (Baker and Byers 1977). Trays are held at a 60° angle and incubated at 24° C, 70% RH and a 14-hr photoperiod. Nutrient solution is poured into each bag every day and bags are replaced every 7 days. After 20 days, data are collected on curculio larval survival and growth.

FIGURE 2.39. A growth pouch for measuring growth of larvae of rootworms on roots of corn seedlings. (From Ortman, E. E. and Branson, T. F., *J. Econ. Entomol.,* 69, 380, 1976. With permission of the Entomological Society of America.)

2.2.5. Adult Longevity and Fecundity

After an insect completes its growth on a host plant, its longevity and fecundity determines its establishment on that plant. The longevity of an insect on test plants is generally recorded as the period of adult survival. Insect fecundity or ability to produce eggs on different host plants is determined as the number of eggs produced and laid.

The longevity and fecundity of phytophagous insect adults is determined by confining newly emerged male and female adults on resistant and susceptible test plants. The newly emerged adults were reared either on their respective susceptible and resistant test cultivars (Teetes et al. 1974, Webster and Inayatullah 1984, Webster et al. 1987, Webster 1990, McCauley, Jr. et al. 1990) or on a common susceptible test cultivar (Hsu and Robinson 1962, Saxena 1969, Cheng and Pathak 1972, Saxena et al. 1974, Heinrichs and Medrano 1984, Khan and Saxena 1985b, Wu et al. 1986, Shanks, Jr. and Doss 1986). For insects whose adults do not feed on host plants (i.e., Lepidoptera), newly emerged males and females are confined on a susceptible host plant or an artificial substrate for oviposition. These adults are reared either on various susceptible and resistant test plants (Leuck 1970, Ng et al. 1985) or on artificial diets containing foliage from the test plants (Pencoe and Martin 1982, Pantoja et al. 1987). Although it is highly desirable to use test insects of a known age, some researchers have occasionally used field-collected insects of unknown ages (Benedict et al. 1981a).

Adult longevity and fecundity of *N. virescens* (Cheng and Pathak 1972), *A. biguttula biguttula* (Saxena et al. 1974), *S. furcifera* (Khan and Saxena 1985b), *S. frugiperda* (Ng et al. 1985), *N. lugens* (Wu et al. 1986), and *N. ribisnigri* (Van Helden et al. 1993) is greater on susceptible plants compared to that on nonhost and resistant plants.

Fecundity of those insects which deposit eggs in egg masses may be determined following the method described by Lynch et al. (1983) and Ng et al. (1985) for *S. frugiperda*. Egg masses deposited by each female are removed from the wax paper and weighed. The weight of a single egg is estimated by counting the number of eggs in each weighed mass and dividing the weight of each egg mass by the total number of eggs in the mass.

The fecundity of those insects which lay their eggs within plant tissue may be determined following the methods described by Heinrichs and Medrano (1984), Khan and Saxena (1985b), and Wu et al. (1986). Newly emerged males and females are enclosed in pairs on uninfested plants. The total number of nymphs appearing on the plants is counted and represented by the number of viable eggs produced by the test females during their lives. At the end of nymphal emergence, unhatched eggs are counted by dissecting plant tissues under a binocular microscope or by obtaining the eggs following the method of Khan and Saxena (1986) (see under oviposition).

Saxena (1969), Saxena et al. (1974), and Shanks, Jr. and Doss (1986) studied insect oviposition on the basis of (a) the preoviposition period (the period required by newly emerged females to develop and lay eggs), and (b) the number of eggs produced and laid. Saxena (1969) used these measurements to calculate a fecundity index expressed as preoviposition period/number of eggs laid.

Bintcliffe and Wratten (1982) assessed potential fecundity of *M. persicae,* by counting the number of embryos within adults. Single aphids are placed in a drop of glycerol on a microscope slide and covered with a coverslip. The embryos are then extruded by gently pressing the coverslip using a mounted coverslip, and the number of embryos are counted.

Using insect fecundity, Lynch et al. (1981) and Pencoe and Martin (1982) determined host suitability index (HSI) for *S. frugiperda* as

$$\text{HSI} = \frac{\times \text{ of eggs laid per female}}{\text{larval duration} \times \text{consumption}} \times \% \text{ survival}.$$

Based on this index, goosegrass was shown to be the most suitable host for *S. frugiperda* reproduction (Pencoe and Martin 1982).

2.2.6. Oviposition

For many phytophagous insects, the selection of an oviposition site is a critical stage in their choice of a host (Singer 1986). Understanding the details of insect oviposition preference is valuable when attempting to identify resistant germplasm in a plant breeding program (Gould 1983).

In order to understand the role of an insect's response to plant ovipositional stimuli in its establishment on susceptible and resistant plants, it is important to eliminate differences in egg production (fecundity) caused by differences in food intake and assimilation. Therefore, ovipositing females reared on a susceptible host plant are allowed to oviposit on different host plants for an identical period and then their ovipositional preference for different plants is compared on the basis of the number of eggs laid during an identical period. Ovipositional bioassays may be conducted in the field, greenhouse, or laboratory, either as ovipositional choice (preference) tests or as ovipositional response (no-choice) tests. Ovipositional responses quantified by one technique may be different when quantified by another technique (Hoffmann 1985).

2.2.6.1. Measurement of Egg Distribution in the Field

Although egg distributions are usually correlated with the presence of host plants in an open field trial, they are likely to be influenced by physical features of the microhabitat (Thomas 1985). Therefore, ovipositional preferences may not be determined in the field from a simple count of the proportion of eggs found on various plant species because such counts can be the result of oviposition by more than one female and not all plants are of equal abundance and availability (Stanton 1982, Singer 1986, Thompson 1988). Field results may give different results when conducted in different sizes of host plant patches (Singer 1986). Although there have been objections to the measurement of preference from egg distribution in the field, this technique has been widely used for the evaluation of plant resistance toward insects where oviposition is strongly correlated with the presence of host plants. From a field experiment Pathak (1969a) reported very strong preference of the *C. suppressalis* for certain rice varieties. In these experiments, even under low borer populations, certain varieties received much larger numbers of eggs. Many resistant varieties, however, had only a few egg masses even under heavy infestations. Ampofo (1985), also reported differences in field oviposition of *C. partellus* on susceptible and resistant maize plants.

To avoid differences in oviposition on different test cultivars due to non-plant factors such as female population, fecundity, and physiological stage several researchers have used large oviposition cages in field plots. Known numbers of ovipositing females are released into each cage to oviposit on test plants within the cages (Masuzawa et al. 1983, Beach and Todd 1988, Ng et al. 1990).

Saxena (1987, 1990) developed and used a three-compartment chamber to evaluate the ovipositional response of *C. partellus* in a field (Figure 2.40). Tests were conducted in a field with a constant number of females in the chamber (210 cm × 80 cm × 80 cm). The oviposition chamber has a control sector between two equal-sized end sectors on either side. The chamber's roof and two vertical end-walls are of glass but open below, the floor being formed by the test arena. The front and rear walls of the central sector are of glass and those of the two end sectors are of removable screen (6 meshes/cm) which allows wind to pass through. In a field the chamber is aligned with its long axis at a right angle to the wind direction. Three to five test plants, 3 to 4 weeks old, are arranged inside one end compartment in a row along the end wall. The opposite end compartment has a similar row of plants of another cultivar in a two-choice test, or contains no plants, but has wax paper sheets glued to its end wall to serve as a "blank" non-plant ovipositional substrate. Gravid females of test insects are released in the central compartment and the eggs laid on the plants and on the wax paper sheets are counted. The difference in numbers of eggs between the two test cultivars or between the test cultivar and wax paper reflects the suitability of the cultivars for oviposition by the insect. Saxena (1987, 1990) demonstrated that both susceptible and resistant sorghum cultivars tested elicited a stronger *C. partellus* ovipositional response than a non-plant substrate and that susceptible sorghum cultivars elicited a stronger ovipositional response than resistant cultivars.

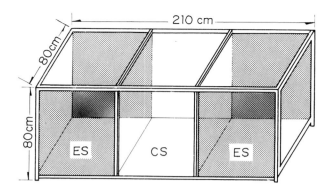

FIGURE 2.40. Three-compartment chamber for studying ovipositional response of *Chilo partellus* to sorghum plants in a field. The plain surfaces are glass, and the strippled areas are finewire mesh. ES, end compartment; CS, central compartment. (From Saxena, K. N., *Entomol. Exp. Appl.,* 55, 91, 1990. With permission of Kluwer Academic Publishers.)

2.2.6.2. Measurement of Egg Distribution in the Greenhouse and Laboratory

Ovipositional preference in a greenhouse or laboratory experiment is measured in choice tests where plants of equal mass of several species or varieties are offered simultaneously. Ovipositional preference is expressed as the proportion of eggs laid on the plants of different species tested and is evaluated using intact or excised plants arranged equidistantly from each other inside a greenhouse compartment or under a Mylar® or a screen cage. Known numbers of ovipositing females of same physiological stage are allowed to oviposit for a certain period of time. For experiments continuing for more than 1 day, plants are inspected daily for eggs.

Using a preference test, Stadelbacher and Scales (1973) reported significant differences in egg deposition among cotton varieties for *H. zea* and *H. virescens.* Pitre et al. (1983) evaluated eight crop plant species for oviposition egg-laying preference by *S. frugiperda.* Ryegrass, wheat, corn, and sorghum were the most preferred, whereas cotton and soybean were least preferred. Pantoja et al. (1986b) reported significant differences in the oviposition preference of *S. frugiperda* on 25 rice varieties. Ovipositional preference by *D. saccharalis* differed among four sugarcane clones, but was significantly more than on maize, sorghum, and rice (Sosa, Jr. 1990). Other researchers have noted similar results (Stephens 1959, Campbell and Dudley 1965, Ibrahim and Hower, Jr. 1979, McCarty, Jr. et al. 1982, Kim et al. 1985, Fatzinger and Merkel 1985, Wilson et al. 1988, Nottingham et al. 1989, Stadler and Schoni 1990, Severson et al. 1991, Yoshida and Parrella 1991, Ramachandran and Khan 1991).

Ovipositional response to a plant in the greenhouse or laboratory is measured in no-choice trials where insects are allowed contact with only one test species and the rate of oviposition is recorded. This technique measures degree of host acceptance for oviposition. Since an insect's oviposition preference under a choice situation could be influenced by a more attractive plant, the relative nonpreference of a host plant could often be misconstrued for true or genetic resistance (Macfoy et al. 1983). To ascertain the presence of true resistance in a cultivar and not the relative preference existing only in a choice situation, insect ovipositional response in a no-choice situation is often measured. Equal numbers of ovipositing females of the same age are confined under a nylon mesh cage or a Mylar® film cage on each test plant for similar duration in a no-choice situation. After the termination of the experiment, the number of eggs are recorded on each test plant.

Saxena et al. (1974) reported significant differences in ovipositional response of cotton leafhopper, *A. biguttula biguttula,* and a caster leafhopper, *Empoasca kerri motti,* on different varieties of cotton and other plant species. On the other hand, the oviposition response of whitebacked planthoppers was similar on various susceptible and resistant rice varieties (Khan and Saxena 1985b). Schultz and Coffelt (1987) demonstrated the role of leaf pubescence in reducing the ovipositional response of hawthorn lace bug, *Corythucha cydoniae* (Fitch), on certain species of *Cotoneaster* cultivars. Ovipositing gravid females of rice leaffolder, *C. medinalis,* showed less ovipositional response for wild rice plants than for cultivated rice plants (Khan et al. 1989). Ovipositional responses of rice thrip, *Stenchaetothrips biformis* (Bagnall), also differed distinctly on rice varieties with diverse insect-resistant genes (Velusamy and Saxena 1991). Other researchers (Dreyer et al. 1987, Tabashnik 1987, Ramaswamy et al. 1987) have reported similar results.

Comparisons of ovipositional responses in free-choice vs. no-choice tests have revealed varying results from insect to insect and from plant to plant. For the plant bug, *Lygus hesperus* Knight, the number of eggs laid on susceptible glandless and resistant glanded cotton genotypes were essentially the same, whether *L. hesperus* females were given a genotype choice or not (Benedict et al. 1981b). However, *M. testulalis,* which showed a clear oviposition preference for the susceptible cowpea cultivar Vita 1 in a choice test, laid equal numbers of eggs on susceptible and resistant cowpea cultivars in a no-choice test (Macfoy et al. 1983). Sosa Jr. (1988) reported that in a no-choice oviposition test, *D. saccharalis* oviposited significantly fewer eggs on pubescent sugarcane clones compared to clones with glabrous leaves, whereas in a free-choice test, more eggs were laid on pubescent leaves than on glabrous leaves.

Several other examples illustrate the differences in results between different techniques used. Hoffman (1985) was able to duplicate Jaenike's (1982) results with a similar technique, but reported that the results were reversed if a larger testing area was used. Therefore, the technique should depend on the specific question being asked and the biology of the test insect.

2.2.6.3. Detection of Eggs Laid in Plant Tissue

Direct measurements of oviposition and determining the site of egg deposition is difficult in insects that deposit eggs within plant tissue (Khan and Saxena 1986). Generally, a count of sealed oviposition punctures or the numbers of emerged first-instar are considered adequate to reflect the numbers of egg laid (Monteith and Hollowell 1929, Poos and Johnson 1936, Everett and Ray 1962). However, such indirect estimation of ovipositional response can be misleading. More accurate data can be obtained by dissecting the plant part exposed to oviposition under a binocular microscope and counting the deposited eggs. This method is not only laborious and time-consuming but eggs may be inadvertently damaged during dissection. Several techniques developed to make direct counts of eggs in plant tissues are described below.

2.2.6.3.1. Lactophenol Clearing

A clearing lactophenol solution consisting of one part each of 85% lactic acid, phenol, distilled water, and two parts of glycerin is brought to boiling (Carlson and Hibbs 1962). Plant tissue containing insect eggs is then immersed in the boiling solution and held for 3 min. The clearing solution renders plant tissue transparent and egg protein is coagulated, revealing the outline of the egg in situ. Plant tissues submerged in cold lactophenol solution are then viewed under a binocular microscope. Lactophenol clearing has been used extensively to detect oviposition of various homopterans, *A. biguttula biguttula* (Khan and Agarwal 1984), including *E. fabae* (Carlson and Hibbs 1962), and the southern garden leafhopper, *Empoasca solana* DeLong (Moffitt and Reynolds 1972).

2.2.6.3.2. Bleaching

Everett and Trahan (1967) used a simpler technique for detecting the eggs of rice water weevil, *Lissorhoptrus oryzophilus* Kuschel. Rice plants are blanched in hot water for about 5 min, transferred to hot ethyl alcohol and bleached for 24 hr. Plants are then transferred to fresh alcohol for final bleaching and preservation. This method removes chlorophyll from rice plant tissues but prolonged storage of the tissue in alcohol tints weevil eggs with chlorophyll, which makes their differentiation from surrounding plant tissue difficult (Gifford and Trahan 1969).

2.2.6.3.3. Modified Lactophenol Method

A modified lactophenol method which uses acid fuchsin for staining *L. oryzophilus* eggs within rice leaves was developed by Gifford and Trahan (1969). Rice plants are blanched in boiling water for 3 to 5 min and then placed in 95% ethyl alcohol for 1 to 3 days. When the alcohol becomes colored with chlorophyll, it is poured off and the seedling plants are resubmerged in fresh alcohol and held until all the natural pigmentation has been extracted. Placing rice plants in boiling alcohol and storing them in direct sunlight or under high-intensity artificial light speeds up the extraction of chlorophyll and provides lighter colored seedlings.

Bleached plants are washed under tap water and placed in a hot or a cold solution of a stain (1 part each of phenol, lactic acid, and distilled water plus 2 parts glycerine with enough acid fuchsin added to produce a deep violet color). In hot boiling stain, eggs are stained in 15 min, while cold staining requires 24 hr to stain the eggs. Stained plants are washed in tap water to remove excess stain and placed in a 3% KOH solution for clearing. Complete clearing takes 24 to 30 hr depending upon the size of the plant culm and the intensity of stain. After clearing is complete, the plants are washed in tap water and placed in 2% acetic solution for storing. Eggs are observed within plant tissue as dark violet-red objects.

The high cost of lactic acid and phenol, their strong corrosive nature, and the hazards of being exposed to the fumes of phenol are major drawbacks of this method and prohibit large numbers of plant samples from being processed.

2.2.6.3.4. Bleaching and Staining

Khan and Saxena (1986) developed a simple bleaching and staining technique to detect eggs of rice leafhoppers and planthoppers in situ. Plants containing eggs are boiled in water for 5 to 7 min to coagulate the egg yolks and also partially bleach the plant. For further bleaching, boiled plants are kept in 95% ethyl alcohol for 3 days. They are then rinsed in water and immensed in 1% aqueous acid fuchsin solution. After 2 days, the plants are washed under running water until the stained eggs are differentiated from the unstained plant tissue (Figure 2.41). Plant samples can be stored in glycerine for an indefinite period without affecting the color of the stained eggs.

Using this technique, Khan and Saxena (1986) reported that indiscriminate oviposition by *N. lugens, S. furcifera,* and *N. virescens* rendered all test varieties equally susceptible for oviposition but significantly fewer eggs hatched on resistant rice varieties than on susceptible ones.

2.2.7. Egg Hatchability

Adverse effects of resistant cultivars on the hatchability of eggs from insects that oviposit within plant tissue are also important in determining the level of resistance against these pests (Pablo 1977, Saxena and Pathak 1977, Rezaul Karim 1978, Khan and Saxena 1985b, 1986). Measuring egg hatchability is, therefore, necessary for a critical evaluation of complementary resistance factors in the host plant.

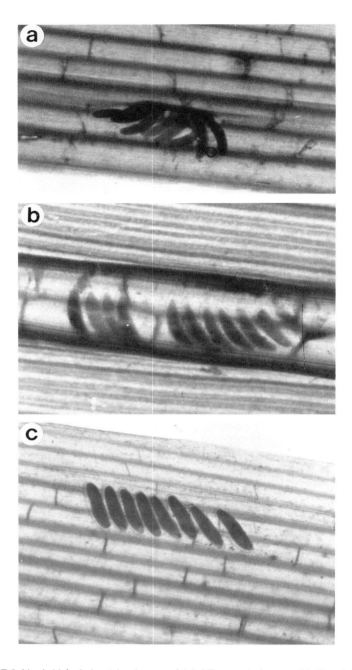

FIGURE 2.41. Acid fuchsin-stained eggs of (a) *Nilaparvata lugens,* (b) *Sogatella fur-cifera,* and (c) *Nephotettix virescens,* within rice plant tissues. (From Khan, Z. R. and Saxena, R. C., *J. Econ. Entomol.,* 79, 271, 1986. With permission of the Entomological Society of America.)

The direct measurement of oviposition and egg hatchability is difficult for those insects that lay their eggs within plant tissues. Generally, after the emergence of first-instar nymphs or larvae, the plant part used by ovipositing females is dissected under the lens of a bionocular microscope and unhatched eggs are counted. However, this method is not always reliable because occasionally eggs are inadvertently damaged during dissection and detecting them subsequently becomes difficult (Khan and Saxena 1986). Egg hatchability can be measured following the method described by Khan and Saxena (1986) for leafhoppers and planthoppers of rice.

For each insect species, plants or plant parts of susceptible and resistant cultivars are exposed to gravid female test insects. The females are allowed to oviposit for 24 hr, after which the insects are removed. The plants are left covered under cages for a period longer than the insect's normal incubation period. The total numbers of nymphs that emerge on the test plants are recorded. After emergence, the plant tissues possibly containing unhatched eggs are bleached and the eggs stained following methods described by Carlson and Hibbs (1962), Gifford and Trahan (1969), Moffitt and Reynolds (1972), or Khan and Saxena (1986) (for detailed methodology, see the section on oviposition). Unhatched eggs can be easily counted under a microscope using this technique.

Rodriguez-Rivera (1972) and Khan and Saxena (1985b) reported that despite indiscriminate egg laying by *S. furcifera* females on both resistant and susceptible rice cultivars, significantly fewer eggs hatched on resistant cultivars. Similarly, Saxena and Pathak (1977) and Rezaul Karim (1978) reported adverse effects of resistant varieties on the egg hatchability of *N. lugens* and *N. virescens,* respectively. On two resistant wild rices, *Oryza punctata* Kotxchy ex Steud. and *Oryza rufipogon* Griff., the egg hatchability of *N. lugens* was significantly reduced (Wu et al. 1986).

2.2.8. Population Increase

The increase of a pest insect population is also an important criterion for assessing the level of insect resistance in a test plant. Insect population increases represent the cumulative effect of an insect's settling, feeding, utilization of ingested food, growth, adult survival, ovipositional rate, and egg hatchability. To determine population increase, known numbers of male and female adults are confined on a test cultivar in the greenhouse or field. The numbers of individual insects recovered from caged plants at the termination of an experiment (30 to 120 days) are considered to be an indication of the overall susceptibility or resistance of each test cultivar. The recovery of a high proportion of adults from a particular cultivar indicates an enhanced developmental rate and, therefore, a form of susceptibility. In contrast, a high proportion of immature forms indicates either delayed oviposition or a reduced developmental rate, which suggests resistance.

To investigate the cumulative effects of varietal resistance in an insect population, the method described by Pathak (1972) for stem borers can be followed. Identical numbers of newly emerged males and females are confined

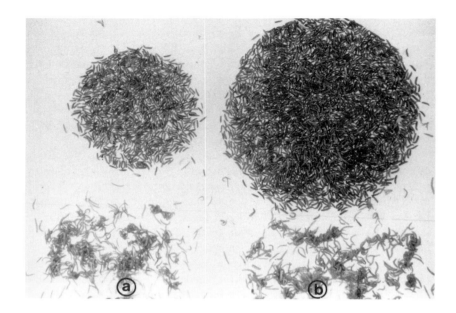

FIGURE 2.42. Population increase of a rice stem borer on resistant (a) and susceptible (b) rice varieties 60 days after infestation. (Photograph provided by M. D. Pathak.)

for several generations on susceptible and resistant rice varieties. The number and emergence of moths and the number of eggs laid on these cultivars are recorded periodically. The plants in each cage are replaced with uninfested healthy plants of the same variety at regular intervals. At 120 days after infestation, Pathak (1972) recorded significantly fewer number of larvae on the resistant Chinan2 rice variety than on the susceptible Sapan Kwai cultivar (Figure 2.42).

Population increase has been used as a criterion for assessing levels of resistance for the *N. virescens* (Cheng and Pathak 1972), *S. furcifera* (Khan and Saxena 1985b, Ye and Saxena 1990), rice thrips, *S. biformis* (Velusamy and Saxena 1991), *N. lugens* on rice (Panda and Heinrichs 1983, Wu et al. 1986), *L. hesperus* on cotton (Leigh et al. 1985), and *Calocoris angustatus* Lethiery on sorghum (Sharma and Lopez 1991). Leigh et al (1985) recorded numbers of adult and immature *L. hesperus* on 52 glandless cotton breeder lines in comparison with a standard glanded variety. Significantly greater numbers of adult bugs were obtained from susceptible lines as compared to resistant lines.

2.3. CONCLUSIONS

As a result of efficient screening techniques, progress in the development of insect resistant cultivars of several crop plants has recently occurred at a much faster than normal rate. The development of numerical damage rating scales has greatly increased the degree of accuracy involved in defining the levels of insect resistance in several crops. Although much progress has been made in developing accurate and efficient techniques to screen and assess plant resistance to insects, more new and improved techniques are still needed. One of the common problems is the lack of artificial diets for several important insect pests. Improved insect infestation techniques and devices that consume less time and minimize damage to insects from handling are also needed. With the increasing importance of resistant cultivars, there is a greater need to develop varieties with multiple resistance. However, success of such a program largely depends upon the availability of sources of resistance, well-planned evaluation techniques, and liberal exchange of cultivars among researchers. International cooperation is also essential in collecting and evaluating germplasm, studying insect biotypes, and identifying the diverse genes for insect resistance.

This chapter also provides a basis for considering the role of the different types of responses that determine the establishment of an insect on different plants. Since certain plants may be more suitable than others for some insect responses and less suitable for other responses, the interaction of these responses will determine the net susceptibility or resistance of a cultivar. According to Saxena et al. (1974), in order to understand the role and interaction of these responses, it is helpful to express their relative intensities as the ratio between the insect response to a test plant and to a standard susceptible plant. Several bioassay techniques for evaluating insect behavior have been discussed. Selection of an appropriate technique will depend on the specific insect and the host plant. These bioassay techniques can be employed on whole plants, plant parts, or fractions of plant tissues.

REFERENCES

Ampofo, J. K. O., *Chilo partellus* (Swinhoe) oviposition on susceptible and resistant maize genotypes, *Insect Sci. Applic.,* 6, 323, 1985.

Ampofo, J. K. O., Effect of resistant maize cultivars on larval dispersal and establishment of *Chilo partellus* (Lepidoptera: Pyralidae), *Insect Sci. Applic.,* 7, 103, 1986.

Ampofo, J. K. O. and Kidiavai, E. L., *Chilo partellus* (Swinhoe) (Lepid., Pyralidae) larval movement and growth on maize plants in relation to plant age and resistance or susceptibility, *J. Appl. Entomol.,* 103, 483, 1987.

Ampofo, J. K. O., Saxena, K. N., Kibuka, J. G. and Nyangiri, E. O., Evaluation of some maize cultivars for resistance of the stemborer *Chilo partellus* (Swinhoe) in Western Kenya, *Maydica,* XXXI, 379, 1986.

Anderson, M. D. and Brewer, G. J., Mechanisms of hybrid sunflower resistance to the sunflower midge (Diptera: Cecidomyiidae), *J. Econ. Entomol.,* 84, 1060, 1991.

Anderson, T. E. and Leppla, N. C., *Advances in Insect Rearing for Research and Pest Management,* Westview Press, 519, 1992.

Armstrong, A. M., Field evaluations of pigeon pea genotypes for resistance against pod borers, *J. Agric. Univ. P. R.,* 75, 73, 1991.

Armstrong, G. and Mordue, W., Oxygen consumption of flying locusts, *Physiol. Entomol.,* 10, 353, 1985.

Auclair, J. L., Honeydew excretion in the pea aphid, *Acyrthosiphon pisum* (Harr.) (Homoptera: Aphididae), *J. Insect Physiol.,* 2, 330, 1958.

Auclair, J. L., Feeding and excretion by the pea aphid, *Acyrthosiphon pisum* (Harr.) (Homoptera: Aphididae), reared on different varieties of peas, *Entomol. Exp. Appl.,* 2, 279, 1959.

Auclair, J. L., Maltais, J. B. and Cartier, J. J., Factors in resistance of peas to the pea aphid, *Acyrthosiphon pisum* (Harr.) (Homoptera: Aphididae). II. Amino acids, *Can. Entomol.,* 89, 457, 1957.

Auclair, J. L., Baldos, E. and Heinrichs, E. A., Biochemical evidence for the feeding sites of the leafhopper *Nephotettix virescens* within susceptible and resistant rice plants, *Insect Sci. Applic.,* 3, 29, 1982.

Backus, E. A., Hunter, W. B. and Arne, C. N., Technique for staining leafhopper (Homoptera: Cicadellidae) salivary sheaths and eggs within unsectioned plant tissue, *J. Econ. Entomol.,* 81, 1819, 1988.

Baker, P. B. and Byers, R. A., A laboratory technique for rearing the clover root curculio, *Melsheimer Entomol. Ser.,* 23, 8, 1977.

Barnes, D. K. and Ratcliffe, R. H., Leaf disk method of testing alfalfa plants for resistance to feeding by adult alfalfa weevils, *J. Econ. Entomol.,* 60, 1561, 1967.

Barnes, D. K., Ratcliffe, R. H. and Hanson, C. H., Interrelationship of three laboratory screening procedures for breeding alfalfa resistant to the alfalfa weevil, *Crop Sci.,* 9, 77, 1969.

Barney, W. P. and Rock, G. C., Consumption and utilization by the Mexican bean beetle of soybean plants varying in levels of resistance, *J. Econ. Entomol.,* 68, 497, 1975.

Bautista, R. C., Heinrichs, E. A. and Rejesus, R. S., Economic injury levels for the rice leaffolder *Cnaphalocrocis medinalis* (Lepidoptera: Pyralidae): insect infestation and artificial leaf removal, *Environ. Entomol.,* 13, 439, 1984.

Beach, R. M. and Todd, J. W., Oviposition preference of the soybean looper (Lepidoptera: Noctuidae) among four soybean genotypes differing in larval resistance, *J. Econ. Entomol.,* 81, 344, 1988.

Beach, R. M., Todd, J. W. and Baker, S. H., Antibiosis of four insect-resistant soybean genotypes to the soybean looper (Lepidoptera: Noctuidae), *Environ. Entomol.* 14, 531, 1985.

Bellotti, A. and Kawano, K., Breeding approaches in cassava, in *Breeding Plants Resistant to Insects,* F. G. Maxwell and P. R. Jennings, eds., John Wiley & Sons, New York, 1980, 683.

Benedict, J. H., Leigh, J. F., Hyer, A. H. and Wynholds, P. F., Nectariless cotton: effect on growth, survival, and fecundity of lygus bugs, *Crop Sci.,* 21, 28, 1981a.

Benedict, J. H., Leigh, T. F., Frazier, J. L. and Hyer, A. H., Ovipositional behavior of *Lygus hesperus* on two cotton genotypes, *Ann. Entomol. Soc. Am.,* 74, 392, 1981b.

Berenbum, M., Postingestive effects of phytochemicals on insects: on *paracelsus* and plant products, in *Insect-Plant Interaction,* J. R. Miller and T. A. Miller, eds., Springer-Verlag, New York, 1986, 342.

Bernays, E. A., Chapman, R. F., Leather, E. M., McCaffery, A. R. and Modder, W. W. D., The relationship of *Zonocerus variegatus* (L.) (Acridoidea: Pyrgomorphidae) with cassava (*Manihot esculenta*), *Bull. Entomol. Res.* 67, 391, 1977.

Bhattacharya, A. K. and Waldbauer, G. P., Quantitative determination of uric acid in feces by lithium carbonate extraction and the enzymatic-spectrophotometric method, *Ann. Entomol. Soc. Am.,* 62, 925, 1969a.

Bhattacharya, A. K. and Waldbauer, G. P., Faecal uric acid as an indicator in the determination of food utilization, *J. Insect Physiol.,* 15, 1129, 1969b.

Bhattacharya, A. K. and Waldbauer, G. P., Use of faecal uric acid method in measuring the utilization of food by *Tribolium confusum, J. Insect Physiol.,* 16, 1983, 1970.

Bhattacharya, A. K. and Waldbauer, G. P., The effect of diet on the nitrogenous end products excreted by larval *Tribolium confusum* with notes on correction of A.D. and E.C.D. for fecal urine, *Entomol. Exp. Appl.,* 15, 238, 1972.

Bing, J. W., Dicke, F. F. and Guthrie, W. D., Genetics of resistance in maize to a complex of three species of thrips (Thysanoptera: Thripidae), *J. Econ. Entomol.,* 83, 621, 1990.

Bintcliffe, E. J. B. and Wratten, S. D., Antibiotic resistance in potato cultivars to the aphid *Myzus persicae, Ann. Appl. Bio.,* 100, 383, 1982.

Birch, A. N. E., A field cage method for assessing resistance to turnip root fly in brassicas, *Ann. Appl. Biol.,* 115, 321, 1989.

Blust, M. H. and Hopkins, T. L., Feeding patterns of a specialist and a generalist grasshopper: electronic monitoring on their host plants, *Physiol. Entomol.,* 15, 261, 1990.

Bowdan, E., An apparatus for the continuous monitoring of feeding by caterpillars in choice, or non-choice tests (automated cafeteria test), *Entomol. Exp. Appl.,* 36, 13, 1984.

Bowdan, E., Microstructure of feeding by tobacco hornworm caterpillars, *Manduca sexta, Entomol. Exp. Appl.,* 47, 127, 1988.

Branson, T. F., Fisher, J. R., Kahler, A. L. and Sutter, G. R., Host plant resistance to corn rootworms, *17th Ann. Illinois Corn Breeders School,* 76, 1981, 110.

Brier, H. B. and Rogers, D. J., Leaf-feeding resistance to six Australian noctuids in soybean, *Crop Prot.,* 10, 320, 1991.

Brown, C. M. and Holbrook, F. R., An improved electronic system for monitoring feeding of aphids, *Am. Potato J.,* 53, 457, 1976.

Burton, R. L., *Mass Rearing the Corn Earworm in the Laboratory,* United States Department of Agriculture, Agriculture Research Service, 134, 1969, 8.

Buscarlet, L. A., The use of ^{22}Na for determining the food intake of the migratory locust, *Oikos,* 25, 204, 1974.

Byers, R. A. and Kendall, W. A., Effects of plant genotypes and root nodulation on growth and survival of *Sitona* spp. larvae, *Environ. Entomol.* 11, 440, 1982.

Cammen, L. M., On the use of liquid scintillation counting of ^{51}Cr and ^{14}C in the twin tracer method of measuring assimilation efficiency, *Oecologia* (Berl.), 30, 249, 1977.

Campbell, W. V. and Dudley, J. W., Differences among *Medicago* species in resistance to oviposition by the alfalfa weevil, *J. Econ. Entomol.,* 58, 245, 1965.

Campbell, B. C., McLean, D. L., Kinsey, M. G., Jones, K. C. and Dreyer, D. L., Probing behavior of the greenbug (*Schizaphis graminum,* biotype C) on resistant and susceptible varieties of sorghum, *Entomol. Exp. Appl.,* 31, 140, 1982.

Cantelo, W. W., Douglass, L. W., Sanford, L. L., Sinden, S. L. and Deahl, K. L., Measuring resistance to the Colorado potato beetle (Coleoptera: Chrysomelidae) in potato, *J. Entomol. Sci.,* 22, 245, 1987.

Carlé, P. and Moutous, G., Observations sur le mode de nutrition sur vigne de quatre espèces de Cicadelles, *Ann. Epiphyt.,* 16, 333, 1965.

Carlson, O. V. and Hibbs, E. T., Direct counts of potato leafhopper, *Empoasca fabae* eggs in *Solanum* leaves, *Ann. Entomol. Soc. Am.,* 55, 512, 1962.

Chalfant, R. B. and Mitchell, E. R., Resistance of peanut varieties to the southern corn rootworm in the field, *J. Econ. Entomol.,* 63, 1825, 1970.

Chang, H. C., Lathoam, R. and Meyer, M. P., Aerial photography: use in detecting simulated insect defoliation in corn, *J. Econ. Entomol.,* 66, 779, 1973.

Chang, N. T., Wiseman, B. R., Lynch, R. E. and Habeck, D. H., Fall armyworm (Lepidoptera: Noctuidae) orientation and preference for selected grasses, *Fla. Entomol.,* 68, 296, 1985.

Chang, V. C. S. and Ota, A. K., Feeding activities of *Perkinsiella* leafhoppers and Fiji disease resistance of sugarcane, *J. Econ. Entomol.,* 71, 297, 1978.

Chang, V. C. S., Feeding activities of the sugarcane leafhopper: identification of electronically recorded waveforms, *Ann. Entomol. Soc. Am.,* 71, 31, 1978.

Chatterji, S. M., Siddiqui, K. H., Panwar, C. G., Sharma, C. G. and Young, W. L., Rearing of the maize stem borer, *Chilo zonellus* (Swinhoe) on artificial diet, *Indian J. Entomol.,* 30, 8, 1968.

Chelliah, S. and Heinrichs, E. A., Factors affecting insecticide-induced resurgence of the brown planthopper, *Nilaparvata lugens* on rice, *Environ. Entomol.,* 9, 773, 1980.

Cheng, C. H. and Pathak, M. D., Resistance to *Nephotettix virescens* in rice varieties, *J. Econ. Entomol.,* 65, 1148, 1972.

Cheung, W. W. K. and Marshall, A. T., Water and ion regulation in cicadas in relation to xylem feeding, *J. Insect Physiol.,* 19, 1801, 1973.

Chou, Y. M., Rock, G. C. and Hodgson, E., Consumption and utilization of chemically defined diets by *Argyrotaenia velutinana* and *Heliothis virescens, Ann. Entomol. Soc. Am.,* 66, 627, 1973.

Dahal G., Hibino, H. and Saxena, R. C., Association of leafhopper feeding behavior with transmission of rice tungro to susceptible and resistant rice cultivars, *Phytopathology,* 80, 371, 1990.

Dahlman, D. L., Effect of L-canavanine on consumption and utilization of artificial diet by tobacco hornworm, *Manduca sexta, Entomol. Exp. Appl.,* 22, 123, 1977.

Dang, K., Anand, M. and Jotwani, M. G., A simple improved diet for mass rearing of sorghum stem borer, *Chilo zonellus* Swinhoe, *Indian J. Entomol.,* 32, 130, 1970.

Daum, R. J., McKibben, G. H., Davich, T. B. and McLaughlin, R., Development of the bait principle for boll weevil control: calco oil red N-1700 dye for measuring ingestion, *J. Econ. Entomol.,* 62, 370, 1969.

Davis, F. M., *Production and Handling of Eggs of the Southwestern Corn Borer for Host Plant Resistance Studies,* United States Department of Agriculture Technical Bulletin, Vol. 74, 1976, 10.

Davis, F. M., A larval dispenser-capper machine for mass rearing the southwestern corn borer, *J. Econ. Entomol.,* 73, 692, 1980a.

Davis, F. M., Fall armyworm resistance program, *Fla. Entomol.,* 63, 420, 1980b.

Davis, F. M., Mechanically removing southwestern corn borer pupae from plastic rearing cups, *J. Econ. Entomol.,* 75, 393, 1982.

Davis, F. M., Simple technique for storing diapausing southwestern corn borers (Lepidoptera: Pyralidae), *J. Econ. Entomol.,* 71, 1171, 1983.

Davis, F. M., Entomological techniques and methodologies used in research programmes on plant resistance to insects, *Insect Sci. Applic.,* 6, 391, 1985.

Davis, F. M. and Oswalt, T. G., Hand inoculator for dispensing lepidopterous larvae, USDA-ARS, ATT-S-9, 1979, 5.

Davis, F. M. and Williams, W. P., Southwestern corn borer: comparison of techniques for infesting corn for plant resistance studies, *J. Econ. Entomol.,* 73, 704, 1980.

Davis, F. M. and Williams, W. P., Survival, growth, and development of southwestern corn borer (Lepidoptera: Pyralidae) on resistant and susceptible maize hybrids, *J. Econ. Entomol.,* 79, 847, 1986.

Davis, F. M., Oswalt, T. G. and Ng, S. S., Improved oviposition and egg collection system for the fall armyworm (Lepidoptera: Noctuidae), *J. Econ. Entomol.,* 78, 725, 1985.

Davis, F. M., Ng, S. S. and Williams, W. P., *Visual Rating Scales for Screening Whorl Stage Corn for Resistance to Fall Armyworm,* Mississippi Agric. and Fores. Exp. Sta. Tech. Bull. 186, 1992, 9.

Diawara, M. M., Wiseman, B. R. and Isenhour, D. J., Mechanism of whorl feeding resistance to fall armyworm among converted sorghum accessions, *Entomol. Exp. Appl.,* 60, 225, 1991.

Diawara, M. M., Wiseman, B. R., Isenhour, D. J. and Hill, N. S., Sorghum resistance to whorl feeding by larvae of the fall armyworm (Lepidoptera: Noctuidae), *J. Agric. Entomol.,* 9, 41, 1992.

Dickson, M. D., Shelton, A. M., Eigenbrode, S. D., Vamosy, M. L. and Mora, M., Selection for resistance to diamondback moth (*Plutella xylostella*) in cabbage, *HortScience,* 25, 1643, 1990.

Dilday, R. H., Methods of screening cotton for resistance to *Heliothis* spp, in *Host Plant Resistance Research Methods for Insects, Diseases, Nematodes and Spider Mites in Cotton,* Southern Cooperative Series Bulletin 280, Mississippi Agricultural and Forestry Experiment Station, 1983, 61.

Dimock, M. B., LaPointe, S. L. and Tingey, W. M., *Solanum neocardensaii:* a new source of potato resistance to the Colorado potato beetle (Coleoptera: Chrysomelidae), *J. Econ. Entomol.,* 79, 1269, 1986.

Dixon, A. G. O., Bramel-Cox, P. J. and Reese, J. C., Feeding behavior of biotype E greenbug (Homoptera: Aphididae) and its relationship to resistance in sorghum, *J. Econ. Entomol.,* 83, 241, 1990a.

Dixon, A. G. O., Bramel-Cox, P. J., Reese, J. C. and Harvey, T. L., Mechanisms of resistance and their interactions in twelve sources of resistance to biotype E greenbug (Homoptera: Aphididae) in sorghum, *J. Econ. Entomol.,* 83, 234, 1990b.

Doggett, H., Starks, K. J. and Eberhart, S. A., Breeding for resistance to the sorghum shootfly, *Crop Sci.,* 10, 528, 1970.

Donovan, L. S., Magee, A. I. and Kalbfleisch, W., A photoelectric device for measurement of leaf areas, *Can. J. Plant Sci.,* 38, 490, 1958.

Downing, N. and Unwin, D. M., A new method for cutting the mouthparts of feeding aphids, and for collecting plant sap, *Physiol. Entomol.,* 21, 275, 1977.

Dreyer, D. L., Jones, K. C. and Jurd, L., Lack of chemical factors in host plant resistance of alfalfa towards alfalfa weevil: Ovicidal activity of some coumarin derivations, *J. Chem. Ecol.,* 13, 917, 1987.

Echendu, T. N. C. and Akingbohungbe, A. E., Intensive free-choice and no-choice cohort tests for evaluating resistance to *Maruca testulalis* (Lepidoptera: Pyralidae) in cowpea, *Bull. Entomol. Res.,* 80, 289, 1990.

Eenink, A. H., Dieleman, F. L., Groenwald, R., Aarts, P. and Clerkx, B., An instant bioassay for resistance of lettuce to the leaf aphid, *Myzus persicae,* *Euphytica,* 33, 825, 1984.

Ellsbury, M. M. and Davis, F. M., Front mounted motorcycle net for mass collection of clover insects, *J. Econ. Entomol.,* 75, 251, 1982.

Escoubas, P., Lajide, L. and Mitzutani, J., An improved leaf-disk antifeedant bioassay and its application for screening of Hokkaids plants, *Entomol. Exp. Appl.,* 66, 99, 1993.

Everett, T. R. and Trahan, G. B., Oviposition by rice water weevils in Louisiana, *J. Econ. Entomol.,* 60, 305, 1967.

Everett, T. R. and Ray, J. O., The utility of sealed punctures for studying fecundity and egg-laying by the boll weevil, *J. Econ. Entomol.,* 55, 634, 1962.

Fatzinger, C. W. and Merkel, E. P., Oviposition and feeding preferences of the southern pine coneworm (Lepidoptera: Pyralidae) for different host-plant materials and observations on monoterpenes as an oviposition stimulants, *J. Chem. Ecol.,* 11, 689, 1985.

Fery, R. L., Cuthbert, F. P. and Perkins, W. D., Artificial infestation of the tomato with eggs of the tomato fruitworm, *J. Econ. Entomol.,* 72, 329, 1979.

Fraenkel, G., Fallil, F. and Kumarasinghe, K. S., The feeding behavior of the rice leaffolder, *Cnaphalocrocis medinalis, Entomol. Exp. Appl.,* 29, 147, 1981.

Frar, D. E. H., Photoelectric apparatus for measuring leaf areas, *Plant Physiol.,* 10, 569, 1935.

Gahukar, R. T., Reaction of locally improved pearl millets to three insect pests and two diseases in Senegal, *J. Econ. Entomol.,* 83, 2102, 1990.

George, B. W., Wilson, F. D. and Wilson, R. L., Methods of evaluating cotton for resistance to pink bollworm, cotton leaf perforator, and lygus bugs, in *Host Plant Resistance Research Methods for Insects, Diseases, Nematodes and Spider Mites in Cotton,* Southern Cooperative Series Bulletin 280, Mississippi Agricultural and Forestry Experiment Station, 1983, 61.

Gifford, J. R. and Trahan, G. B., Staining technique for eggs of rice water weevils oviposited intracellularly in the tissue of the leaf sheaths of rice, *J. Econ. Entomol.,* 62, 740, 1969.

Gilson, W. E., Differential respirometer of simplified and improved design, *Science,* 141, 531, 1963.

Goonewardene, H. F., Kwolek, W. F., Mouzin, T. E. and Williams, E. B., A "no choice" study for evaluating resistance of apple fruits to four insect pests, *Hortscience,* 14, 165, 1979.

Gould, F., Genetics of plant-herbivore systems: interactions between applied and basic study, in *Variable Plants and Herbivores in Natural and Managed Systems,* R. R. Denno and M. S. McClure, eds., Academic Press, New York, 1983, 717.

Guthrie, W. D., Dicke, F. F. and Neiswander, C. R., *Leaf and Sheath Feeding Resistance to the European Corn Borer in Eight Inbred Lines of Dent Corn,* Ohio Agric. Exp. Sta. Res. Bull. 860, 1960, 38.

Guthrie, W. D., Raun, E. S., Dicke, F. F., Pesho, G. R. and Carter, S. W., Laboratory production of European corn borer egg masses, *Iowa State J. Sci.,* 40, 65, 1965.

Guthrie, W. D., Russell, W. A. and Jenning, C. W., Resistance of maize to second brood European corn borers, *Hybrid Corn and Sorghum Conf. Proc.,* 26, 165, 1971.

Guthrie, W. D., Russell, W. A., Reed, G. L., Halbauer, A. R. and Cox, D. F., Methods of evaluating maize for sheath-collar-feeding resistance to the European corn borer, *Maydica,* 23, 45, 1978.

Hagley, E. A. C. and Blackman, J. A., Site of feeding of the sugar froghopper, *Aeneolamia varia saccharia* (Homoptera: Cercopidae), *Ann. Entomol. Soc. Am.,* 59, 1289, 1966.

Haniotakis, G. E. and Lange, W. H., Beet yellow virus resistance in sugar beets: mechanisms of resistance, *J. Econ. Entomol.,* 67, 25, 1974.

Hansen, J. D., Asay, K. H. and Nielson, D. C., Screening range grasses for resistance to black grass bug *Labops hesperius* and *Irbisia pacifica* (Hemiptera: Miridae), *J. Range Manage.,* 38, 254, 1985.

Hansen, J. D., Differential feeding on range grass seedlings by *Irbisia pacifica* (Hemiptera: Miridae), *J. Kansas Entomol. Soc.,* 59, 199, 1986.

Hardie, J., Holyoak, M., Taylor, N. J. and Griffiths, D. C., The combination of electronic monitoring and video-assisted observations of plant penetration by aphids and behavioral effects of polygodial, *Entomol. Exp. Appl.,* 62, 233, 1992.

Hargrove, W. W., A photographic technique for tracking herbivory on individual leaves through time, *Ecol. Entomol.,* 13, 359, 1988.

Hargrove, W. W. and Crossley, Jr., D. A., Video digitizer for the rapid measurement of leaf area lost to herbivorous insects, *Ann. Entomol. Soc. Am.,* 81, 593, 1988.

Harris, G. and MacWilliam, I. C., A dipping technique for revealing sugars on paper chromatograms, *Chem. Ind.,* Feb. 27, 1954.

Hart, S. V., Burton, J. W. and Campbell, W. V., Comparison of three techniques to evaluate advanced breeding lines of soybean for leaf-feeding resistance to corn earworm (Lepidoptera: Noctuidae), *J. Econ. Entomol.,* 81, 615, 1988.

Heinrichs, E. A. and Medrano, F. G., *Leersia hexandra,* a weed host of the rice brown planthopper, *Nilaparvata lugens* (Stål), *Crop Prot.,* 3, 77, 1984.

Heinrichs, E. A. and Pruess, K. P., Chromatogen-ratio method for determining digestibility of plants by grasshoppers, *J. Econ. Entomol.,* 59, 550, 1966.

Heinrichs, E. A., Medrano, F. G. and Rapusas, H. R., *Genetic Evaluation for Insect Resistance in Rice,* International Rice Research Institute, Los Baños, Philippines, 1985, 356.

Hoffman, A. A., Effects of experience on oviposition and attraction in *Drosophila*: comparing apples and oranges, *Am. Nat.,* 126, 41, 1985.

Hollay, M. E., Smith, C. M. and Robinson, J. F., Structure and formation of feeding sheaths of the rice stink bug (Homoptera: Pentatomidae) on rice grains and their association with fungi, *Ann. Entomol. Soc. Am.,* 80, 212, 1987.

Hopkins, R. M., A method for demonstrating sap uptake in the rice brown planthopper, *Nilaparvata lugens,* using a radioactive tracer technique, *Symp. Biol. Hung.,* 39, 479, 1990.

Hsu, S. J. and Robinson, A. G., Resistance of barley varieties to the aphid *Rhopalosiphum padi* (L.), *Can. J. Plant Sci.,* 42, 247, 1962.

Ibrahim, Y. B. and Hower, Jr., A. A., Oviposition preference of the seed corn maggot for various developmental stages of soybeans, *J. Econ. Entomol.,* 72, 64, 1979.

Inayatullah, C., Webster, J. A. and Fargo, W. S., Index for measuring resistance to insects, *The Entomologist,* 109, 146, 1990.

IRRI (International Rice Research Institute), *Standard Evaluation System for Rice,* International Rice Testing Program, IRRI, Los Baños, Laguna Philippines, 1988, 54.

Jackai, L. E. N., A field screening technique for resistance of cowpea (*Vigna unguiculata*) to the pod-borer *Maruca testulalis* (Geyer) (Lepidoptera: Pyralidae), *Bull. Entomol. Res.,* 72, 145, 1982.

Jackai, L. E. N., Laboratory and screenhouse assays for evaluating cowpea resistance to legume pod borer, *Crop Prot.,* 10, 48, 1991.

Jaenike, J., Environmental modifications of oviposition behavior in *Drosophila, Am. Nat.,* 119, 784–802, 1982.

Jensen, R. L., Newsom, L. D., Herzog, D. C., Thomas, Jr., J. W., Farthing, B. R. and Martin, F. A., A method of estimating insect defoliation of soybean, *J. Econ. Entomol.,* 70, 240, 1977.

Johnson, J. W. and Teetes, G. L. , Breeding for arthropod resistance in sorghum, in *Biology and Breeding for Resistance to Arthropods and Pathogens on Agricultural Plants,* M. K. Harris, ed., TAMU Publ., 1979.

Jones, C. G. and Coleman, J. S., Leaf disc size and insect feeding preferences: implications for assays and studies on induction of plant defense, *Entomol. Exp. Appl.,* 47, 167, 1988.

Jones, Jr., W. A. and Sullivan, M. J., Soybean resistance to the southern green stink bug, *Nezara viridula, J. Econ. Entomol.,* 72, 628, 1979.

Kabrick, L. R. and Bakus, E. A., Salivary deposits and plant damage associated with specific probing behaviors of potato leafhopper, *Empoasca fabae,* on alfalfa stems, *Entomol. Exp. Appl.,* 56, 287, 1990.

Kashin, P. and Wakeley, H. G., An insect "biometer", *Nature* (*London*), 208, 462, 1965.

Kasting, R. and McGinnis, A. J., Measuring consumption of food by an insect with carbon-14 labeled compounds, *J. Insect Physiol.,* 11, 1253, 1965a.

Kasting, R. and McGinnis, A. J., Radioisotopes and the determination of nutrient requirement, *Ann. N.Y. Acad. Sci.,* 139, 98, 1965b.

Kawabe, S., Mechanisms of varietal resistance to the rice green leafhopper (*Nephotettix cincticeps* Uhler), *JARQ,* 19, 115, 1985.

Kawabe, S., Fukumorita, T. and Chino, M., Collection of rice phloem sap from stylets of homopterous insects severed by YAG laser, *Plant and Cell Physiol.,* 21, 1319, 1980.

Keister, M. and Buck, J., Respiration: Some exogenous and endogenous effects on rate of respiration, in *The Physiology of Insects,* Vol. 6, 2nd edition, M. Rockstein, ed., Academic Press, New York, 1974, 469.

Kendall, W. A. and Leath, K. T., Slant-board culture methods for root observations of red clover, *Crop Sci.,* 14, 317, 1974.

Kennedy, G. G. and Kishaba, A. N., Response of alate aphids to resistant and susceptible muskmelon lines, *J. Econ. Entomol.,* 70, 407, 1977.

Kennedy, G. G. and Schaefers, G. A., Evidence for nonpreference and antibiosis in aphid resistant red raspberry cultivars, *Environ. Entomol.,* 3, 773, 1974.

Kennedy, G. G., McLean, D. L. and Kinsey, M. G., Probing behavior of *Aphis gossyppi* on resistant and susceptible muskmelon, *J. Econ. Entomol.,* 71, 13, 1978.

Khan, Z. R., Artificial diet for rearing rice leaffolder (LF), *Int. Rice Res. Newsl.,* 12(6), 30, 1987.

Khan, Z. R. and Agarwal, R. A., Ovipositional preference of jassid, *Amrasca biguttula biguttula* Ishida on cotton, *J. Entomol. Res.,* 8, 78, 1984.

Khan, Z. R. and Saxena, R. C., Techniques for demonstrating phloem or xylem feeding by leafhoppers (Homoptera: Cicadellidae) and planthoppers (Homoptera: Delphacidae) in rice plants, *J. Econ. Entomol.,* 77, 550, 1984a.

Khan, Z. R. and Saxena, R. C., Electronically recorded waveforms associated with the feeding behavior of *Sogatella furcifera* (Homoptera: Delphacidae) on susceptible and resistant rice varieties, *J. Econ. Entomol.,* 77, 1479, 1984b.

Khan, Z. R. and Saxena, R. C., Mode of feeding and growth of *Nephotettrix virescens* (Homoptera: Cicadellidae) on selected resistant and susceptible rice varieties, *J. Econ. Entomol.,* 78, 583, 1985a.

Khan, Z. R. and Saxena, R. C., Behavioral and physiological responses of *Sogatella furcifera* (Homoptera: Delphacidae) to selected resistant and susceptible rice cultivars, *J. Econ. Entomol.,* 78, 1280, 1985b.

Khan, Z. R. and Saxena, R. C., Behavior and biology of *Nephotettix virescens* (Homoptera: Cicadellidae) on tungro virus-infected rice plants: epidemiology implications, *Environ. Entomol.,* 14, 297, 1985c.

Khan, Z. R. and Saxena, R. C., Techniques for locating planthopper (Homoptera: Delphacidae) and leafhopper (Homoptera: Cicadellidae) eggs in rice plants, *J. Econ. Entomol.,* 79, 271, 1986.

Khan, Z. R. and Saxena, R. C., Probing behavior of three biotypes of *Nilaparvata lugens* (Homoptera: Delphacidae) on different resistant and susceptible rice varieties, *J. Econ. Entomol.,* 8, 1338, 1988.

Khan, Z. R. and Joshi, R. C., Varietal resistance to *Cnaphalocrocis medinalis* (Guenee) in rice, *Crop Prot.,* 9, 243, 1990.

Khan, Z. R., Ward, J. T. and Norris, D. M., Role of trichomes in soybean resistance to cabbage looper, *Trichoplusia ni, Entomol. Exp. Appl.,* 42, 109, 1986.

Khan, Z. R., Rueda, B. P. and Caballero, P., Behavioral and physiological responses of rice leaffolder, *Cnaphalocrocis medinalis* to selected wild rices, *Entomol. Exp. Appl.,* 52, 7, 1989.

Kim, T. H., Eckenrode, C. J. and Dickson, M. H., Resistance in beans to bean seed maggots (Diptera: Anthomyiidae), *J. Econ. Entomol.,* 78, 133, 1985.

Kogan, M., Fluorescence photography in the quantative evaluation of feeding by phytophagous insects, *Ann. Entomol. Soc. Am.,* 65, 277, 1972a.

Kogan, M., Feeding and nutrition of insects associated with soybeans. 2. Soybean resistance and host preferences of the Mexican bean beetle, *Epilachna varivestis, Ann. Entomol. Soc. Am.,* 65, 675, 1972b.

Kogan, M. and Goeden, D., The host-plant range of *Lema trilineata dauturaphila* (Coleoptera: Chrysomelidae), *Ann. Entomol. Soc. Am.,* 63, 1175, 1970.

Kogan, M. and Ortman, E. E., Antixenosis—a new term proposed to replace Painter's "non-preference" modality of resistance, *Bull. Entomol. Soc. Am.,* 24, 175, 1978.

Kogan, M. and Parra, J. R. P., Techniques and applications of measurement of consumption and utilization of food by phytophagous insects, in *Current Topics in Insect Endocrinology and Nutrition,* G. Bhaskaran, S. Friedman and J. G. Rodriguez, eds., Plenum Press, New York, 1979, 362.

Kogan, M. and Paxton, J., Natural inducers of plant resistance to insects, *ACS Symp. Ser.,* 208, 153, 1983.

Kramer, P. J., An improved photoelectric apparatus for measuring leaf areas, *Am. J. Botany,* 24, 375, 1937.

Krishna, S. S. and Saxena, K. N., Measurement of the quantity of food ingested by insects infesting stored food material, *Naturwissenschaften,* 49, 309, 1962.

Kugler, J. L. and Ratcliffe, R. H., Resistance in alfalfa to a red form of the pea aphid (Homoptera: Aphididae), *J. Econ. Entomol.,* 76, 74, 1983.

Kushwaha, K. S. and Singh, R., Field evaluation of rices for whitebacked planthopper (WBPH) and leaffolder (LF) resistance, *Int. Rice Res. Newsl.,* 11(1), 8, 1986.

Lambert, L., An improved screen-cage design for use in plant and insect research, *Agron. J.,* 76, 168, 1984.

Lambert, L., Jenkins, J. N., Parrott, W. L. and McCarty, J. C., Greenhouse technique for evaluating resistance to the banded winged whitefly (Homoptera: Aleyrodidae) used to evaluate thirty-five foreign cotton cultivars, *J. Econ. Entomol.,* 75, 1166, 1982.

Laster, M. L. and Meredith, M. R., Evaluating the response of cotton cultivars to tarnished plant bug injury, *J. Econ. Entomol.,* 67, 686, 1974.

Lee, J. O., Kim, Y. H., Park, J. S., Seok, S. J. and Goh, H. G., Screening method of varietal resistance to planthoppers labeled with radioisotope ^{32}P(I), *Korean J. Plant Prot.,* 20, 117, 1981a (in Korean).

Lee, J. O., Kim, Y. H., Park, J. S. and Lippold, P. C., Screening method of varietal resistance to planthoppers labeled with radioisotope ^{32}P(II), *Korean J. Plant Prot.,* 20, 173, 1981b (in Korean).

Leigh, T. F., Hyer, A. H., Benedict, J. H. and Wynholds, P. F., Observed population increase, nymphal weight gain, and oviposition nonpreference as indicators of *Lygus hesperus* Knight (Heteroptera: Miridae) resistance in glandless cotton, *J. Econ. Entomol.,* 78, 1109, 1985.

Leuck, D. B., The role of resistance in pearl millet in control of the fall armyworm, *J. Econ. Entomol.,* 63, 1679, 1970.

Linduska, J. J. and Harrison, F. P., Corn earworm: artificial infestation as a means of evaluating resistance in sweet corn, *J. Econ. Entomol.,* 70, 565, 1977.

Lukefahr, M. J., Noble, L. W. and Houghtaling, J. E., Growth and infestation of bollworms and other insects on glanded and glandless strains of cottons, *J. Econ. Entomol.,* 59, 817, 1966.

Lye, B. H. and Smith, C. M., Evaluation of rice cultivars for antibiosis and tolerance resistance to fall armyworm (Lepidoptera: Noctuidae), *Fla. Entomol.,* 71, 254, 1988.

Lyman, J. M. and Cardona, C., Resistance in lima beans to a leafhopper, *Empoasca Kraemeri, J. Econ. Entomol.,* 75, 281, 1982.

Lynch, R. E., Pair, S. D. and Johnson, R., Fall armyworm fecundity: relationship of egg mass weight to number of eggs, *J. Georgia Entomol. Soc.,* 18, 507, 1983.

Lynch, R. E., Branch, W. D. and Garner, J. W., Resistance of *Arachis* species to fall armyworm, *Spodoptera frugiperda, Peanut Sci.,* 8, 106, 1981.

Mansour, M. H., Aboul-Nasr, A. E. and Amr, E. M., Dietary influence of two allelochemicals on larval growth of the spiny bollworm, *Earias insulana* Boisd., *J. Plant Dis. Prot.,* 97, 580, 1990.

Macfarlane, J. H., Damage assessment and yield losses in sorghum due to the stem borer *Busseola fusca* (Fuller) (Lepidoptera: Noctuidae) in northern Nigeria, *Trop. Pest Manage.,* 36, 131, 1990.

Macfoy, C. A., Dabrowski, Z. T. and Okech, S., Studies on the legume pod-borer, *Maruca testulalis* (Geyer)-VI. Cowpea resistance to oviposition and larval feeding, *Insect Sci. Applic.,* 4, 147, 1983.

Manuwoto, S. and Scriber, J. M., Consumption and utilization of three maize genotypes by the southern armyworm, *J. Econ. Entomol.,* 75, 163, 1982.

Masuzawa, T., Swa, H. and Nakasuji, F., Differences of oviposition preference and survival rate of two skipper butterflies *Parnara guttata* and *Pelopidas mathias* (Lepidoptera: Hesperiidae) on rice plant and cogon grass, *New Entomol.,* 32(3), 1, 1983.

McBride, M. C., A method of demonstrating rust hyphae and haustoria in unsectioned leaf tissue, *Am. J. Bot.,* 23, 686, 1936.

McCarty, Jr., J. C., Jenkins, J. N. and Parrott, W. L., Partial suppression of boll weevil oviposition by a primitive cotton, *Crop. Sci.,* 22, 490, 1982.

McCauley, Jr., G. W., Margolies, D. C. and Reese, J. C., Feeding behavior, fecundity and weight of sorghum- and corn-reared greenbugs on corn, *Entomol. Exp. Appl.,* 55, 183, 1990.

McGinnis, A. J. and Kasting, R., Colorimetric analysis of chromic oxide used to study food utilization by phytophagous insects, *Agric. Food Chem.,* 12, 259, 1964a.

McGinnis, A. J. and Kasting, R., Comparison of gravimetric and chromic oxide methods for measuring percentage utilization and consumption of food by phytophagous insects, *J. Insect Physiol.,* 10, 989, 1964b.

McLean, D. L. and Kinsey, M. G., A technique for electronically recording aphid feeding and salivation, *Nature (London),* 202, 1358, 1964.

McLean, D. L. and Kinsey, M. G., Identification of electrically recorded curve patterns associated with aphid salivation and ingestion, *Nature (London),* 205, 1130, 1965.

McLean, D. L. and Weight, Jr., W. A., An electronic measuring system to record aphid salivation and ingestion, *Ann. Entomol. Soc. Am.,* 61, 180, 1969.

Meehan, M. and Wilde, G., Screening for sorghum line and hybrid resistance to chinch bug (Hemiptera: Lygaeidae) in the greenhouse and growth chamber, *J. Econ. Entomol.,* 82, 616, 1989.

Mihm, J. A., Peairs, F. B. and Ortega, A., New procedures for efficient mass production and artificial infestation with lepidopterous pests of maize, *CIMMYT Review,* International Maize and Wheat Improvement Center, El Batan, Mexico, 1978, 138.

Mihm, J. A., Evaluating maize for resistance to tropical stem borers, armyworms, and earworms, in *Toward Insect Resistant Maize for the Third World,* Proceedings of the International Symposium on Methodologies for Developing Host Plant Resistance to Maize Insects, Mexico, D. F. CIMMYT, 1989, 327.

Mihm, J. A., *Techniques for Efficient Mass Rearing and Infestation in Screening for Host Plant Resistance to Corn Earworm,* Heliothis zea, International Maize and Wheat Improvement Center, El Batan, Mexico, 1983a, 16.

Mihm, J. A., *Efficient Mass-Rearing and Infestation Techniques to Screen for Host Plant Resistance for Fall Armyworm,* Spodoptera frugiperda, International Maize and Wheat Improvement Center, El Batan, Mexico, 1983b, 16.

Mihm, J. A., *Efficient Mass Rearing and Infestation Techniques to Screen for Host Plant Resistance to Maize Stem Borers,* Diatraea sp., International Maize and Wheat Improvement Center, El Batan, Mexico, 1983c, 23.

Mitchell, J. W., Measurement of the leaf of attached and detached leaves, *Science,* 83, 334, 1936.

Mize, T. W. and Wilde, G., New grain sorghum sources of antibiosis to the chinch bug (Heteroptera: Lygaeidae), *J. Econ. Entomol.,* 79, 176, 1986.

Moffitt, H. R. and Reynolds, H. T., Bionomics of *Empoasca solana* Delong on cotton in southern California, *Hilgardia,* 41, 247, 1972.

Montandon, R., Stipanovic, R. D., Williams, H. J., Sterling, W. L. and Vinson, S. B., Nutritional indices and excretion of gossypol by *Alabama argillacea* (Hübner) and *Heliothis virescens* (F.) (Lepidoptera: Noctuidae) fed glanded and glandless cotyledonary cotton leaves, *J. Econ. Entomol.,* 80, 32, 1987.

Monteith, J. E. and Hollowell, Pathological symptoms in legumes caused by the potato leafhopper, *J. Agric. Res.,* 38, 649, 1929.

Morgan, C. L., Bluebook of assembly routines for the IBM PC and XT, *New American Library,* New York, 1984.

Mulrooney, J. E., Parrott, W. L. and Jenkins, J. N., Nutritional indices of second-instar tobacco budworm larvae (Lepidoptera: Noctuidae) fed different cotton strains, *J. Econ. Entomol.,* 78, 757, 1985.

Ng, S. S., Davis, F. M. and Williams, W. P., Ovipositional response of southwestern corn borer (Lepidoptera: Pyralidae) and fall armyworm (Lepidoptera: Noctuidae) to selected maize hybrids, *J. Econ. Entomol.,* 83, 1575, 1990.

Ng, S. S., Davis, F. M. and Williams, W. P., Survival, growth, and reproduction of the fall armyworm (Lepidoptera: Noctuidae) as affected by resistant corn genotypes, *J. Econ. Entomol.,* 78, 967, 1985.

Nielson, M. W. and Don, H., Probing behavior of biotypes of the spotted alfalfa aphid on resistant and susceptible alfalfa clones, *Entomol. Exp. Appl.,* 17, 477, 1974.

Nolting, S. P. and Edwards, C. R., Defoliation assessment using video imagery and a microcomputer, *Bull. Entomol. Soc. Am.,* 31, 38, 1985.

Nottingham, S. F., Son, K. C., Wilson, D. D., Severson, R. F. and Kays, S. J., Feeding and oviposition preferences of sweet potato weevil, *Cylas formicarius elegantulus* (Summers), on storage roots of sweet potato cultivars with differing surface chemistries, *J. Chem. Ecol.,* 15, 895, 1989.

Nwanze, K. F. and Reddy, Y. V. R., A rapid method for screening sorghum for resistance to *Chilo partellus* (Swinhoe) (Lepidoptera: Pyralidae), *J. Agric. Entomol.,* 8, 41, 1991.

Okech, S. H. O. and Saxena, K. N., Responses of *Maruca testulalis* (Lepidoptera: Pyralidae) larvae to variably resistant cowpea cultivars, *Environ. Entomol.,* 19, 1792, 1990.

Oliver, B. F. and Gifford, J. R., Weight differences among stalk borer larvae collected from rice lines showing resistance in field studies, *J. Econ. Entomol.,* 68, 134, 1975.

Orchard, B., Measurement of leaf area, *Rep. Rothamsted Exp. Sta.,* 1958, 81.

Ortman, E. E. and Branson, T. F., Growth pouches for studies of host plant resistance to larvae of corn rootworms, *J. Econ. Entomol.,* 69, 380, 1976.

Overman, J. L. and MacCarter, L. E., Evaluating seedlings of cantaloupe for varietal non preference-type resistance to *Diabrotica* spp., *J. Econ. Entomol.,* 65, 1140, 1972.

Pablo, S., Resistance to Whitebacked Planthopper, *Sogatella furcifera (Horváth),* in Rice Varieties, Ph.D. Thesis, Indian Agricultural Research Institute, New Delhi, 1977.

Padgham, D. E. and Woodhead, S., Variety-related feeding patterns in the brown planthopper *Nilaparvata lugens* (Stål) (Hemiptera: Delphacidae), on its host, the rice plant, *Bull. Entomol. Res.,* 78, 339, 1988.

Paguia, P., Pathak, M. D. and Heinrichs, E. S., Honeydew excretion measurement techniques for determining differential feeding activity of biotypes of *Nilaparvata lugens* on rice varieties, *J. Econ. Entomol.,* 73, 35, 1980.

Painter, R. H., *Insect Resistance in Crop Plants,* Macmillan, New York, 1951, 520.

Painter, R. H. and Grandfield, C. O., Preliminary report on resistance of alfalfa varieties to pea aphids (*Illinoia pisi* [Kalt.]), *Am. Soc. Agron. J.,* 27, 671, 1935.

Panda, N., *Principles of Host-Plant Resistance to Insect Pests,* Allanheld, Osmun and Co. and Universe Books, New York, 1979, 386.

Panda, N. and Heinrichs, E. A., Levels of tolerance and antibiosis in rice varieties having moderate resistance to the brown planthopper, *Nilaparvata lugens* (Stål) (Hemiptera: Delphacidae), *Environ. Entomol.,* 12, 1204, 1983.

Pantoja, A., Smith, C. M. and Robinson, J. F., Development of fall armyworm, *Spodeptera frugiperda* (J.E. Smith) (Lepidoptera: Noctuidae), strains from Louisiana and Puerto Rico, *Environ. Entomol.,* 16, 116, 1987.

Pantoja, A., Smith, C. M. and Robinson, J. F., Evaluation of rice plant material for resistance to the fall armyworm (Lepidoptera: Noctuidae), *J. Econ. Entomol.,* 79, 1319, 1986a.

Pantoja, A., Smith, C. M. and Robinson, J. F., Fall armyworm oviposition and egg distribution on rice, *J. Agric. Entomol.,* 3, 114, 1986b.

Parra, J. R. and Kogan, M., Comparative analysis of methods for measurements of food intake and utilization using the soybean looper *Pseudoplusia includens* and artificial media, *Entomol. Exp. Appl.,* 30, 45, 1981.

Pathak, M. D., Stemborer and leafhopper-planthopper resistance in rice varieties, *Entomol. Exp. Appl.,* 12, 789, 1969a.

Pathak, M. D., Resistance to *Nephotettix cincticeps* and *Nilaparvata lugens* in varieties of rice, *Nature,* 223, 502, 1969b.

Pathak, M. D., Resistance to insect pests in rice varieties, in *Rice Breeding,* International Rice Research Institute, Los Baños, Philippines, 1972, 738.

Pathak, M. D. and Saxena, R. C., Breeding approaches in rice, in *Breeding Plants Resistant to Insects,* F. G. Maxwell and P. R. Jennings, eds., John Wiley & Sons, New York, 1980, 683.

Pathak, P. K. and Heinrichs, E. A., Bromocresol green indicator for measuring feeding activity of *Nilaparvata lugens* on rice varieties, *Philipp. Entomol.,* 11, 85, 1982.

Pathak, P. K., Saxena, R. C. and Heinrichs, E. A., Parafilm sachet for measuring honeydew excretion by *Nilaparvata lugens* on rice, *J. Econ. Entomol.,* 75, 194, 1982.

Pedigo, L. P., Stone, J. D. and Clemen, R. B., Photometric device for measuring foliage loss caused by insects, *Ann. Entomol. Soc. Am.,* 63, 815, 1970.

Pencoe, N. L. and Martin, P. B., Fall armyworm (Lepidoptera: Noctuidae) larval development and adult fecundity on five grass hosts, *Environ. Entomol.,* 11, 720, 1982.

Peterson, G. C., Ali, A. E. B., Teetes, G. L., Jones, J. W. and Schaefer, K., Grain sorghum resistance to midge by yield loss vs. visual scores, *Crop Sci.,* 29, 1136, 1989.

Pfannenstiel, R. S. and Meagher, Jr., R. L., Sugarcane resistance to stalkborers (Lepidoptera: Pyralidae) in South Texas, *Fla. Entomol.,* 74, 300, 1991.

Pitre, H. N., Mulrooney, J. E. and Hogg, D. B., Fall armyworm (Lepidoptera: Noctuidae) oviposition: crop preferences and egg distribution on plants, *J. Econ. Entomol.,* 76, 463, 1983.

Poos, F. W. and Johnson, H. W., Injury to alfalfa and red clover by the potato leafhopper, *J. Econ. Entomol.,* 29, 325, 1936.

Powell, G. S., Campbell, W. V., Cope, W. A. and Chamblee, D. S., Ladino clover resistance to the clover root curculio (Coleoptera: Curculionidae), *J. Econ. Entomol.,* 76, 264, 1983.

Prakasa Rao, P. S., Testing for field resistance in rice under induced brown planthopper (BPH) outbreaks, *Int. Rice Res. News.,* 10(4), 5, 1985.

Raina, A. K., Benepal, P. S. and Sheikh, A. Q., Effects of excised and intact leaf methods, leaf size, and plant age on Mexican bean beetle feeding, *Entomol. Exp. Appl.,* 27, 303, 1980.

Ramachandran, R. and Khan, Z. R., Mechanisms of resistance in wild rice, *Oryza brachyantha* to rice leaffolder, *Cnaphalocrocis medinalis* (Guenée) (Lepidoptera: Pyralidae), *J. Chem. Ecol.,* 17, 41, 1991.

Ramaswamy, S. B., Ma, W. K. and Baker, G. T., Sensory cues and receptors for oviposition by *Heliothis virescens, Entomol. Exp. Appl.,* 43, 159, 1987.

Reese, J. C. and Beck, S. D., Effects of plant allelochemics on the black cutworm, *Agrotis ipsilon*; effects of p-benzoquinone, hydroquinone, and duroquinone on larval growth, development, and utilization of food, *Ann. Entomol. Soc. Am.,* 69, 59, 1976a.

Reese, J. C. and Beck, S. D., Effects of plant allelochemics on the black cutworm, *Agrotis ipsilon*; effects of resorcinol, phloroglucinol, and gallic acid on larval growth, development, and utilization of food, *Ann. Entomol. Soc. Am.,* 69, 999, 1976b.

Reese, J. C. and Beck, S. D., Effects of certain allelochemics on the growth and development of the black cutworm, *Symp. Biol. Hung.,* 16, 217, 1976c.

Reynolds, G. W. and Smith, C. M., Effects of leaf position, leaf wounding, and plant age of two soybean genotypes on soybean looper (Lepidoptera: Noctuidae) growth, *Environ. Entomol.,* 14, 475, 1985.

Reynolds, G. W., Smith, C. M. and Kester, K. M., Reduction in consumption, utilization and growth rate of soybean looper (Lepidoptera: Noctuidae) larvae fed foliage of soybean genotype PI227687, *J. Econ. Entomol.,* 77, 1371, 1984.

Rezaul Karim, A. M. N., Varietal Resistance to the Green Leafhopper *Nephotettix virescens* (Distant): Sources, Mechanisms, and Genetics of Resistance, Ph.D. Thesis, University of the Philippines, Los Baños, Philippines, 1978.

Rezaul Karim, A. N. M. and Saxena, R. C., Resistance of rice to green leafhopper (GLH) *Nephotettix virescens* in free-choice and no-choice tests, *Int. Rice Res. Newsl.,* 15(1), 15, 1990.

Risch, S. J., Effects of induced chemical changes on interpretation of feeding preference tests, *Entomol. Exp. Appl.,* 39, 81, 1985.

Robinson, J. F., Klun, J. A. and Brindley, T. A., European corn borer: a non-preference mechanism of leaf feeding resistance and its relationship to 1,4-benzoxazin-3-one concentration in dent corn tissue, *J. Econ. Entomol.,* 71, 461, 1978.

Rodriquez-Rivera, R., Resistance to the Whitebacked Planthopper *Sogatella furcifera* (Horváth) in Rice Varieties, M.S. Thesis, University of the Philippines, Los Baños, Philippines, 1972.

Roof, M. E., Horber, E. and Sorensen, E. L., Evaluating alfalfa cuttings for resistance to the potato leafhopper, *Environ. Entomol.,* 5, 295, 1976.

Rufener II, G. K., Hammond, R. B., Cooper, R. L. and St. Martin, S. K., Mexican bean beetle (Coleoptera: Coccinellidae) development on resistant and susceptible soybean lines in the laboratory and relationship to field selection, *J. Econ. Entomol.,* 79, 1354, 1986.

Rufener II, G. K., Hammond, R. B., Cooper, R. L. and St. Martin, S. K., Larval antibiosis screening technique for Mexican bean beetle resistance in soybean, *Crop Sci.,* 27, 598, 1987.

Sams, D. W., Lauer, F. I. and Radcliffe, E. B., Excised leaflet test for evaluating resistance to green peach aphid in tuber-bearing *Solanum* plant material, *J. Econ. Entomol.,* 68, 607, 1975.

Saxena, K. N., Patterns of insect-plant relationship determining susceptibility or resistance of different plants to an insect, *Entomol. Exp. Appl.,* 12, 751, 1969.

Saxena, K. N., Ovipositional responses of the stem borer *Chilo partellus* (Swinhoe) to certain sorghum cultivars in relation to their resistance or susceptibility, in *Insects-Plants, Proceedings of the 6th International Symposium on Insect-Plant Relationships,* V. Labeyrie, G. Fabres and D. Lachaise, eds., Dr. W. Junk Publishers, Netherlands, 1987, 459.

Saxena, K. N., Mechanisms of resistance/susceptibility of certain sorghum cultivars to the stem borer *Chilo partellus*: role of behavior and development, *Entomol. Exp. Appl.,* 55, 91, 1990.

Saxena, K. N., Gandhi, J. R. and Saxena, R. C., Patterns of relationship between certain leafhoppers and plants. I. Responses to plants, *Entomol. Exp. Appl.,* 17, 303, 1974.

Saxena, R. C. and Khan, Z. R., Comparison between free-choice and no-choice seedling bulk tests for evaluating resistance of rice cultivars to the whitebacked planthopper, *Crop Sci.,* 24, 1204, 1984.

Saxena, R. C. and Khan, Z. R., Electronic recording of feeding behavior of *Cnaphalocrocis medinalis* (Lepidoptera: Pyralidae) on resistant and susceptible rice cultivars, *Ann. Entomol. Soc. Am.,* 84, 316, 1991.

Saxena, R. C. and Pathak, M. D., Factors affecting resistance of rice varieties to the brown planthopper, *N. lugens,* paper presented at the 8th Conf. Pest Control Council Philipp., Bacolod City, Philippines, 18–20 May, 1977.

Schaefers, G. A., The use of direct current for electronically recording aphid feeding and salivation, *Ann. Entomol. Soc. Am.,* 59, 1022, 1966.

Schalk, J. M. and Jones, A., Methods to evaluate sweet potatoes for resistance to the banded cucumber beetle in the field, *J. Econ. Entomol.,* 75, 76, 1982.

Schalk, J. M. and Stoner, A. K., A bioassay differentiates resistance to the Colorado potato beetle on tomatoes, *J. Am. Soc. Hort. Sci.,* 101, 74, 1976.

Scheller, V. H. and Shukle, R. H., Feeding behavior and transmission of barley yellow dwarf virus by *Sitobion avenae* on oats, *Entomol. Exp. Appl.,* 40, 189, 1986.

Schmidt, D. J. and Reese, J. C., Sources of error in nutritional index studies of insect on artificial diet, *J. Insect Physiol.,* 32, 193, 1986.

Schultz, P. B. and Coffelt, M. A., Oviposition and nymphal survival of the hawthorn lace bug (Hemiptera: Tingidae) on selected species of *Cotoneaster* (Rosaceae), *J. Econ. Entomol.,* 16, 365, 1987.

Sell, C. R., Weiss, M. A., Moffitt, H. R., Jr. and Burditt, A. K., An automated technique for monitoring carbon dioxide respired by insects, *Physiol. Entomol.,* 10, 317, 1985.

Severson, R. F., Jackson, D. M., Johnson, A. W., Sisson, V. A. and Stephenson, M. G., Ovipositional behavior of tobacco budworm and tobacco hornworm, in *Naturally Occurring Pest Bioregulators, ACS Symp. Ser. 449,* P. A. Hedin, ed., American Chemical Society, Washington, DC, 1991, 456.

Shanks, Jr., C. H. and Chase, D., Electrical measurement of feeding by the strawberry aphid on susceptible and resistant strawberries and nonhost plants, *Ann. Entomol. Soc. Am.,* 69, 784, 1976.

Shanks, Jr., C. H. and Doss, R. P., Black vine weevil (Coleoptera: Curculionidae) feeding and oviposition on leaves of weevil-resistant and -susceptible strawberry clones presented in various quantities, *Environ. Entomol.,* 15, 1074, 1986.

Sharma, H. C. and Lopez, V. F., Stability of resistance in sorghum to *Calocoris angustatus* (Hemiptera: Miridae), *J. Econ. Entomol.,* 84, 1088, 1991.

Sharma, H. C., Taneja, S. L., Leuschner, K. and Nwanze, K. F., *Techniques to Screen Sorghums for Resistance to Insect Pests,* International Crops Research Institute for the Semi-Arid Tropics, Information Bulletin no. 32, Patancheru, India, 1992, 48.

Sharma, H. C., Vidyasagar, P. and Leuschner, K., Field screening sorghum for resistance to sorghum midge (Diptera: Cecidomyiidae), *J. Econ. Entomol.,* 81, 327, 1988a.

Sharma, H. C., Vidyasagar, P. and Leuschner, K., No-choice cage technique to screen for resistance to sorghum midge (Diptera: Cecidomyiidae), *J. Econ. Entomol.,* 81, 415, 1988b.

Simpson, S. J. and Simpson, C. L., The mechanisms of nutritional compensation by phytophagous insects, in *Insect-Plant Interactions,* Vol. 1, E. A. Bernays, ed., CRC Press, Boca Raton, FL, 1989, 164.

Simpson, S. J., Change in the efficiency of utilization of food through the fifth-instar nymphs of *Locusta migratoria, Entomol. Exp. Appl.,* 10, 443, 1982.

Singer, M. C., The definition and measurement of oviposition preference in plant-feeding insects, in *Insect-Plant Interactions,* J. R. Miller and T. A. Miller, eds., Springer-Verlag, New York, 1986, 342.

Singh, P., *Artificial Diets for Insects, Mites, and Spiders,* Plenum Press, New York, 1977, 594.

Singh, P. and Moore, P. F., *Handbook of Insect Rearing,* Vol. 1, 2, Elsevier, New York, 1985.

Smith, J. J. B. and Friend, W. G., The application of split-screen television recording and electrical resistance measurement to the study of feeding a blood-sucking insect (*Rhodnius prolixus*), *Can. Entomol.,* 103, 167, 1971.

Smith, C. M., *Plant Resistance to Insects, A Fundamental Approach,* John Wiley & Sons, New York, 1989, 286.

Smith, C. M., Khan, Z. R. and Caballero, P., Techniques and methods to evaluate the chemical bases of insect resistance in the rice plant, in *Rice Insects: Management Strategies,* E. A. Heinrichs and T. A. Miller, eds., Springer-Verlag, New York, 1991, 347.

Smith, C. M., Schotzko, D. J., Zemetra, R. S. and Souza, E. J., Categories of resistance in plant introductions of wheat resistant to Russian wheat aphid (Homoptera: Aphididae), *J. Econ. Entomol.,* 85, 1480, 1992.

Smith, Jr., J. W., Sams, R. L., Agnew, C. W. and Simpson, C. E., Methods for estimating damage and evaluating the reaction of selected peanut cultivars to the potato leafhopper, *Empoasca fabae* (Homoptera: Cicadellidae), *J. Econ. Entomol.,* 78, 1059, 1985.

Smith, C. N., *Insect Colonization and Mass Production,* Academic Press, New York, 1966, 618.

Sogawa, K. and Pathak, M. D., Mechanism of brown planthopper resistance in Mudgo variety of rice (Hemiptera: Delphacidae), *Appl. Entomol. Zool.,* 5, 145, 1970.

Soo Hoo, C. F. and Fraenkel, G., The consumption, digestion and utilization of food plants by a polyphagous insect, *Prodenia eridania* (Cramer), *J. Insect Physiol.,* 12, 711, 1966.

Sosa, Jr., O., Pubescence in sugarcane as a plant resistance character affecting oviposition and mobility by the sugarcane borer (Lepidoptera: Pyralidae), *J. Econ. Entomol.,* 81, 663, 1988.

Sosa, Jr., O., Oviposition preference by the sugarcane borer (Lepidoptera: Pyralidae), *J. Econ. Entomol.,* 83, 866, 1990.

Stadelbacher, E. A. and Scales, A. L., Techniques for determining oviposition preference of the bollworm and tobacco budworm for varieties and experimental stocks of cotton, *J. Econ. Entomol.,* 66, 418, 1973.

Städler, E. and Schöni, R., Oviposition behavior of the cabbage root fly, *Delia radicum* (L.), influenced by host plant extracts, *J. Insect Behavior,* 3, 195, 1990.

Stanton, M. L., Searching in a patchy environment: food plant selection by *Colias philodice* butterflies, *Oecologia,* 60, 365, 1982.

Starks, K. J., Increasing infestation of the sorghum shootfly in experimental plots, *J. Econ. Entomol.,* 63, 1715, 1970.

Starks, K. J. and Burton, R. L., *Greenbugs: Determining Biotypes, Culturing, and Screening for Plant Resistance,* USDA, Agric. Res. Service Tech. Bull. 1556, 1977, 12.

Starks, K. J. and Doggett, H., Resistance to a spotted stem borer in sorghum and maize, *J. Econ. Entomol.,* 63, 1790, 1970.

Starks, K. J. and Mirkes, K. S., Yellow sugarcane aphid: plant resistance in cereal crops, *J. Econ. Entomol.,* 72, 486, 1979.

Stephens, S. G., Laboratory studies of feeding and oviposition preference of *Anthonomus grandis* Boh., *J. Econ. Entomol.,* 52, 390, 1959.

Sutter, G. R. and Branson, T. F., A procedure for artificially infesting field plots with corn rootworm eggs, *J. Econ. Entomol.,* 73, 135, 1980.

Tabashnik, B. E., Plant secondary compounds as oviposition deterrents for cabbage butterfly, *Pieris rapae* (Lepidoptera: Pieridae), *J. Chem. Ecol.,* 13, 309, 1987.

Teetes, G. L., Schaefer, C. A., Johnson, J. W. and Rosenow, D. T., Resistance in sorghums to the greenbug field evaluation, *Crop Sci.,* 14, 706, 1974.

Thomas, J. G., Sorensen, E. L. and Painter, R. H., Attached vs. excised trifoliates for evaluation of resistance in alfalfa to the spotted aphid, *J. Econ. Entomol.,* 59, 444, 1966.

Thomas, C. D., Specializations and polyphagy of *Plebejus arqus* (Lep.: Lycaenidas) in Northwest Britain, *Ecol. Entomol.,* 10, 325, 1985.

Thomas, J. G., Mechanisms of Spotted Alfalfa Aphid, *Therioaphis maculata* (Buckton), Resistance in Selected Alfalfa (*Medicago Sativa* L.) Clones, Ph.D. Thesis, Kansas State University, Manhattan, KS, 1970.

Thompson, J. N., Evolutionary ecology of the relationship between oviposition preference and performance of offsprings in phytophagous insects, *Entomol. Exp. Appl.,* 47, 3, 1988.

Tingey, W. M., Techniques for evaluating plant resistance to insects, in *Insect-Plant Interactions,* J. R. Miller and T. A. Miller, eds., Springer-Verlag, New York, 1986, 342.

Tjallingii, W. F., Electronic recording of penetration behavior by aphids, *Entomol. Exp. Appl.,* 24, 521, 1978.

Tjallingii, W. F., Electrical recording of stylet penetration activities, in *Aphids, Their Biology, Natural Enemies and Control,* Vol. 213, A. K. Minks and P. Harrewijn, eds., Elsevier, Amsterdam, 1988, 364.

Traynier, R. M. M. and Hines, E. R., Probes by aphids indicated by stain induced fluorescence in leaves, *Entomol. Exp. Appl.,* 45, 198, 1987.

van Emden, H. F., Vidyasagar, P. and Kazemi, M. H., Use of systemic insecticide to measure antixenosis to aphids in plant choice experiments, *Entomol. Exp. Appl.*, 58, 69, 1991.

van Helden, M., Tjallingii, W. F. and Dieleman, F. L., The resistance of lettuce (*Lactuca sativa* L.) to *Nasonovia ribisnigri*: bionomics of *N. ribisnigri* on near-isogenic lettuce lines, *Entomol. Exp. Appl.*, 66, 53, 1993.

Van Loon, J. J. A., A flow-through respirometer for leaf chewing insects, *Entomol. Exp. Appl.*, 49, 265, 1988.

Van Loon, J. J. A., Measuring food utilization in plant-feeding insects toward a metabolic and dynamic approach, in *Insect-Plant Interactions, Volume III*, E. Bernays, ed., CRC Press, Boca Raton, FL, 1991, 258.

Velusamy, R. and Heinrichs, E. A., Electronic monitoring of feeding behavior of *Nilaparvata lugens* (Homoptera: Delphacidae) on resistant and susceptible rice cultivars, *J. Econ. Entomol.*, 15, 678, 1986.

Velusamy, R. and Saxena, R. C., Genetic evaluation for resistance to rice thrips (Thysanoptera: Thripidae) in leafhopper and planthopper-resistant rice varieties, *J. Econ. Entomol.*, 84, 664, 1991.

Velusamy, R., Heinrichs, E. A. and Medrano, F. G., Greenhouse techniques to identify field resistance to the brown planthopper, *Nilaparvata lugens* (Stål) (Homoptera: Delphacidae), in rice cultivars, *Crop Prot.*, 5, 328, 1986.

Villani, M. G. and Gould, F., Use of radiographs for movement analysis of the corn wireworm, *Melanotus communis* (Coleoptera: Elateridae), *Environ. Entomol.*, 15, 462, 1986.

Visser, J. H., Differential sensory perceptions of plant compounds by insects, in *Plant Resistance to Insects*, P. A. Hedin, ed., *ACS Symp. Ser. 208*, American Chemical Society, Washington, DC, 1983, 375.

Voisey, P. W. and Mason, W. J., Note on an improved device for measuring leaf areas, *Can. J. Plant Sci.*, 43, 247, 1963.

Waldbauer, G. P., The consumption, digestion and utilization of solanaceous and non-solanaceous plants by larvae of the tobacco hornworm, *Protoparce sexta* (Johan) (Lepidoptera: Sphingidae), *Entomol. Exp. Appl.*, 7, 253, 1964.

Waldbauer, G. P., The consumption and utilization of food by insects, in *Advances in Insect Physiology, Vol. 5*, J. W. L. Beament, J. E. Treherne and V. B. Wigglesworth, eds., Academic Press, New York, 1968, 357.

Webster, J. A., Starks, K. J. and Burton, R. L., Plant resistance studies with *Diuraphis noxia* (Homoptera: Aphididae), a new United States wheat pest, *J. Econ. Entomol.*, 80, 944, 1987.

Webster, J. A. and Inayatullah, C., Greenbug (Homoptera: Aphididae) resistance in *Triticale, Environ. Entomol.*, 13, 444, 1984.

Webster, J. A. and Inayatullah, C., Assessment of experimental design for greenbug (Homoptera: Aphididae) antixenosis tests, *J. Econ. Entomol.*, 81, 1246, 1988.

Webster, J. A., Yellow sugarcane aphid (Homoptera: Aphididae): detection and mechanisms of resistance among Ethiopian sorghum lines, *J. Econ. Entomol.*, 83, 1053, 1990.

Widstorm, N. W. and Burton, R. L., Artificial infestation of corn with suspensions of corn earworm eggs, *J. Econ. Entomol.*, 63, 443, 1970.

Wiegert, R. G., The ingestion of xylem sap by meadow spittlebug, *Philaenus spumarius* (L.), *American Midland Naturalist*, 71, 442, 1964.

Wilde, G. and Apostol, R., Armyworm resistance in rice, *Environ. Entomol.*, 12, 376, 1983.

Wilson, D. D., Severson, R. F., Son, K-C and Kays, S. J., Oviposition stimulant in sweet potato periderm for the sweet potato weevil, *Cylas formicarius elegantulus* (Coleoptera: Curculionidae), *Environ. Entomol.*, 17, 691, 1988.

Wiseman, B. R., Williams, W. P. and Davis, F. M., Fall armyworm: resistance mechanisms in selected corns, *J. Econ. Entomol.*, 74, 622, 1981.

Wiseman, B. R., Gueldner, R. C. and Lynch, R. E., Resistance in common centipedegrass to the fall armyworm, *J. Econ. Entomol.*, 75, 245, 1982.

Wiseman, B. R., Davis, F. M. and Williams, W. P., Fall armyworm: larval density and movement as an indication of nonpreference in resistant corn, *Prot. Ecol.*, 5, 135, 1983.

Wiseman, B. R., Hall, C. V. and Painter, R. H., Interactions among cucurbit varieties and feeding responses of the striped and spotted cucumber beetles, *Proc. Am. Soc. Hort. Sci.*, 68, 379, 1961.

Wiseman, B. R. and Widstrom, N. W., Fall armyworm damage ratings on corn at various infestation levels and plant development stages, *J. Agric. Entomol.*, 1, 115, 1984.

Wiseman, B. R., Davis, F. M. and Campbell, J. E., Mechanical infestation device used in fall armyworm plant resistance programs, *Fla. Entomol.*, 63, 425, 1980.

Wiseman, B. R. and Gourley, L., Fall armyworm (Lepidoptera: Noctuidae): infestation procedures and sorghum resistance evaluations, *J. Econ. Entomol.*, 75, 1048, 1982.

Wiseman, B. R. and Widstrom, N. W., Comparison of methods of infesting whorl stage corn with fall armyworm larvae, *J. Econ. Entomol.*, 73, 440, 1980.

Wiseman, B. R. and Davis, F. M., Plant resistance to the fall armyworm, *Fla. Entomol.*, 62, 123, 1979.

Wood, Jr., E. A., Description and results of a new greenhouse technique for evaluating tolerance of small grains to the greenbug, *J. Econ. Entomol.*, 54, 303, 1961.

Wu, J. T., Heinrichs, E. A. and Medrano, F. G., Resistance of wild rices, *Oryza* spp., to the brown planthopper *Nilaparvata lugens* (Homoptera: Delphacidae), *Environ. Entomol.*, 15, 648, 1986.

Ye, Z. H. and Saxena, R. C., Resistance to whitebacked planthopper in elite lines of cultivated x wild rice crosses, *Crop Sci.*, 30, 1178, 1990.

Yoshida, H. A. and Parrella, M. P., Chrysanthemum cultivar preferences exhibited by *Spodoptera exigua* (Lepidoptera: Noctuidae), *Environ. Entomol.*, 20, 160, 1991.

Zemetra, R. S., Schotzko, D. J., Smith, C. M. and Lauver, M., In vitro selection for Russian wheat aphid (*Diuraphis noxia*) resistance in wheat (*Triticum aestivum*), *Plant Cell Rep.*, 12, 312, 1993.

Zhang, Z. T., Heinrichs, E. A. and Medrano, F. G., Seedbox screening tests to determine resistance to brown planthopper (BPH), *Int. Rice Res. Newsl.* 11(2), 10, 1986.

CHAPTER 3

Techniques to Determine Categories
of Plant Resistance

The categories of insect resistance in plants generally accepted and used frequently by researchers in this area of science are those proposed by Painter (1951) of antibiosis and tolerance. Painter's third resistance category, non-preference, has been replaced by the term antixenosis (Kogan and Ortman 1978) (see below).

Much of the effort in a varietal screening program involves the elimination of susceptible plant material. Therefore, large-scale varietal evaluations where insects are offered a free choice of plant materials, either in field plot or greenhouse experiments, are often conducted initially. Materials identified from these tests as potentially resistant are then reevaluated in a smaller group. After resistance in a cultivar is confirmed, the category(s) of resistance is determined in order to fully understand and utilize the resistant cultivar in an integrated pest management program.

To determine which category or combinations of categories are operating in a resistant plant, experiments must be carefully designed that prove or disprove the involvement of each of the three categories. Different experimental test procedures are necessary to differentiate between the antixenosis, antibiosis, and tolerance categories of plant resistance to insects. These experiments may be conducted in a laboratory, in a greenhouse, or in the field. For studying antixenosis and antibiosis, plants require testing under choice and no-choice conditions respectively; for tolerance, they require infested and uninfested conditions (Davis 1985).

3.1. ANTIXENOSIS

Kogan and Ortman (1978) proposed the term antixenosis to describe more accurately the term of nonpreference (Painter 1951) of insects for a resistant plant. Antixenosis, a term derived from the Greek word *xenos* (guest), describes the inability of a plant to serve as a host to an insect herbivore. As a result, insect pests are forced to select an alternate host plant. Both antixenosis and nonpreference denote the presence of morphological or chemical plant factors that adversely alter insect behavior, resulting in selection of an alternate host plant.

To confirm antixenosis resistance, test germplasm is planted together and infested within each experimental replication. In a greenhouse test, different cultivars under evaluation are often planted in a circular arrangement in pots and test insects are released in the center of the test plants. Test insect populations are left on plants until the susceptible control has acquired a heavy population accumulation. Plants are then evaluated for insect population levels, insect feeding damage, and/or oviposition. By identifying resistant plants in a choice test, the researcher is assured the potentially resistant material possesses antixenosis.

Antixenosis tests to measure the settling response of insect larvae or adults are designed to evaluate an insect's choice for intact or excised plant parts. The settling response of a test insect may be observed either after a fixed period ranging from 18 to 24 hr (Wiseman et al. 1982, Khan et al. 1989, Dixon et al. 1990) to 10 days (Soroka and Mackay 1991), or at various intervals ranging from 2 hr to 14 days (Cheng and Pathak 1972, Mize and Wilde 1986, Salifu et al. 1988, Webster et al. 1991). Wiseman et al. (1983) designed tests to evaluate antixenosis in corn plants to the fall armyworm, *Spodoptera frugiperda* (J. E. Smith), by recording larval density on resistant and susceptible maize genotypes at varying intervals after infestation and by determining the degree of larval movement off the resistant genotypes to surrounding susceptible plants (please see Figure 2.15). Ampofo (1986) also studied larval dispersal as an indicator of the degree of antixenosis in corn genotypes to *Chilo partellus* (Swinhoe).

Various feeding choice tests (Connin et al. 1958, Khan et al. 1989, Bodnaryk and Lamb 1991, Yoshida and Parrella 1991) and ovipositional choice tests (Cheng and Pathak 1972, Ellis and Hardman 1975, Olatunde et al. 1991) have been widely used for identifying antixenosis resistance to insects in plants. Techniques for measuring plant resistance in relation to insect orientation, settling, feeding, and oviposition are discussed in detail in Chapter 2.2.

3.2. ANTIBIOSIS

Antibiosis is the category of resistance that includes those adverse effects on the insect life history which result after a resistant host plant is used for food (Painter 1951). Both chemical and morphological plant defenses mediate antibiosis, and antibiotic effects of these resistant plants range from mild to lethal. The effects of antibiosis are measured as the death of early instars, reduced size or low weight, prolonged periods of development of the immature stages, reduced adult longevity and fecundity, death in the prepupal or pupal stage, and abnormal (wandering or restless) behavior.

To identify antibiosis, plant materials are planted, caged, and infested individually in a no-choice situation (Davis 1985). Test insects have no choice but to feed on the plants of the cultivar upon which they are caged. During the course of the development of the insects, antibiosis measurements are recorded in relation to larval survival (Kennedy and Schaefers 1974, Teetes et al. 1974a, Wiseman et al. 1982, Srivastava and Srivastava 1990, Diawara et al. 1992), metabolic utilization of ingested food (Khan and Saxena 1985, Reynolds and Smith 1985, Wu et al. 1986, Khan et al. 1989), larval development (Leuck 1970, Kennedy and Schaefers 1974, Wiseman et al. 1982, Panda and Heinrichs 1983, Mize and Wilde 1986, Salifu et al. 1988, Khan et al. 1989, Diawara et al. 1992), weight increase (Schuster and Starks 1973, Parrott et al. 1978, Raman et al. 1980, Lye and Smith 1988, Srivastava and Srivastava 1990), adult longevity (Cheng and Pathak 1972, Khan and Saxena 1985, Robinson et al. 1991), fecundity (Sotherton and van Emden 1982, Dixon et al. 1990, White 1990, Scott et al. 1991, Soroka and Mackay 1991, Webster et al. 1991), egg hatchability (Khan and Saxena 1985, Wu et al. 1986), and population increase (Teetes et al. 1974b, Bintcliffe and Wratten 1982, Panda and Heinrichs 1983, Wilson and George 1984, Khan and Saxena 1985, Wu et al. 1986).

Inayatullah et al. (1990) calculated the numbers of adult and nymph greenbug, *Schizaphis graminum* (Rondani), on wheat plants daily for 15 days. An antibiosis index was measured as follows:

$$\frac{\text{Number of aphids at greatest density}}{\text{Days required to reach greatest density}}$$

Techniques for measuring plant resistance in relation to insect feeding, metabolic utilization of ingested food, growth, longevity, fecundity, egg hatchability, and population increase on plants, callus tissue and artificial diets are discussed in detail in Chapters 2.2.2 and 2.2.7.

3.3. TOLERANCE

Plants may also be resistant to insect attack by possessing the ability to withstand or recover from damage caused by insect populations equal to those on a susceptible cultivar. According to Painter (1951) tolerance is a "basis of resistance in which the plant possesses an ability to grow and reproduce or to repair injury to a marked degree in spite of supporting a population approximately equal to that damaging a susceptible host." The expression of tolerance is determined by the inherent genetic ability of a plant. Unlike antixenosis and antibiosis, tolerance involves only plant characteristics and is not part of an insect-plant interaction. However, tolerance often occurs in combination with antibiosis and antixenosis. Because of its unique nature in plant resistance to insects, the quantitative measurement of tolerance is accomplished by using entirely different experimental procedures from those used to study antibiosis or antixenosis. Normally, the existence of tolerance is determined by comparing damage, plant stand, and the production of plant biomass or yield in infested and noninfested plants of the same cultivar (Table 3.1). For further reading see the review of Velusamy and Heinrichs (1986). Several different techniques have been used for the quantitative measurement of insect tolerance in crop plants. These techniques have been developed in an attempt to separate tolerance from antibiosis and antixenosis.

3.3.1. Seedling Survival Test

Tolerance among sorghum, wheat, barley, and rye seedlings can be commonly measured by the degree of seedling survival after aphid infestation following a method developed by Wood (1961). Test cultivars are sown in a wooden flat (50 × 35 × 8 cm) with a susceptible and a resistant check variety. Test plants are infested at emergence by uniformly sprinkling approximately 2000 aphids over each flat. The number of days the plants survive is used as a measure of tolerance. Plants destroyed within 2 weeks are considered to be susceptible while others are rated as tolerant. Any cultivar that approaches the level of tolerance of the resistant check is tested again to assure the presence of tolerance. Using the seedling survival test, Wood (1961) reported that several *Triticum vulgare* and *T. durum* varieties were highly tolerant to *S. graminum.*

Mize and Wilde (1986) effectively used a no-choice seedling mortality test to evaluate sorghum lines for tolerance to the chinch bug, *Blissus leucopterus leucopterus* (Say). Ten days after planting, 20 seedlings are exposed to 75 fifth-instar nymphs of *B. leucopterus leucopterus.* Plant mortality counts are made at daily intervals to obtain damage ratings.

3.3.2. Recovery Resistance Test

A recovery resistance tolerance test measures the ability of an injured plant to successfully recover from insect attack and still produce a satisfactory grain yield. Doggett et al. (1970) screened sorghum cultivars for seedling

Table 3.1. Selected Examples of Methods Used to Measure Insect Tolerance in Crop Plants

Crop	Pest	Method(s) of tolerance measurement	Ref.
Alfalfa	*Hypera postica* (Gyllenhal)	Yield	Showalter et al. (1975), Stern et al. (1980)
		Plant damage	Dogger and Hanson (1963)
	Therioaphis maculata (Buckton)	Plant survival	Howe and Pesho (1960)
		Yield	Kindler et al. (1971)
		Plant damage	Berberet et al. (1991)
		Damage index	Jimenez et al. (1988)
	Acyrthosiphon kondoi Shinji	Damage threshold	Bishop et al. (1982)
	Acyrthosiphon pisum (Harris)	Yield	Kindler et al. (1971)
Barley	*Delia flavibasis* Stein	Seedling survival	Macharia and Mueke (1986)
	Schizaphis graminum (Rondani)	Seedling survival	Dahms (1948), Webster and Starks (1984)
		Damage	Starks et al. (1983)
	Diuraphis noxia (Mordvilko)	Tolerance index	Robinson et al. (1991)
		Seedling survival	Webster et al. (1991)
		Damage	Webster et al. (1987)
	Sipha flava (Forbes)	Damage, height	Starks and Mirkes (1979)
	Rhopalosiphum padi (L.)	Seedling survival	Hsu and Robinson (1962)
Corn	*Diabrotica virgifera* LeConte and *Diabrotica longicornis* (Say)	Root pull resistance	Ortman et al. (1968), Rogers et al. (1976), Zuber et al. (1971)
		Root volume	Zuber et al. (1971)
	Blissus leucopterus leucopterus (Say)	Seedling survival	Painter et al. (1935)
	Sipha flava (Forbes)	Damage, height	Starks and Mirkes (1979)
Cotton	*Lygus lineolaris* (Palisot de Beauvois)	Yield	Meredith, Jr. and Laster (1975), Meredith, Jr. and Schuster (1979)
	Earias vittela F.	Plant growth	Sharma and Agarwal (1984)
Cowpea	*Empoasca dolichi* (Paoli)	Plant damage	Raman et al. (1980)
Muskmelon	*Aphis gossyppi* Glover	Plant damage	Bohn et al. (1973)
Mustard	*Phyllotreta cruciferae* (Goeze)	Plant growth	Bodnaryk and Lamb (1991)
Oats	*Diuraphis noxia* (Mordvilko)	Damage	Webster et al. (1987)
	Sipha flava (Forbes)	Height	Starks and Mirkes (1979)
	Schizaphis graminum (Rondani)	Damage	Starks et al. (1983)
Okra	*Amrasca biguttula biguttula* (Ishida)	Plant damage	Teli and Dalaya (1981)
Rice	Stem borers	Plant damage	Das (1976)
	Spodoptera frugiperda (J. E. Smith)	Plant dry weight reduction, plant growth, root volume, root weight	Lye and Smith (1988)

continued

Table 3.1. Continued

Crop	Pest	Method(s) of tolerance measurement	Ref.
	Nilaparvata lugens (Stål)	Yield	Panda and Heinrichs (1983), Ho et al. (1982)
		Plant damage	Wu et al. (1986)
		FPLI	Panda and Heinrichs (1983), Wu et al. (1986)
		Photosynthetic activity	Ho et al. (1982)
		Root volume	Panda and Heinrichs (1983)
	Lissorhoptrus oryzophilus Kuschel	Yield	Grigarick (1984), Tseng et al. (1987)
		Root volume	Oliver et al. (1972)
Sorghum	*Atherigona varia* Meigen	Seedling survival	Doggett et al. (1970)
	Schizaphis graminum (Rondani)	Seedling height	Schuster and Starks (1973)
		Tolerance index	Dixon et al. (1990), Inayatullah et al. (1990)
		Functional plant loss index (FPLI)	Morgan et al. (1980)
		Damage	Wood et al. (1969), Starks et al. (1983)
		Foliage loss	Schweissing and Wilde (1978)
	Sipha flava (Forbes)	Seedling height	Webster (1990), Starks and Mirkes (1979)
	Diuraphis noxia (Mordvilko)	Seedling height	Webster et al. (1991)
	Blissus leucopterus leucopterus (Say)	Seedling survival	Mize and Wilde (1986)
	Chilo partellus (Swinhoe)	Yield	Dabrowski and Kidiavai (1983)
Sugarcane	*Sipha flava* (Forbes)	Leaf discoloration and photosynthetic activity	White (1990)
Triticale	*Diuraphis noxia* (Mordvilko)	Plant height	Scott et al. (1991)
	Schizaphis graminum (Rondani)	Plant damage	Webster and Inayatullah (1984)
	Sipha flava (Forbes)	Damage, height	Starks and Mirkes (1979)
Wheat	*Schizaphis graminum* (Rondani)	Presence of plant growth hormones in honeydew	Maxwell and Painter (1962c)
	Sipha flava (Forbes)	Damage, height	Starks and Mirkes (1979)
	Diuraphis noxia (Mordvilko)	Damage	Webster et al. (1987)
Wheat grass	*Labops hesperius* Uhler	Yield	Hewitt (1980)
	Diuraphis noxia Mordvilko	Plant growth	Smith et al. (1992)
Small grains	*Diuraphis noxia* Mordvilko	Plant growth	Webster et al. (1987)
		Seedling survival	Wood, Jr. (1961)

tolerance to the sorghum shoot fly, *Atherigona varia* Meigen. Doggett et al. (1970) reported that screening sorghum cultivars for seedling resistance to *A. varia* gave inconsistent results, but that screening older plants for recovery resistance was much more reliable. Tolerance was identified in the cultivars Serena and Nematare in which more than 70% of the infested plants recovered and yielded normally.

3.3.3. Plant Height Test

Schuster and Starks (1973), Morgan et al. (1980), Webster (1990), and Webster et al. (1991) measured tolerance to *S. graminum,* to the Russian wheat aphid, *Diuraphis noxia* (Mordvilko), and yellow sugarcane aphid, *Sipha flava* (Forbes), as differences in the heights of infested and uninfested plants of each sorghum cultivar. To measure tolerance, entries are planted separately in pots and 4 days after emergence, plant height from the soil surface to the tip of the longest leaf is measured and each plant is infested with 10 adult aphids. Another set of plants of each entry are left as uninfested controls. Every 2 days, all nymphs are removed and 10 aphids are maintained on each plant. Ten days after infestation, the heights of infested and uninfested plants are measured and the percent differences in plant heights are calculated. Using this method, Schuster and Starks (1973) and Webster (1990) successfully identified sorghums tolerant to *S. graminum* and *S. flava,* respectively.

3.3.4. Visual Ratings

Webster and Starks (1984) developed a technique based on a visual rating scale (1 to 6, where 1 = no damage and 6 = a dead or dying plant) to detect differences among barley seedlings for tolerance to *S. graminum* damage. Young seedlings (50 to 60 cm in height) are infested with 20 aphids per plant and the plants are checked daily to remove or add aphids as needed to maintain 20 per plant. Plants are rated for damage 20 days later. Webster and Starks (1984) used this technique to identify barley with tolerance to *S. graminum* biotype C and biotype E.

Webster and Inayatullah (1984) developed an expanded rating scale, ranging from 1 = no damage to 9 = dead or dying plant, to identify *S. graminum* tolerance in triticale. Webster et al. (1987) also used this rating scale to identify tolerance in wheat, barley, oats, and rye to *D. noxia.* Inayatullah et al. (1990) rated barley plants daily using a 0 to 9 damage scale (0 = no damage, 9 = dead) for tolerance to *S. graminum.* The daily damage ratings are regressed against time, and the slope of the regression line indicates the amount of damage per day and is considered a tolerance index.

Dogger and Hanson (1963) developed a tolerance rating for alfalfa weevil, *Hypera postica* (Gyllenhal), damage to alfalfa in the field, using a numerical grade of 1 to 5, where 1 = no damage and 5 = severe damage. A tolerance rating, based on the relationship between *H. postica* larval populations and stem damage is calculated as:

$$\frac{\text{Number of larvae/10 stems}}{\text{Stem damage rating}}$$

Dogger and Hanson (1963) successfully identified alfalfa cultivars with a relatively low damage rating in spite of having high *H. postica* infestations.

3.3.5. Functional Plant Loss Index (FPLI)

Schweissing and Wilde (1978, 1979), Morgan et al. (1980), Panda and Heinrichs (1983), and Wu et al. (1986) developed and used formulas to assess sorghum and rice tolerance to insect damage. Schweissing and Wilde (1978, 1979) determined the average sorghum foliar loss by obtaining the difference between paired infested and uninfested plants. The effects of different *S. graminum* populations on plant loss of different genotypes, was used to develop the following tolerance formula:

$$\text{mg of dry foliage lost per mg of dry aphid produced} = \frac{\text{mg of dry foliage lost}}{\text{mg of dry aphids}}$$

Panda and Heinrichs (1983) also followed the method of Schweissing and Wilde (1978) to measure the level of tolerance to the brown planthopper, *Nilaparvata lugens* (Stål), by calculating the rice plant dry weight loss per mg of *N. lugens* dry weight produced. The calculation was made as follows: plant dry weight loss/mg of insect dry weight produced = (dry weight of uninfested plant (mg) – dry weight of infested plant) per dry weight of insects produced on infested plants (mg).

Morgan et al. (1980) devised a further measure of sorghum tolerance, the Functional Plant Loss Index (FPLI), that combines measurements of leaf area loss due to stunting with visual *S. graminum* damage ratings. The FPLI is calculated as:

$$1 - \left(\frac{\text{Leaf area of control plant} - \text{leaf area of infested plant}}{\text{Leaf area of control plant}} \right) \times (1 - \text{Average visual damage rating}) \times 100$$

During a short duration test when damage is not severe enough to warrant a damage rating, only leaf area is measured to calculate FPLI, i.e.:

$$\left(\frac{\text{Leaf area of uninfested plant} - \text{leaf area of infested plant}}{\text{Leaf area of control plant}} \right) \times 100$$

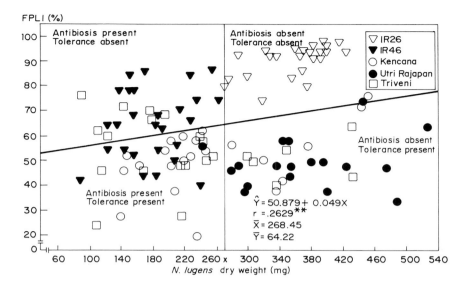

FIGURE 3.1. Identification of the components of resistance to *Nilaparvata lugens* in rice varieties, using *N. lugens* dry weight as the antibiosis indicator and the functional plant loss index (FPLI) as tolerance indicator. (From Panda, N. and Heinrichs, E. A., *Environ. Entomol.,* 12, 1204, 1983. With permission of the Entomological Society of America.)

Panda and Heinrichs (1983) and Wu et al. (1986) determined the levels of tolerance in rice cultivars to *N. lugens* based on plant weight loss and plant height reduction, respectively, using the formulae:

$$\underset{\text{(Plant weight)}}{\textbf{FPLI}} = 1 - \left(\frac{\text{Dry weight of infested plant}}{\text{Dry weight of uninfested plant}} \right) \times \left(1 - \frac{\text{Damage rating}}{9} \right) \times 100$$

$$\underset{\text{(Plant growth)}}{\textbf{FPLI}} = 1 - \left(\frac{\text{Increased height of infested plant}}{\text{Increased height of uninfested plant}} \right) \times \left(1 - \frac{\text{Damage rating}}{9} \right) \times 100$$

Additional detailed methodology for evaluating for tolerance to *N. lugens* under greenhouse conditions is described by Heinrichs et al. (1985).

In addition, Panda and Heinrichs (1983) calculated the regression of FPLI (Y) on *N. lugens* dry weight (X) for each cultivar tested and a pooled regression over all cultivars. The mean value of the independent variable, *N. lugens* dry weight, is then computed and a vertical line is drawn through the graphed data. The intersection of the mean and the pooled regression line is used to separate the components of resistance in the tested cultivars into four categories (Figure 3.1). Utilizing FPLI regression values, Panda and Heinrichs (1983) identified rice tolerance to *N. lugens* biotype 2.

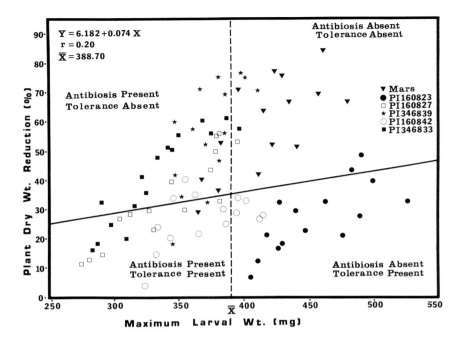

FIGURE 3.2. Relationship between percent rice plant dry weight reduction (tolerance) and maximum larval weight of fall armyworm (antibiosis) fed foliage of five rice plant introductions (PI) and the variety 'Mars' at 5-leaf stage of development. (From Lye, B. H. and Smith, C. M., *Fla. Entomol.*, 71, 254, 1988. With permission.)

The regression analysis technique was also used by Lye and Smith (1988) for partitioning the tolerance and antibiosis resistance components in rice to *S. frugiperda*. Plant dry weight values are plotted against larval weight and the intersection of a line marking the mean maximum larval weight and the regression line forms four quadrants indicating different combinations of tolerance and antibiosis (Figure 3.2). Lye and Smith (1988) reported the rice plant introduction (PI) 160823 to be tolerant to *S. frugiperda* feeding damage.

3.3.6. Tolerance Index

Panda and Heinrichs (1983) partitioned tolerance and antibiosis in rice varieties to *N. lugens* based on *N. lugens* dry weight produced on the test varieties. In plants with tolerance but no antibiosis, increased *N. lugens* weights are as high as on susceptible plants. If resistance is due to antibiosis, *N. lugens* weights will not increase. Tolerance and antibiosis indices are calculated as follows:

$$\textbf{Tolerance index} = \frac{\text{Insect dry weight on test variety}}{\text{Insect dry weight on susceptible check variety}}$$

$$\textbf{Antibiosis index} = 1 - \text{tolerance index}$$

Using the tolerance and antibiosis indices, Panda and Heinrichs (1983) reported that *N. lugens* dry weights on Utri Rajapan, a tolerant variety, are equal to those on susceptible IR26, and have a tolerance index of 1.0. Other tolerant varieties (Kencana and Triveni) had tolerance indices of 0.75 and 0.65 respectively, which are higher than that of resistant IR46 (0.56).

Dixon et al. (1990) measured the tolerance of sorghum cultivars to *S. graminum* based on a weight index. Each of the test entries was planted in two polycast tubes. At 9 days after planting (approximately the 2-leaf stage), 1 of the 2 plants of similar height per entry was caged with 5 adult aphids and the other (control) plant was caged without aphids. The total number of aphids on the infested plants and the above ground dry weights of the control and infested plants were determined 12 days after infestation. The weight index was calculated as:

$$\frac{(CP - IP)/CP \times 100}{TGB}$$

where CP = above ground dry weight of control plant, IP = above ground dry weight of infested plant, and TGB = total number of *S. graminum.* Using this method, Dixon et al. (1990) reported that the sorghum cultivar PI229825 was highly tolerant to *S. graminum.*

Similarly, Robinson et al. (1991) measured tolerance indices in barley genotypes to *D. noxia,* using procedures similar to those of Schweissing and Wilde (1979) and Dixon et al. (1990). Tolerance indices are calculated based on heights and dry weights of infested and uninfested plants as:

Tolerance Index (weight) =

$$\frac{\text{Mean dry weight of control plants (mg)} - \text{Mean dry weight of infested plants (mg)}}{\text{Weight of aphids (mg)} \times \text{Mean dry weight of control plants (mg)}} \times 100$$

Tolerance Index (height) =

$$\frac{\text{Mean height of control plants (cm)} - \text{Mean height of infested plants (cm)}}{\text{Weight of aphids (mg)} \times \text{Mean height of control plants (cm)}} \times 100$$

Based on these indices, Robinson et al. (1991) reported the S12 and S13 barley genotypes to be more tolerant to *D. noxia* than other genotypes tested.

3.3.7. Photosynthetic Activity Test

Ho et al. (1982) compared the carbon dioxide uptake of rice seedlings to determine the reduction of photosynthetic activity in infested tolerant and susceptible rice plants damaged by *N. lugens.* Ten-day-old rice seedlings, grown individually in culture solution within test tubes, are infested with 10 male and 10 female *N. lugens* adults. Dead insects are replaced every 12 hr to maintain a consistent density. Three days after infestation, seedlings showing severe damage symptoms (chlorotic and withered leaves and curled leaf tips) are selected for experiments. Individual infested plants, maintained with culture solution in small glass containers and infested with five *N. lugens* females in a small plastic cage, are transferred to CO_2 leaf chambers (Figure 3.3). Uninfested seedlings with five females in leaf chambers are held as controls. Plants are illuminated with two banks of lights, each composed of nine incandescent 75-W bulbs. Lights are placed 3 m from the plants and left on for 24 hr.

Radioactive $^{14}CO_2$, generated through a reaction between lactic acid and sodium-bi-^{14}carbonate in a generating cylinder, is circulated in a closed system. Honeydew excreted by *N. lugens* females is collected on a filter paper placed on the bottom of the insect cage. After 24 hr, the plants are removed from the leaf chamber and autoradiographed on x-ray film to measure the intensity of ^{14}C incorporation. Radioactivity of ^{14}C in plants, insects, and excreta are read with a scintillation spectrometer.

Ho et al. (1982) reported that although the seedlings suffered severely from damage caused by *N. lugens,* plants of the tolerant cultivar Triveni exhibited a higher CO_2 uptake, indicating higher photosynthetic activity than the susceptible TN1 cultivar. When healthy and damaged plants are compared, the reduction in CO_2 uptake is less in Triveni than in TN1.

White (1990) determined sugarcane tolerance to *S. flava* by the extent of leaf discoloration and reduction in photosynthetic activity. Clip-on cages are placed on sugarcane leaves and five nymphs (4 to 5 days old) are placed in each cage. Cages are removed after 10 days and the degree of leaf discoloration is estimated visually on a scale of 1 (0 to 20% discolored) to 5 (80 to 100% discolored). Reduction in photosynthetic activity as a result of aphid feeding is estimated using a fluorometer, which absorbs a small portion of the light energy intercepted by plant photosynthetic pigments and measures changes in chlorophyll fluorescence due to plant injury. White (1990) reported that leaf discoloration ratings and fluorescence measurement readings were negatively correlated, suggesting that *S. flava* infestation and subsequent discoloration is not a good indicator of reduced photosynthesis.

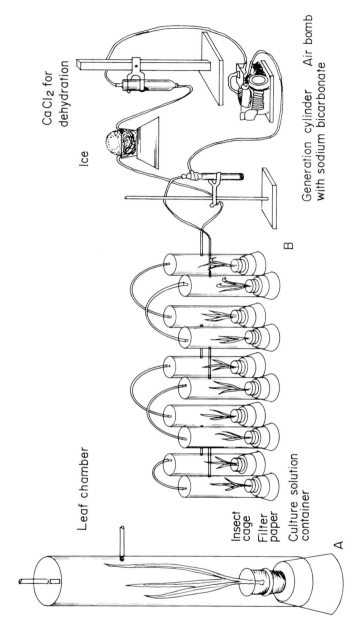

FIGURE 3.3. Schematic diagram of the apparatus used to determine the photosynthetic rate of rice seedlings using $^{14}CO_2$. (From Ho, D. T. et al., *Environ. Entomol.*, 11, 598, 1982. With permission of the Entomological Society of America.)

3.3.8. Auxins Test

Auxins (plant growth hormones) in the honeydew of aphids are found to be correlated with the degree of tolerance of a host plant (Maxwell and Painter 1962c). Aphids can effectively reduce free acid auxin levels in susceptible barley, wheat, and alfalfa plants, but fewer auxins are found in the honeydew of the same alphid species feeding on tolerant plants (Maxwell and Painter 1962c). In auxin tests, honeydew of aphids feeding on susceptible and tolerant plants is collected on aluminum foil strips, washed with 70% ethyl alcohol and filtered. The alcohol filtrate containing the honeydew is then analyzed for the presence of free auxins using paper chromatography following the methods described by Maxwell and Painter (1962a,b).

Maxwell and Painter (1962c) reported a striking relationship between auxins present in the honeydew of aphids feeding on susceptible and tolerant plants. The presence of more auxins in the honeydew of aphids feeding on susceptible plants explains why aphids stunt plant growth, while the presence of fewer auxins in the honeydew of aphids feeding on tolerant plants indicates that growth hormones are not removed by aphids with the sap of tolerant plants.

3.3.9. Vertical Root-Pull Technique

Measurements to identify plants tolerant to root-feeding insects, which compensate for the differences between infested and uninfested plants, can be undertaken using a vertical root-pull technique developed by Ortman et al. (1968), Zuber et al. (1971), and Rogers et al. (1976). The resistance of susceptible and tolerant plant roots is estimated by pulling plants vertically out of the ground, using a lever and clamp device and recording the tension (kg) required to remove the plant from the soil with a dynamometer (Figure 3.4). Before the plants are pulled, they are cut 30 to 40 cm above ground and the leaves are removed from the cut stub. The tripod is positioned over the stub and a grip is attached to the stub. A steadily increasing force is applied to the end of the lever until the root system is lifted from the soil. The maximum force required to remove the root mass from the soil is considered the vertical pull resistance.

Ortman et al. (1968), Zuber et al. (1971), and Rogers et al. (1976) reported that the vertical root pull technique is efficient and useful in obtaining quantitative measurements of tolerance in maize root systems to the corn rootworms, *Diabrotica virgifera* LeConte, *D. longicornis* (Say) and *D. undecimpunctata howardi* Barber.

3.3.10. Root Volume Measurement

Root volume measurements have also been used for evaluating corn and rice for tolerance to *D. virgifera* and the rice water weevil, *Lissorhoptrus oryzophilus* Kuschel (Zuber et al. 1971, Oliver et al. 1972). Plants are dug or pulled from field plots, the soil is washed from the roots, and the root volume

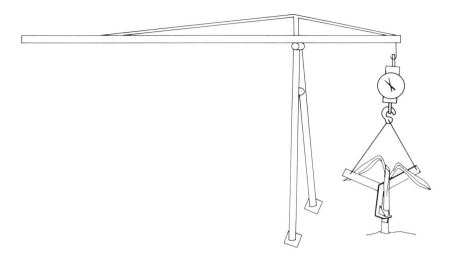

FIGURE 3.4. A dynamometer device used to extract root system from the soil. (From Rogers, R. R. et al., *Crop Sci.,* 16, 591, 1976. With permission of the Crop Science Society of America.)

is determined by the quantity of water displaced by the root mass in a graduated cylinder. The procedure consists of placing excised root masses in a 1000-ml graduated cylinder, adding 500 ml of water, and recording the amount of water displaced. Zuber et al. (1971) and Oliver et al. (1972) reported significant differences in root volumes of infested and control maize and rice plants, respectively. Panda and Heinrichs (1983) reported that the root growth of the susceptible rice cultivar IR26 was adversely affected by low *N. lugens* populations, but high *N. lugens* populations had little effect on the tolerant cultivar Utri Rajapan (Figure 3.5).

3.3.11. Yield

A tolerant cultivar has the ability to produce satisfactory yield without suppressing insect pest populations. This phenomenon of producing a yield higher than a susceptible cultivar is generally cumulative and may be due to general plant vigor, inter- and intra-plant compensatory growth, wound compensation, mechanical strength of tissues and organs, and/or nutrient and growth regulator partitioning (Velusamy and Heinrichs 1986).

Insect tolerance based on crop yield is generally measured by comparing yields under infested and uninfested conditions. Each test cultivar is exposed to similar levels of insect infestation in screenhouses or in field plots under cages. At plant maturity, plots are harvested and the yield losses due to insect infestation are compared.

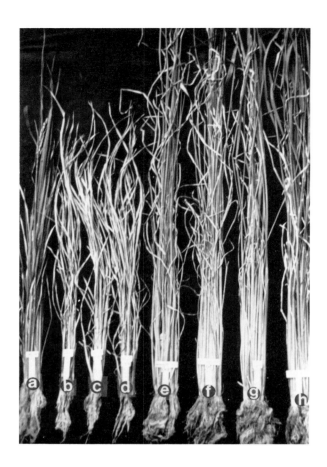

FIGURE 3.5. Roots of IR26 (susceptible) (a–d) and Utri Rajapan (tolerant) (e–g) rice varieties 32 days after initial infestation with 0 (a, h), 25 (b, g), 50 (c, f), and 100 (d, e) *Nilaparvata lugens* nymphs per two plants. (From Panda, N. and Heinrichs, E. A., *Environ. Entomol.,* 12, 1204, 1983. With permission of the Entomological Society of America. Photo courtesy of N. Panda.)

Panda and Heinrichs (1983) evaluated seven rice varieties for their tolerance to *N. lugens.* Thirty-five 21-day-old seedlings are transplanted at a spacing of 15 by 20 cm in 1-m^2 plots at a rate of 1 seedling per hill. Each plot is covered with a fiber glass mesh cage (1.5 × 1.5 × 2.0 m high) (Figure 3.6), and 20 days after transplanting, each cultivar is infested with 1 pair per 2 hills, 1 pair per hill, or 2 pairs per hill of *N. lugens* adults. The *N. lugens* population and plant damage rating are recorded as indicators of antibiosis and

FIGURE 3.6. Tolerant Utri Rajapan and susceptible IR26 rice plants in a field test after initial infestation at a rate of 2 pairs per plant and 0.5 pairs per plant, respectively. IR26 plants were completely hopperburned while Utri Rajapan plants remained healthy. (From Panda, N. and Heinrichs, E. A., *Environ. Entomol.*, 12, 1204, 1983. With permission of the Entomological Society of America. Photo courtesy of N. Panda.)

tolerance, respectively, at 30, 60, and 75 days after infestation. At plant maturity, entire plots are harvested and grain weights are recorded. Percent grain yield reduction due to insect infestation is calculated as:

$$\frac{\text{Yield in uninfested plot} - \text{yield in infested plot}}{\text{Yield in uninfested plot}} \times 100$$

Panda and Heinrichs (1983) reported that at 50 days after infestation, susceptible IR26 plants were completely hopperburned resulting in 100% yield loss, whereas tolerant Utri Rajapan plants had a very low yield reduction in spite of high *N. lugens* populations (Figure 3.6). Similarly Ho et al. (1982) reported only a 40% yield reduction of Triveni, a tolerant rice variety, compared to a 100% reduction on the susceptible variety TN1. Tseng et al. (1987) reported that the rice genotype PI50623, which is tolerant to *L. oryzophilus,* showed only 27% yield reduction due to insect infestation compared to a 67% yield reduction of the susceptible cultivar, M9.

3.4. CONCLUSIONS

A good understanding of the category(s) of resistance must be attained before intelligent decisions can be made about how to use a resistant cultivar in various integrated pest-management schemes. Although it is easy to identify tolerance in a cultivar, there is often an overlap between the antibiosis and antixenosis categories of resistance because of the difficulty involved in designing experiments to delineate between the two. This is especially evident when experiments are conducted with early stages of immature insects (Horber 1980). In these types of experiments, the death of test insects may result from either the toxic factor(s) involved in antibiosis or the strong deterrent factors involved in antixenosis. Quite often, very detailed sets of experiments are required to determine the actual roles of plant factors in each category that contribute to insect resistance. However, it is not unusual to find that combinations of each category contribute to insect resistance. From a practical standpoint, the absolute contribution of a given resistance mechanism may never be fully elucidated.

REFERENCES

Ampofo, J. K. O., Effect of resistant maize cultivars on larval dispersal and establishment of *Chilo partellus* (Lepidoptera: Pyralidae), *Insect Sci. Applic.,* 7, 103, 1986.

Berberet, R. C., McNew, R. W., Dillwith, J. W. and Caddel, J. L., Within-plant patterns of *Therioaphis maculate* on resistant, tolerant, and susceptible alfalfa plants, *Environ. Entomol.,* 20, 551, 1991.

Bintcliffe, E. J. B. and Wratten, S. D., Antibiotic resistance in potato cultivars to the aphid *Myzus persicae, Ann. Appl. Biol.,* 100, 383, 1982.

Bishop, A. L., Walters, P. J., Holtkamp, R. H. and Dominiak, B. C., Relationships between *Acyrthosiphon kondoi* and damage in three varieties of alfalfa, *J. Econ. Entomol.,* 75, 118, 1982.

Bodnaryk, R. P. and Lamb, R. J. Mechanisms of resistance to the flea beetle, *Phyllotreta cruciferae* (Goeze), in mustard seedlings, *Sinapis alba* L., *Can. J. Plant Sci.,* 71, 13, 1991.

Bohn, G. W., Kishaba, A. N., Principe, J. A. and Toba, H. H., Tolerance to melon aphid in *Cucumis melo* L., *J. Am. Soc. Hort. Sci.,* 98, 37, 1973.

Cheng, C. H. and Pathak, M. D., Resistance to *Nephotettix virescens* in rice varieties, *J. Econ. Entomol.,* 65, 1148, 1972.

Connin, R. V., Gorz, H. J. and Gardner, C. O., Greenhouse technique for evaluating sweetclover weevil preference for seedling clover plants, *J. Econ. Entomol.,* 51, 190, 1958.

Dabrowski, Z. T. and Kidiavai, E. L., Resistance of some sorghum lines to the spotted stalkborer *Chilo partellus* under Western Kenya conditions, *Insect Sci. Applic.,* 4, 109, 1983.

Dahms, R. G., Comparative tolerance of small grains to greenbugs from Oklahoma and Mississippi, *J. Econ. Entomol.,* 41, 825, 1948.

Das, Y. T., Cross resistance to stemborers in rice varieties, *J. Econ. Entomol.,* 69, 41, 1976.

Davis, F. M., Entomological techniques and methodologies used in research programmes on plant resistance to insects, *Insect Sci. Appl.,* 6, 391, 1985.

Diawara, M. M., Wiseman, B. R., Isenhour, D. J. and Hill, N. S., Sorghum resistance to whorl feeding by larvae of the fall armyworm (Lepidoptera: Noctuidae), *J. Agric. Entomol.,* 9, 41, 1992.

Dixon, A. G. O., Bramel-Cox, P. J., Reese, J. C. and Harvey, T. L., Mechanisms of resistance and their interactions in twelve sources of resistance to biotype E greenbug (Homoptera: Aphididae) in sorghum, *J. Econ. Entomol.,* 83, 234, 1990.

Dogger, J. R. and Hanson, C. H., Reaction of alfalfa varieties and strains to alfalfa weevils, *J. Econ. Entomol.,* 56, 192, 1963.

Doggett, H., Starks, K. J. and Eberhart, S. A., Breeding for resistance to the sorghum shootfly, *Crop Sci.,* 10, 528, 1970.

Ellis, P. R. and Hardman, J. A., Laboratory methods for studying non-preference resistance to cabbage root fly in cruciferous crops, *Ann. Appl. Biol.,* 79, 253, 1975.

Grigarick, A. A., General control problems with rice invertebrate pests and their control in the United States, *Prot. Ecol.,* 7, 105, 1984.

Heinrichs, E. A., Medrano, F. G. and Rapusas, H. R., *Genetic Evaluation for Insect Resistance in Rice,* International Rice Research Institute, Los Baños, Philippines, 1985, 356.

Hewitt, G. B., Tolerance of ten species of *Agropyron* to feeding by *Labops hesperius, J. Econ. Entomol.,* 73, 779, 1980.

Ho, D. T., Heinrichs, E. A. and Medrano, F., Tolerance of the rice variety Triveni to the brown planthopper, *Nilaparvata lugens, Environ. Entomol.,* 11, 598, 1982.

Horber, E., Types and classification of resistance, in *Breeding Plants Resistant to Insects,* F. G. Maxwell and P. R. Jennings, eds., John Wiley & Sons, New York, 1980, 683.

Howe, W. L. and Pesho, G. R., Influence of plant age on the survival of alfalfa varieties differing in resistance to the alfalfa aphid, *J. Econ. Entomol.,* 53, 142, 1960.

Hsu, S. J. and Robinson, A. G., Resistance of barley varieties to the aphid *Rhopalosiphum padi* (L.,), *Can J. Plant Sci.,* 42, 247, 1962.

Inayatullah, C., Webster, J. A. and Fargo, W. S., Index for measuring plant resistance to insects, *The Entomologist,* 109, 146, 1990.

Jimenez, H. O., Caddel, J. L., and Berberet, R. C., Selection and characterization of tolerance to the spotted alfalfa aphid (Homoptera: Aphididae) in alfalfa, *J. Econ. Entomol.,* 1, 1768, 1988.

Kennedy, G. G. and Shaefers, G. A., Evidence for nonpreference and antibiosis in aphid-resistant red raspberry cultivars, *Environ. Entomol.,* 3, 773, 1974.

Khan, Z. R. and Saxena, R. C., Behavioral and physiological responses of *Sogatella furcifera* (Homoptera: Delphacidae) to selected resistant and susceptible rice cultivars, *J. Econ. Entomol.,* 78, 1280, 1985.

Khan, Z. R., Rueda, B. R. and Caballero, P., Behavioral and physiological responses of rice leaffolder *Cnaphalocrocis medinalis* to selected wild rices, *Entomol. Exp. Appl.,* 2, 7, 1989.

Kindler, S. D., Kehr, W. R. and Ogden, R. L., Influence of pea aphids and spotted alfalfa aphids on the stand, yield of dry matter, and chemical composition of resistant and susceptible varieties of alfalfa, *J. Econ. Entomol.,* 64, 653, 1971.

Kogan, M. and Ortman, E. E., Antixenosis—a new term proposed to replace Painter's "non-preference" modality of resistance, *Bull. Entomol. Soc. Am.,* 24, 175, 1978.

Leuck, D. B., The role of resistance in pearl millet in control of the fall armyworm, *J. Econ. Entomol.,* 63, 1679, 1970.

Lye, B. H. and Smith, C. M., Evaluation of rice cultivars for antibiosis and tolerance resistance to fall armyworm (Lepidoptera: Noctuidae), *Fla. Entomol.,* 71, 254, 1988.

Macharia, M. and Mueke, J. M., Resistance of barley varieties to barley fly *Delia flavibasis* Stein (Diptera: Anthomyiidae), *Insect Sci. Applic.,* 7, 75, 1986.

Maxwell, F. G. and Painter, R. H., Auxin content of extracts of certain tolerant and susceptible host plants to *Toxoptera graminum, Macrosiphum pisi,* and *Therioaphis maculata* and relation to host plant resistance, *J. Econ. Entomol.,* 55, 46, 1962a.

Maxwell, F. G. and Painter, R. H., Plant growth hormones in either extracts of the greenbug, *Toxoptera graminum,* and the pea aphid, *Macrosiphum pisi,* fed on selected tolerant and susceptible host plants, *J. Econ. Entomol.,* 55, 57, 1962b.

Maxwell, F. G. and Painter, R. H., Auxins in honeydew of *Toxoptera graminum, Therioaphis maculata,* and *Macrosiphum pisi,* and their relation to degree of tolerance in host plants, *J. Econ. Entomol.,* 55, 229, 1962c.

Meredith, W. R. Jr. and Laster, M. L., Agronomic and genetic analysis of tarnished plant bug tolerance in cotton, *Crop Sci.,* 15, 535, 1975.

Meredith, W. R. Jr. and Schuster, M. F., Tolerance of glamorous and pubescent cottons to tarnished plant bug, *Crop Sci.,* 19, 484, 1979.

Mize, T. W. and Wilde, G., New resistant germplasm to the chinch bug (Homoptera: Lygaeidae) in grain sorghum: contribution of tolerance and antixenosis as resistance mechanisms, *J. Econ. Entomol.,* 79, 42, 1986.

Morgan, J., Wilde, G. and Johnson, D., Greenbug resistance in commercial sorghum hybrids in the seedling stage, *J. Econ. Entomol.,* 73, 510, 1980.

Olatunide, G. O., Odebiyi, J. A. and Jackai, L. E. N., Cowpea antixenosis to the pod sucking bug, *Clavigralla tomentosicollis* Stål (Hemiptera: Coreidae), *Insect Sci. Applic.,* 12, 449, 1991.

Oliver, B. F., Gifford, J. R. and Trahan, G. B., Studies of the differences in root volume and dry root weight of rice lines where rice water weevil were controlled and not controlled, *Ann. Prog. Rep. Rice Res. Sta.,* LAES, LSU Agric. Ctr., 64, 212, 1972.

Ortman, E. E., Peters, D. C. and Fitzgerald, P. J., Vertical-pull technique for evaluating tolerance of corn root systems to northern and western corn rootworms, *J. Econ. Entomol.,* 61, 373, 1968.

Painter, R. H., *Insect Resistance in Crop Plants,* Macmillan, New York, 1951, 520.

Painter, R. H., Snelling, R. O. and Brunson, A. M., Hybrid vigor and other factors in relation to chinch bug resistance in corn, *J. Econ. Entomol.,* 28, 1025, 1935.

Panda, N. and Heinrichs, E. A., Levels of tolerance and antibiosis in rice, cultivars having moderate resistance to the brown planthopper, *Nilaparvata lugens* (Stål) (Hemiptera: Delphacidae), *Environ. Entomol.,* 12, 1204, 1983.

Parrott, W. L., Jenkins, J. N., McCarthy, J. C., Jr. and Lambert, L., A procedure to evaluate antibiosis in cotton to the tobacco budworm, *J. Econ. Entomol.,* 71, 310, 1978.

Raman, K. V., Singh, S. S. and van Emden, H. F., Mechanisms of resistance to leafhopper damage in cowpea, *J. Econ. Entomol.,* 73, 484, 1980.

Reynolds, G. W. and Smith, C. M., Effects of leaf position, leaf wounding, and plant age off two soybean genotypes on soybean looper (Lepidoptera: Noctuidae) growth, *Environ. Entomol.,* 14, 475, 1985.

Robinson, J., Vivar, H. E., Burnett, P. A. and Calhoun, D. S., Resistance to Russian wheat aphid (Homoptera: Aphididae) in barley genotypes, *J. Econ. Entomol.,* 84, 674, 1991.

Rogers, R. R., Russell, W. A. and Owens, J. C., Evaluation of a vertical-pull technique in population improvement of maize for corn rootworm tolerance, *Crop Sci.,* 16, 591, 1976.

Salifu, A. B., Singh, S. R. and Hodgson, C. J., Mechanism of resistance in cowpea (*Vigna unguiculata* (L.) Walp.) genotype, TVX3236, to bean flower thrips, *Megalurothrips sjostedts* (Trybom) (Thysanoptera: Thripidae). X. Non preference and antibiosis, *Trop. Pest Manage.,* 34, 185, 1988.

Schuster, D. J. and Starks, K. J., Greenbugs: Components of host-plant resistance in sorghum, *J. Econ. Entomol.,* 66, 1131, 1973.

Schweissing, F. C. and Wilde, G., Temperature influence on greenbug resistance of crops in the seedling stage, *Environ. Entomol.,* 7, 831, 1978.

Schweissing, F. C. and Wilde, G., Temperature and plant nutrient effects on resistance of seedling sorghum to the greenbug, *J. Econ. Entomol.,* 172, 20, 1979.

Scott, R. A., Worrall, W. D. and Frank, W. A., Screening for resistance to Russian wheat aphid in triticale, *Crop Sci.,* 31, 32, 1991.

Sharma, H. C. and Agarwal, R. A., Factors imparting resistance to stem damage by *Earias vittella* F. (Lepidoptera: Noctuidae) in some cotton phenotypes, *Prot. Ecol.,* 6, 35, 1984.

Showalter, A. H., Pienkowski, R. L. and Wolf, D. D., Alfalfa weevil: host response to larval feeding, *J. Econ. Entomol.,* 68, 619, 1975.

Smith, C. M., Schotzko, D. J., Zemetra, R. S. and Souza, E. J., Categories of resistance in plant introductions of wheat resistant to the Russian wheat aphid (Homoptera: Aphididae), *J. Econ. Entomol.,* 85, 1480, 1992.

Soroka, J. J. and Mackay, P. A., Antibiosis and antixenosis to pea aphid (Homoptera: Aphididae) in cultivars of field peas, *J. Econ. Entomol.,* 84, 1951, 1991.

Sotherton, N. W. and van Emden, H. F., Laboratory assessments of resistance to the aphids *Sitobion avenae* and *Metopolophium dirhodum* in three *Triticum* species and two modern wheat cultivars, *Ann. Appl. Biol.,* 101, 99, 1982.

Srivastava, C. P. and Srivastava, R. P., Antibiosis in chickpea (*Cicer arietinum* L.) to gram pod borer, *Heliothis armigera* (Hübner) (Noctruidae: Lepidoptera) in India, *Entomon,* 15, 89, 1990.

Starks, K. J. and Mirkes, K. A., Yellow sugarcane aphid: plant resistance in cereal crops, *J. Econ. Entomol.,* 72, 486, 1979.

Starks, K. J., Burton, R. L. and Merkle, O. G., Greenbugs (Homoptera: Aphididae) plant resistance in small grains and sorghum to biotype E, *J. Econ. Entomol.,* 76, 877, 1983.

Stern, V. M., Sharma, R. and Summers, C., Alfalfa damage from *Acyrthosiphon kondoi* and economic threshold studies in southern California, *J. Econ. Entomol.,* 73, 145, 1980.

Teetes, G. L., Schaefer, C. A. and Johnson, J. W., Resistance in sorghums to the greenbug; laboratory determination of mechanisms of resistance, *J. Econ. Entomol.,* 67, 393, 1974a.

Teetes, G. L., Schaefer, C. A., Johnson, J. W. and Rosenow, D. T., Resistance in sorghums to the greenbug: field evaluation, *Crop Sci.,* 14, 706, 1974b.

Teli, V. S. and Dalaya, V. P., Varietal resistance in okra to *Amrasca biguttula biguttula* (Ishida), *Indian J. Agric. Sci.,* 51, 729, 1981.

Tseng, S. T., Johnson, C. W., Grigarick, A. A., Rutger, J. N. and Carnahan, H. L., Registration of short stature, early maturing, and water weevil tolerant plant material lines of rice, *Crop Sci.,* 27, 1320, 1987.

Velusamy, R. and Heinrichs, E. A., Tolerance in crop plants to insect pests, *Insect Sci. Applic.,* 7, 689, 1986.

Webster, J. A., Yellow sugarcane aphid (Homoptera: Aphididae): detection and mechanisms of resistance among Ethiopian sorghum lines, *J. Econ. Entomol.,* 83, 1053, 1990.

Webster, J. A. and Inayatullah, C., Greenbug (Homoptera: Aphididae) resistance in triticale, *Environ. Entomol.,* 13, 444, 1984.

Webster, J. A. and Starks, K. J., Sources of resistance in barley to two biotypes of the greenbug *Schizaphis graminum* (Rondani), Homoptera: Aphididae, *Prot. Ecol.,* 6, 51, 1984.

Webster, J. A., Baker, C. A. and Porter, D. R., Detection and mechanisms of Russian wheat aphid (Homoptera: Aphididae) resistance in barley, *J. Econ. Entomol.,* 84, 669, 1991.

Webster, J. A., Starks, K. J. and Burton, R. L., Plant resistance studies with *Diuraphis noxia* (Homoptera: Aphididae), a new United States wheat pest, *J. Econ. Entomol.,* 80, 944, 1987.

White, W. H., Yellow sugarcane aphid (Homoptera: Aphididae) resistance mechanisms in selected sugarcane cultivars, *J. Econ. Entomol.,* 83, 2111, 1990.

Wilson, F. D. and George, B. W., Pink bollworm (Lepidoptera; Gelechiidae): selecting for antibiosis in artificially and naturally infested cotton plants, *J. Econ. Entomol.,* 77, 720, 1984.

Wiseman, B. R., Gueldner, R. C. and Lynch, R. E., Resistance in common centipedegrass to fall armyworm, *J. Econ. Entomol.,* 75, 245, 1982.

Wiseman, B. R., Davis, F. M. and Williams, W. P., Fall armyworm: larval density and movement as an indication of nonpreference in resistant corn, *Prot. Ecol.,* 5, 135, 1983.

Wood, E. A. Jr., Description and results of a new greenhouse technique for evaluating tolerance of small grains to the greenbug, *J. Econ. Entomol.,* 54, 303, 1961.

Wood, E. A. Jr., Chada, H. L., Weibel, D. E. and Davies, F. F., *A Sorghum Variety Highly Tolerant to the Greenbug, Schizaphis graminum* (Rond.), Progress Report P-614, Agricultural Research, Oklahoma State University, 1969, 7.

Wu, J. T., Heinrichs, E. A. and Medrano, F. G., Resistance of wild rices, *Oryza* spp., to the brown planthopper, *Nilaparvata lugens* (Homoptera: Delphacidae), *Environ. Entomol.,* 15, 648, 1986.

Yoshida, H. A. and Parrella, M. P., Chrysanthemum cultivar preferences exhibited by *Spodoptera exigua* (Lepidoptera: Noctuidae), *Environ. Entomol.,* 20, 160, 1991.

Zuber, M. S., Musick, G. J. and Fairchild, M. L., A method of evaluating corn strains for tolerance to the western corn rootworm, *J. Econ. Entomol.,* 64, 1514, 1971.

CHAPTER 4

Techniques and Methods to Evaluate the Allelochemical Bases of Insect Resistance

Plant allelochemicals greatly influence the behavior and physiology of insects and are very important in imparting resistance or susceptibility to insects in plants. Allelochemicals are classified as "allomones" and "kairomones." The allomones are associated with repellence, feeding deterrence, toxicity, or other adverse effects on insects. The kairomones may serve as attractants and feeding or ovipositional stimulants. It is therefore important to clearly understand the roles of these chemicals in imparting resistance or susceptibility to insects and to use them as cues in breeding resistant crop varieties. However, determination of these factors is often difficult to comprehend because adequate bioassays are necessary before chemical differences can be identified as the cause of resistance.

This chapter has been prepared to serve as a comprehensive overview of the entomological techniques developed during the past 40 years and to demonstrate how each has been utilized to understand the roles of both volatile and nonvolatile allelochemicals in imparting insect resistance in plants.

4.1. ISOLATION AND CHARACTERIZATION OF ALLELOCHEMICALS

Allelochemicals are largely involved in plant resistance to insect attack. Any study of the chemical basis of resistance requires a hand-in-hand evaluation of biological activity toward an insect and sequential chemical isolation and purification. Initial experiments are conducted with crude extracts that are rather complex in composition. As isolation and purification methods are followed and the biological activity of plant extract fractions is monitored, the quantity and complexity of fractions are reduced. Very frequently, natural product organic chemists face the challenging problem of being able to develop methodologies that purify only minute amounts of active allelochemicals.

4.1.1. Purification

Although there are many methods for the separation and purification of organic compounds based on their chromatographic properties, only gas chromatography and liquid chromatography will be briefly discussed here.

4.1.1.1. Gas Chromatography

Gas chromatography (GC) is the technique most widely used for the purification and identification of stable and volatile compounds (Littlewood 1970, Crippen 1973). Crude purification can be accomplished (mg scale) with the use of prepacked columns. Fractions are normally obtained with relatively high purity and good reproducibility is frequently achieved. This methodology has been extensively used in the field of insect pheromones (Heath and Tumlinson 1984), drugs and metabolites (Gudzinowicz et al. 1976, 1977), essential oils (Masada 1976), steroid hormones (Wotiz and Clark 1966), and polymers (Berezkin et al. 1977).

With improved capillary columns and the technological advances in the design of equipment, gas chromatography–mass spectrometry (GC-MS) has been most widely used to characterize volatile organic compounds (Pierce et al. 1978, Visser et al. 1979, Withycombe et al. 1978, Stubbs et al. 1985, Maeshima et al. 1985, Phelan and Lin 1991, Lin and Phelan 1991, Haynes et al. 1991). For a more comprehensive methodology on the application of GC-MS, readers are referred to Masada (1976), Gudzinowicz et al. (1976, 1977), Odham et al. (1984), and Massage (1984).

The addition of fourier transform infrared spectroscopy (FTIR) to the GC-MS offers additional information for the characterization of minute volatile compounds (Charpentier et al. 1986, Sebeido et al. 1987, Pu et al. 1988, Chyau and Wu 1989). In general, GC-FTIR-MS provides more accurate results, since the chemical characterization is based on two different spectroscopic techniques that complement each other. For a detailed description of FTIR, the reader is referred to Mackenzie (1988).

4.1.1.2. High Performance Liquid Chromatography

High performance liquid chromatography (HPLC) is the most important technique to isolate labile nonvolatile active compounds (Altman et al. 1989, Benincasa et al. 1990, King et al. 1990). New developments in this area are being made to isolate biologically active compounds at the preparative scale. As in the case of GC, the coupling of HPLC with MS provides a powerful technique to characterize active nonvolatiles in small amounts (Voyksner et al. 1990). Moreover, new developments in soft ionization techniques for MS, such as field desorption (FD) and fast atom bombardment (FAB) (Busch and Cooks 1982), are making possible the analysis of unstable high molecular weight compounds in the form of positive or negative ions. In general, applications of HPLC to studies of insect resistance have been directed to the identification of flavonoids, terpenoids and cyclic hydroxamic acids (Besson et al. 1985, Kim et al. 1985, Stipanovic et al. 1988, Lyons et al. 1988, Altman et al. 1989). For a detailed description of HPLC the reader is referred to Henschen (1985) and Hancock (1990).

4.1.2. Interpretation

Characterization of new compounds is made by interpretation of spectral data. The most common techniques used by phytochemists are ultraviolet spectroscopy (UV) (Jaffe and Orchin 1962, Demchenko 1986), infrared spectroscopy (IR) (Stewart 1970, Messerschmidt and Harthcock 1988), mass spectroscopy (MS) (Roboz 1968, Schlunegger 1980), fourier transform nuclear magnetic resonance (FT-NMR) (Mullen and Pregosin 1976, Shaw 1984), and in cases where the sample is in the form of crystalline solids, X-ray crystallography has been used (Jaffery 1971, Duniz 1979). Of these techniques, FT-NMR provides most of the information needed for structure elucidation. New computer technology has made possible the development of new areas of two-dimensional NMR (2D-NMR), including heteronuclear and homonuclear experiments. Multipulse experiments (Turner 1984) are now possible and have tremendously simplified the assignment of structures to complex molecules; this was almost impossible to accomplish in the past. For a more comprehensive and simple summary on the applications of FT-NMR and other techniques to characterize biologically active compounds, the reader is referred to Kubo and Hanke (1976).

4.2. EXTRACTION OF PLANT ALLELOCHEMICALS

4.2.1. Extraction of Volatile Plant Allelochemicals

Volatile secondary plant chemicals play a decisive role in determining host plant selection by insects. Insects are either lured by these chemicals to a susceptible plant for food, oviposition site, and shelter or are repelled away from a resistant plant. Evaluating and identifying the volatile compounds mediating the interactions between an insect and its host plant may have potential in insect management strategies. Plant volatiles are generally chemicals of low to moderate molecular weights (<250) and low boiling points (20° to 340° C). These include alcohols ($<C_{16}$), aldehydes ($<C_{16}$), ketones ($<C_{16}$), acids ($<C_{16}$), esters ($<C_{14}$), etc. The following section includes both extraction and evaluation methods for volatile plant allelochemicals which impart resistance or susceptibility to insect pests.

FIGURE 4.1. Set up for collecting steam distillates from soybean leaves. a, vapor generator flask; b, flask holding plant sample; c, flask for collection of distillates. (Photograph provided by Z. R. Khan.)

4.2.1.1. Steam Distillation

Plant volatiles can be extracted for bioassay as steam distillates following the methods described by Saxena and Okech (1985), Shaw et al. (1985), Khan and Saxena (1986), Andersen et al. (1987), Muzika et al. (1990), Takeoka et al. (1990), Tatsuka et al. (1990), and Gabel et al. (1992). According to Saxena and Okech (1985) and Khan and Saxena (1986), plant parts are ground with an electric grinder. A 200-g ground sample is placed in a 2000-ml two-neck flask (Figure 4.1). A vapor generator, consisting of a 3000-ml flask, filled with 2000 ml water, is connected to one neck of the flask with the ground sample. The other neck of the flask is attached to the distillate collector. During operation, the sample container temperature is kept at 40° C and the vapor generator temperature at 100° C. The sample is steam distilled for 3 hr, during which 900 ml of distillate is collected. The distillate is extracted 3 times with diethyl ether (300 ml distillate: 100 ml diethyl ether) by thoroughly shaking the mixture in a separatory funnel for 5 min. Diethyl ether absorbs the essential oils and other volatiles and the mixture settles above the water layer in the funnel. The water layer is then discarded and the ether extract is pooled in a 1000-ml Erlenmeyer flask and left to dry for 12 hr with 100 g of anhydrous sodium sulfate. The solution is filtered and the diethyl ether is evaporated under reduced pressure on a rotary evaporator. The resulting yellow oily residue is redissolved in a small amount of diethyl ether, and transferred to a preweighted glass vial. Finally, the ether is evaporated with a gentle stream of purified nitrogen. The vial is reweighed, sealed with nitrogen, and kept at $-10°$ C. The residue is diluted in acetone to desired concentrations for various bioassays. Shaw et al. (1985), Andersen et al. (1987), Muzika et al. (1990), and Takeoka et al. (1990) used reduced pressure steam distillation to collect volatiles from test plants.

4.2.1.2. Molecular Distillation

Steam distillates of plants are obtained at very high temperatures which may destroy volatile compounds. With molecular, or low-temperature-high-vacuum distillation, most volatile compounds are efficiently recovered in sufficient quantity to conduct phytochemical bioassays (Hibbard and Bjostad 1988). However, recovery of compounds for chemical characterization is low because the extract is collected with large amounts of water and sometimes requires an additional extraction with organic solvents.

Molecular distillation involves condensing the air itself, along with any volatile organic compound in the air. Molecular distillation allows nitrogen (B.P. $-196°$ C) and oxygen (B.P. $-183°$ C) to be removed after the collection is completed, while plant volatiles with boiling points higher than $-183°$ C are retained.

Generally, fresh plants or plant parts are cut in pieces and placed in a 500-ml glass container connected to one side of a "U" tube (10 cm long). On the other end, a glass thimble is used to collect the distillate. A high-vacuum valve is connected 5 cm above the thimble. After assembling, the sample container is immersed for 20 min in a dry ice/acetone bath connected to a high-vacuum line and evacuated until the inside pressure reaches 10^{-3} mmHg. The high-vacuum valve is closed, the dry ice/acetone bath transferred to the extract collector, and the apparatus is left to equilibrate at room temperature for 24 hr. Volatile compounds emanating from the foliage sample are collected by convection resulting from the difference in temperature between the sample container and the extract collector. For biological evaluation, the extract is used as it is obtained or diluted to the desired concentration. For chemical analysis, the extract is saturated with sodium chloride and extracted 3 times with 2 ml of HPLC grade dichloromethane. The organic layer is dried with 100 mg of anhydrous sodium sulfate for 2 hr, filtered, concentrated to 5 µl with a stream of purified nitrogen, and analyzed by gas chromoatography-mass spectrometry.

4.2.1.3. Dynamic Headspace Collection

An extremely accurate method to characterize low amounts of volatile allelochemicals involves the use of dynamic headspace collection (DHC). DHC has been used to concentrate and analyze volatiles in polluted air and water and to successfully elucidate chromatographic profiles of potato (Visser 1983), apple (Fein et al. 1982), red clover (Kamm and Buttery 1984), soybean and lima bean (Liu et al. 1988), rice (Hernandez et al. 1989), maize (Hibbard and Bjostad 1990), and various other graminaceous plants (Hopkins and Young 1990).

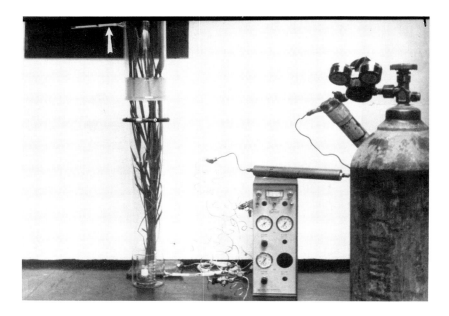

FIGURE 4.2. Set up for collection of volatiles from rice plant by dynamic headspace sampling. The arrow shows a glass column packed with Tenax TA for collection of volatiles. (Photo courtesy P. Caballero.)

The major disadvantage of DHC is that the amount of sample collected is sufficient only for chemical characterization. Thus, the biological activity of crude extracts may not be determined. However, molecular distillation and DHC techniques are complementary to each other and can be used jointly to measure biological activity.

To collect volatiles for DHC analysis, plants or plant parts are placed in a 150-ml beaker containing enough water to moisten the roots. The beaker is placed in a glass chamber (75 cm high × 10 cm inside diameter, I.D.) (Figure 4.2), and purified air or nitrogen gas is passed through the bottom of the chamber at a flow rate of 80 ml to 1 l/min. Volatiles are collected for 24 hr at 25° C in a trap located at the top of the chamber. The trap is made of a glass column the same dimensions of the injector liner (78 mm long × 4 mm I.D. × 6 mm outside diameter, O.D.) of a 5890 Hewlett Packard Gas Chromatograph. The trap is packed with approximately 100 mg of Tenax TA (2,6-diphenyl-p-phenylene oxide polymer) 60–80 Mesh column packing. After collection, the trap is purged with purified nitrogen to remove condensed water. The Tenax trap is then placed in the injector port of a GC-MS at 40° C in the splitless mode. The injector port is heated to 185° C for 25 min and desorbed volatiles are trapped in a 1-m long empty capillary coil immersed in a dry ice/acetone mixture. After desorption, the oven temperature is programmed from 40 to 240° C at the rate of 2° C/min.

Volatiles are chromatographed at the following conditions: carrier gas, He; head pressure, 15 psi; carrier gas velocity, 25 cm/s; carrier gas flow rate, 0.76 ml/min; column temperatures, 40° C for 5 min (initial), 2° C/min (rate), to 175° C for 15 min (final). The chromatographed volatiles were then analyzed by GC-MS with an electron ionization voltage of 70 eV and an electron multiplier voltage of 1800 V.

Trapped volatiles are eluted from the Tenax with HPLC grade hexane or diethyl ether for 2 min for bioassay. The resultant extractables are held at −10° C for 2 hr, filtered, weighed, and stored at −20° C for bioassays (Blust and Hopkins 1987b, Liu et al. 1988).

4.2.2. Extraction of Nonvolatile Plant Allelochemicals

Nonvolatile allelochemicals play a vital role in host selection by insects for feeding and oviposition. These chemicals are designated as feeding stimulants (phagostimulant) or oviposition stimulants; those that inhibit feeding or oviposition are referred to as feeding deterrents, antifeedants or feeding inhibitors, or oviposition deterrents or inhibitors. Some allelochemicals are acute insect toxins (Todd et al. 1971, Holyoke and Reese 1987) while others act as growth inhibitors (Elliger et al. 1981, Binder and Waiss, Jr. 1984, Isman et al. 1989), ovicides (Pouzat 1978), or egg hatching inhibitors (Saxena and Puma 1979, Dreyer et al. 1987a).

Different methods have been used to extract nonvolatile allelochemicals from susceptible and resistant plants. The following are the extraction methods commonly used for nonvolatile allelochemicals.

4.2.2.1. Sequential Solvent Extraction

Plant tissues, weighed upon removal from the plants, are placed directly into an organic solvent. Allelochemicals can be extracted from plant tissue following any of several methods, including those described by Elliger et al. (1981), Binder and Waiss, Jr. (1984), Khan et al. (1986), Huang and Mack (1989), Waladde et al. (1990), Dimock et al. (1991), Ramachandran and Khan (1991), Torto et al. (1991), and Wheeler and Slanksy, Jr. (1991). Plant tissues are often sequentially extracted in a series of solvents, ranging from nonpolar to polar.

Elliger et al. (1981) extracted growth inhibitors in tomato affecting *Heliothis zea* (Boddie). One hundred grams of tomato leaves are ground with 800 ml of acetone for 5 min. The mixture is filtered and the residue is extracted again with acetone. The extracted residue was further extracted with 800 ml methanol and water, respectively.

Binder and Waiss, Jr. (1984) and Khan et al. (1986) extracted allelochemicals of resistant soybean leaves with solvents of increasing polarity such as hexane, isooctane, ethyl acetate, acetone, methanol, and water.

Khan et al. (1986) homogenized soybean leaves with 60% methanol (1:10, wt/vol) in a dry ice-cooled blender for 4 min. Extractables were filtered through Whatman no. 1 filter paper and the plant residue was mechanically

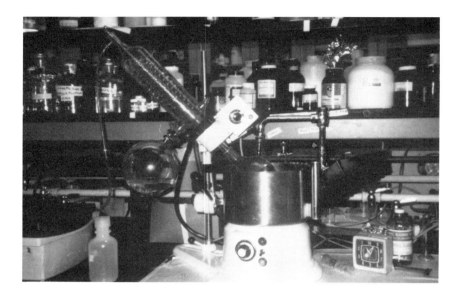

FIGURE 4.3. Set up for rotoevaporating nonvolatile allelochemicals to dryness. (Photograph provided by Z. R. Khan.)

shaken in an excess quantity of 60% methanol for 24 hr with 3 changes of the solvent (i.e., every 8 hr). All methanol extractables from a leaf sample were finally combined and rotoevaporated to dryness (Figure 4.3). Dry methanol extractables were subsequently extracted sequentially in 200 ml each of hexane, diethyl ether, ethyl acetate, 100% methanol, and distilled water. After each extraction, residues were removed by filtration using Whatman no. 1 filter paper in a Buchner funnel. Solubles in each organic solvent were rotoevaporated to dryness and the water-solubles were lyophylized to dryness. Resultant fractions were stored in sealable glass containers under nitrogen in darkness at –20° C. Diethyl ether, ethyl acetate, hexane, and water extractables were redissolved in diethyl ether, ethyl acetate, hexane, and water, respectively, at a known concentration and methanol extractables were redissolved in 80% methanol.

Similarly, phagostimulants for *Elasmopalpus lignosellus* (Zeller) from peanut plants were extracted sequentially (Huang and Mack 1989). Plant materials were homogenized with methanol and water (4:1) and the residue was extracted with ethyl acetate chloroform, chloroform methanol (3:1), and methanol and water.

Diethyl ether, chloroform ethanol (8:2), methanol, and water were used to extract nonvolatile allelochemicals from corn plants to identify antifeedants and feeding stimulants against *Chilo partellus* (Swinhoe) (Waladde et al. 1990). Citrus plants have been extracted in chloroform and isobutanol to identify *Papilio protenor* oviposition stimulants (Honda 1990). Wild crucifers have been extracted in ethanol, hexane, η-butanol and water to identify feeding deterrents against the imported cabbageworm, *Pieris rapae* (L.) (Dimock et al. 1991). Sorghum was extracted in hexane, ethyl acetate, and methanol to identify feeding stimulants in sorghum to *C. partellus* (Torto et al. 1991). Finally, rice has been extracted in η-hexane, methylene chloride, methanol and water to find antifeedants and growth inhibitors against the rice leaffolder, *Cnaphalocrosis medinalis* (Guenée) (Ramachandran and Khan 1991).

4.2.2.2. Water Extraction

Water extraction is accomplished by homogenizing plant parts in distilled water in a blender. The solution is then filtered and the filtrate is quick frozen in a dry ice/acetone bath and lyophilized. McMillan et al. (1970) reported that water extracts from maize silks and kernels of a susceptible maize plant stimulated *H. zea* feeding. Water extracts of several indigenous plants deterred oviposition by *H. virescens* (Tingle and Mitchell 1984).

4.2.2.3. Extraction of Plant Surface Waxes

Surface waxes of plants can be extracted following one of several methods described by Woodhead (1983), Wilson et al. (1988), Woodhead and Padgham (1988), Pelletier and Smilowitz (1990), Harris and Rose (1990), Jackson et al. (1991), and Severson et al. (1991). All begin by immersing intact plants or plant parts in chloroform or methylene chloride for 30 to 60 sec. After concentrating under vacuum to a small volume and finally to dryness under nitrogen, the weight of crude surface wax per plant is calculated from the dry weight of total wax and the number of plants extracted. Samples are redissolved in appropriate volumes of chloroform or methylene chloride in different concentrations for bioassays.

Plant waxes can also be fractionated by column chromatography for bioassays (Woodhead and Padgham 1988). Samples are dissolved in hot chloroform, applied to a silica gel column and eluted with chloroform followed by chloroform:methanol (50:50, V/V) before finally washing with sodium hydroxide. The chloroform fraction is redissolved in petroleum spirits, and separated on a freshly packed column of silica gel by elution with petroleum spirits followed by 3% diethyl ether in petroleum spirits and finally with chloroform. The classes of compounds extracted by this method can be crudely designated as hydrocarbon, carbonyl compounds, alcohols, aldehydes, and polar compounds. For bioassays, hydrocarbons are dissolved in hexane, carbonyls, alcohols and aldehydes in chloroform, and polar compounds are dissolved in diethyl ether.

4.2.2.4. Extraction of Some Specific Groups of Chemicals

Lipid-soluble constituents. One gram of plant tissue is continuously extracted with petroleum ether (b.p. 40° to 60° C) for 3 hr in a Soxhlet extractor. After filtering through celite, the extract is concentrated to dryness under vacuum and redissolved in petroleum ether (Woodhead and Bernays 1978).

Alkaloids and related compounds. One gram of plant tissue is extracted with 10% acetic acid in ethanol for 4 hr at room temperature with continuous stirring. After filtering and concentrating under vacuum to approximately 25% of its original volume, any alkaloids are precipitated from the filtrate by the dropwise addition of concentrated NH_4OH. After centrifuging, the precipitate is washed with 1% NH_4OH and the residue dissolved in ethanol (Woodhead and Bernays 1978).

Hydroxyaromatic acids bound as esters, glycosides, or in the cell wall. 100 mg of plant tissue is hydrolyzed with $2N$ HCl at 100° C for 45 min. After filtering and cooling, the filtrate is extracted twice with ether, once with bicarbonate, and once again with ether after acidification. The final ether extract is taken to dryness and redissolved in methanol (Woodhead and Bernays 1978).

Aglycones of flavones and flavonols. 100 mg of plant tissue is hydrolyzed with $2N$ HCl at 100° C for 45 min. After filtering and cooling, the filtrate is extracted twice with ethyl acetate and the combined extracts taken to dryness and redissolved in 80% methanol (Woodhead and Bernays 1978).

Hydroxyaromatic acids bound as soluble esters. 100 mg of plant tissue is hydrolyzed with $2N$ NaOH in Thunberg tubes under N_2 at room temperature for 4 hr, filtered, acidified with $2N$ H_2SO_4 and extracted with ether, bicarbonate and ether again as stated above, concentrated to dryness and redissolved in 80% methanol (Woodhead and Bernays 1978).

Terpenoids. Plants are dried in an oven at 50° C for 24 hr. Leaves are finely chopped with a food grinder, extracted with dichloromethane (approximately 125 g/l), filtered, and concentrated in a rotary evaporator under reduced pressure. The remaining solvent is removed under a nitrogen stream. The sample is weighed and redissolved in chloroform (Blust and Hopkins 1987a).

Glucosinolates. Plant tissue (300 to 500 mg) is extracted with boiling 70% (V/V) methanol, and re-extracted twice in boiling 70% methanol. Extracts are concentrated under vacuum and the resulting crude aqueous mixture is filtered. The sample is diluted with distilled water (1 g fresh weight plant tissue per milliliter) and stored at −20° C until assay (Reed et al. 1989, Koritsas et al. 1991).

Anthocyanidins. 100 mg of plant tissue is hydrolyzed with $2N$ HCl at 100° C for 45 min. After filtering and cooling, the filtrate is extracted twice with amyl alcohol and the combined extracts are concentrated to dryness and redissolved in 80% methanol plus one drop of $2N$ HCl (Woodhead and Bernays 1978).

4.3. EVALUATION OF VOLATILE PLANT ALLELOCHEMICALS

4.3.1. Orientational Response

The orientational response of an insect to odors of susceptible or resistant plants may be evaluated either in a laboratory test using a suitable olfactometer, a wind tunnel, simple petri dish test arena, or in a field test using a sticky trap or a water trap.

4.3.1.1. Olfactometers

Orientational responses of insects to volatile allelochemicals have been measured since the beginning of this century after Barrows (1907) and McIndoo (1926) developed the first olfactometers. Olfactometers offer an efficient and rapid technique for evaluating insect response to volatile chemicals (Rowlands 1985) and their designs vary widely from simple tubes and boxes relying upon odor diffusion or natural air currents (Painter 1930, Folsom 1931, Khan et al. 1987, Liu et al. 1988) to t-shaped, y-shaped, and trident olfactometers (Barrows 1907, McIndoo 1926, Brewer et al. 1983, Ascoli and Albert 1985, Sakuma and Fukami 1985, Reid and Lampman 1989, Rembold et al. 1989, Hopkins and Young 1990, Cannon, Jr. 1990). Numerous complex arenas complete with controlled airflow, humidity, or temperature have been designed to measure the olfactory responses of insects to plant odors (Howell and Goodhue 1965, Wearing et al. 1973, Payne et al. 1976, Kennedy 1977, Price et al. 1978, White and Richmond 1979, Katsoyannos et al. 1980, Turlings et al. 1989, Huang et al. 1990).

The usual focus of an olfactometer is a chamber in which the test insects are expected to respond to an odor. Any irregularities or barriers along the path of air flow can disturb movement of volatiles and may prevent valid results. Therefore, before attempting to copy any olfactometer, it is essential to thoroughly test the equipment design to ensure satisfactory air flow and a sufficient gradient between control and treatment air currents. A prerequisite in designing an olfactometer is a knowledge of the insect behavior associated with the odors being investigated. In all cases, simple systems are preferable if they provide accurate results. The following are descriptions of some common types of olfactometers.

4.3.1.1.1. Single Tube Olfactometers

In single tube olfactometers, the test chemical is applied on one end of the tube and the other end serves as the control. Insects are released in the center of the tube and their orientation in either direction is recorded.

Khan et al. (1987) and Liu et al. (1988) used a single tube olfactometer to measure orientational responses of cabbage looper, *Trichoplusia ni* (Hübner), larvae and adults to volatiles from susceptible and resistant soybean plants. The orientational responses of *T. ni* adults are measured in a horizontal cylindrical glass chamber (14 cm diameter × 21 cm long), open at both ends with a centered 15-mm sidewall opening for introducing insects (Figure 4.4). The opening at each end of the cylindrical arena is covered by 36-gauge mesh plastic screen. Test chemical is applied to a 9-cm-diameter filter paper disk that is positioned opposite a disk bearing control solvent. Both disks are placed outside the 2 screened ends of the arena and each positioned disk is secured by a 14-cm-diameter plastic petri dish lid. Each dish lid has 4 2-cm-diameter "air holes" positioned equidistantly on a circle 3 cm away from the center of the lid. Ten 1-day-old *T. ni* females are introduced into the middle of each test chamber, and insect responses are observed for 30 min. The percentages of

FIGURE 4.4. A single tube olfactometer chamber for testing orientational responses of *Trichoplusia ni* adults. (From Khan, Z. R. et al., *J. Chem. Ecol.*, 13, 1903, 1987. With permission of Plenum Publishing Corporation.)

adult females arriving on the nylon net wall adjacent to extract-treated and control filter paper disks are calculated.

Orientational responses of *T. ni* larvae were recorded similarly in a horizontal 3 cm diameter × 15 cm long, cylindrical glass chamber with nylon net ends (Figure 4.5) (Khan et al. 1987). Test chemicals are applied to 2.5-cm-diameter filter paper disks, and extract- or solvent-treated disks are attached to the center of the inside bottom of a plastic cup (4 cm high, 3.5 cm diameter) with 0.5-cm-diameter ventilation holes arranged equidistantly in a circle both on the bottom and on the side. One plastic cup with either an extract- or solvent-treated filter paper disk is snug fitted over each end of the assay chamber. Ten third-instar *T. ni* larvae are introduced at the midpoint in the assay chamber and larval responses are observed for 2 hr. The percentages of larvae arriving on the nylon net end adjacent to the extract-treated or control filter paper disks are calculated. Khan et al. (1987) and Liu et al. (1988) reported that soybean plant volatiles trapped on Tenax® or extracted as steam distillate extracts from the susceptible variety Davis attracted *T. ni* larvae and female moths, but volatiles from resistant PI227687 plants repelled *T. ni.*

The orientational response of rice leafhopper and planthopper adults to odors of extracts of susceptible and resistant rice cultivars may be tested following methods described by Obata et al. (1981, 1983) and Khan and Saxena (1986). Orientational responses are tested in a horizontal cylindrical glass chamber (7.5 cm long × 13 cm diameter) with nylon net walls at both ends (Figure 4.6). Eight 40-day-old susceptible rice plants with roots immersed in vials of water are sprayed with 1 ml of a known concentration of the extract

FIGURE 4.5. A single tube olfactometer chamber for testing orientational responses of *Trichoplusia ni* larvae. (From Khan, Z. R. et al., *J. Chem. Ecol.,* 13, 1903, 1987. With permission of Plenum Publishing Corporation.)

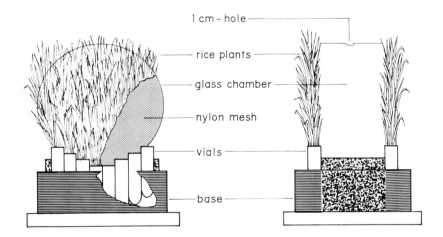

FIGURE 4.6. A chamber for testing orientation response of Sogatella furcifera to rice plant volatiles. (From Khan, Z. R. and Saxena, R. C., J. Econ. Entomol., 79, 928, 1986. With permission of the Entomological Society of America.)

of a test cultivar, while another set of control plants are sprayed with solvent. After the solvent has evaporated, the extract-treated and control plants are placed at opposite ends of the chamber outside the nylon net walls. Ten newly emerged leafhopper or planthopper females (starved for 2 hr but water satiated) are introduced into the middle of the chamber using a glass blowing tube and observed for 30 min. The percentages of arriving test insects on the nylon net walls are calculated as $100A/(A + B)$; where A is the number of ar-

rivals on the wall facing the extract-treated plants and B is the number of arrivals on the walls facing the control plants (Saxena et al. 1974). Each experiment is replicated several times. In a test using the whitebacked planthopper,
Sogatella furcifera (Horváth), Khan and Saxena (1986) reported that the
odors of extracts of the susceptible cultivar TN1 and the moderately susceptible cultivar Podiwi A8 rice cultivars attracted significantly more *S. furcifera*
females than control plants, but the odor of the extract of highly resistant
IR2035-117-3 plants was repellent.

Using a simple 50 × 5.5 cm glass tube olfactometer, Reed et al. (1988)
evaluated the ovipositional behavior of the lesser peachtree borer,
Synanthedon pictipes (Grote and Robinson), to host plant volatiles. One end
of the tube is closed with a cork containing a 0.5-cm-long latex tube leading
to a vacuum pump. The other end of the glass tube is corked with a rubber
stopper modified to hold a screen cage (8.5 × 4 cm) in which the test insect is
confined inside the tube during testing. Air is drawn through a charcoal filter
into a glass sample chamber and finally into the olfactometer. The sample
chamber contains a filter paper strip (2 × 0.5 cm) on which the test chemical is
placed. Air flow is adjusted at 0.1 to 0.3 m/s and temperature is maintained at
25° C. A single borer moth is placed in the cage inserted into the olfactometer
chamber and observed for 1 min. Strong ovipositional thrusts with the
ovipositor are considered positive responses to host volatiles. Reed et al.
(1988) reported that *S. pictipes* females were stimulated by volatile mixtures
of peach tree derived by solvent extraction, steam distillation, and volatile
trapping techniques.

4.3.1.1.2. Y-Tube Olfactometer

Y-tube olfactometers that provide insects with one of two choices are commonly used to measure unit olfactory responses. Odor is passed through one
arm of the Y-tube and clean air through the other (McIndoo 1926, Brewer et
al. 1983, Ascoli and Albert 1985, Reid and Lampman 1989, Cannon, Jr. 1990,
Hopkins and Young 1990) and the proportion of insects entering the arm containing the odor is used as a measure of chemical attractiveness.

Brewer et al. (1983) measured the attractiveness of odors of susceptible and
resistant alfalfa to the alfalfa seed chalcid, *Bruchophagus roddi*
(Gussakovsky), in a Y-tube olfactometer with three attached components
(Figure 4.7). Two test chambers are attached to the Y-tube branches and an introduction chamber is located at the base of the Y-tube. Each test chamber of
the Y-tube contains a glass funnel with an opening of 2 mm in diameter which
traps chalcids moving into each chamber. Filtered compressed air is split into
two streams, calibrated by flowmeters, and passed over distilled water in a
flask to equalize humidity before entering the olfactometer arms. Test plant
material or extract-treated filter paper disks are placed in one test chamber of
the olfactometer and the other (control) chamber is left blank or contains a solvent-treated filter paper. The data is interpreted using a chi-square test with a
null hypothesis that the air-stream choice is random. Brewer et al. (1983)

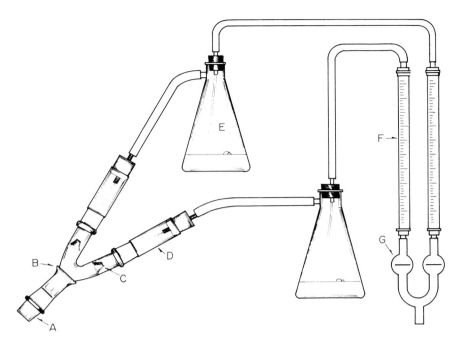

FIGURE 4.7. A Y-tube olfactometer. A, introduction chamber; B, Y-tube; C, funnel; D, test chamber; E, humidifier; F, flowmeter; G, air flow valve. (From Brewer, G. J. et al., *Environ. Entomol.,* 12, 1504, 1983. With permission of the Entomological Society of America.)

demonstrated that alfalfa seed pod aroma stimulated *B. roddi* movement to the seed pods but the movement was not significantly related to resistance. Hopkins and Young (1990) noted that the odors of chopped leaves of wheat, sorghum alfalfa, and ryegrass were significantly attractive to the grasshopper, *Melanoplus sanguinipes* (F.), in a Y-tube olfactometer.

4.3.1.1.3. Dual-Choice Olfactometer

In addition to Y-tube olfactometers, dual-choice olfactometers have been developed. The olfactometer described by Nottingham et al. (1989b) to evaluate sweet potato weevil, *Cylas formicarius elegantulus* (Summers), ovipositional preference consists of a main chamber (A) of 600-ml volume with a fitted lid, 2 gauze-covered holes through which air exits the system, and a central baffle to reduce turbulence (Figure 4.8). Two glass connecting tubes (E) join the main chamber with 2 glass jars (200 mm), one a treatment chamber (B) and the other a control chamber (C). Test chemicals, dried on a filter paper disk, are placed in test tubes (F) with 1.2-cm-diameter gauze-covered holes. A regulated air flow is fed through tubing (G) to each side of the olfactometer. Test insects are placed in the center of the main chamber and after 2 hr the weevils moving into treated and control jars are counted. An

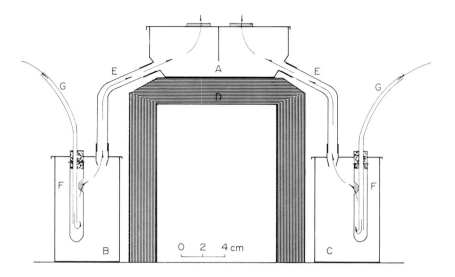

FIGURE 4.8. A dual-choice olfactometer. Arrows indicate direction of air flow. A, main chamber; B and C, treatment and control chambers; D, wooden stand; E, glass connecting tubes; F, test tubes with gauze-covered holes; and G, air input lines. (From Nottingham, S. F. et al., *J. Chem. Ecol.*, 15, 1095, 1989a. With permission of Plenum Publishing Corporation.)

Attraction Index is calculated as $[(T - C)/N] \times 100$; where T is the number of weevils in the treatment chamber, C is the number in the control chamber, and N is the total number of weevils originally in the main chamber.

The dual-choice olfactometer described by Dickens (1986) to study orientational response of the boll weevil, *Anthonomus grandis* Boheman, to plant volatiles consists of a main chamber (8×4 cm) of a crystallizing dish (Figure 4.9). The chamber is covered by plexiglass into which 2 holes of 1.5 cm diameter are cut. Test vials (7.0×2.1 cm) containing test chemicals are positioned over the 2 holes leading to the lower chamber. To facilitate gradation of olfactory stimuli, air is continuously exhausted from the device from a hole in the center of the bottom of the main chamber. Test insects are released from a screw cap vial through a 1-cm-diameter hole located 1 cm from the center of the bottom of the main chamber. The insects are allowed to orient for 20 to 30 min in the dark, and the number of insects in each test vial is counted and recorded. A paired *t* test is used to evaluate differences between various treatments.

4.3.1.1.4. Dual-Choice Pitfall Olfactometer

Various dual-choice still-air pitfall olfactometers have been developed by Phillips and Burkholder (1981), Pierce et al. (1981, 1990, 1991), and Budenberg et al. (1993) to evaluate volatile host food attractants and

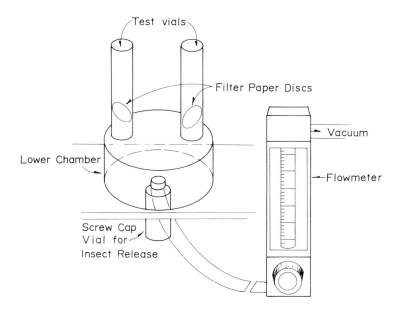

FIGURE 4.9. A dual-choice olfactometer to test orientation responses of *Anthonomus grandis*. (From Dickens, J. C., *J. Chem. Ecol.,* 12, 91, 1986. With permission of Plenum Publishing Corporation.)

pheromones against several species of stored-product beetles. The olfactometers described by Pierce et al. (1981, 1990, 1991) utilize plastic petri plates (14 × 1 cm) with two holes (14 mm diameter) drilled diameteretrically opposite one another, and 65 mm apart in the bottom (Figure 4.10). A glass vial is suspended from each hole. A filter paper disk treated with the extract is put into the bottom of one vial after evaporation of the solvent. A control disk similarly treated with solvent is put into another vial. Test insects are released in the center of the petri dish and the petri dish arena is closed with a lid. The test insects are left in darkness for 2 hr and the numbers of insects in control and treated vials are recorded. Data is analyzed using a *t* test and results are expressed as the mean percent response = 100 [(treated − control)/N], where treated and control are the number of insects in the vial containing the experimental and control stimuli, respectively, and N is the total number of insects released.

4.3.1.1.5. Trident Olfactometer

A trident olfactometer (Figure 4.11) was developed by Rembold et al. (1989) to demonstrate that the volatiles of *Cicer arietinum* L. seed flour elicited a significant positive orientational behavior by first instar larvae of *Heliothis armigera* (Hübner). The principal feature of the trident olfactometer

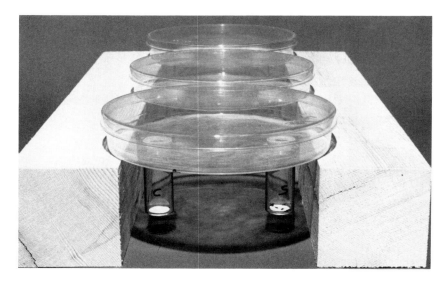

FIGURE 4.10. A two-choice, pitfall olfactometer. (From Pierce, A. M., *Can. J. Zool.*, 59, 1980, 1991. With permission of National Research Council, Canada. Photo courtesy A. M. Pierce.)

is the central tube, which keeps the air streams coming from treated and control arms separated at their cross point. This separation is attained by maintaining a flow of 25 ml/min in the central arm and 11 ml/min in the 2 lateral arms of the olfactometer. Volatiles are adsorbed on a Tenax tube and the tube is placed into the treatment side of the olfactometer and an empty control Tenax tube is placed into the other side. Larvae are released at the base of the trident olfactometer and their migration towards treatment odors (+) or the control arm (–) is recorded. A preference value (PV) is calculated as follows:

$$PV = \frac{\text{Larvae showing + response (no.)} - \text{Larve showing – response (no.)}}{\text{Larvae showing + response (no.)} + \text{Larvae showing – response (no.)}}$$

4.3.1.1.6. Four-Arm Olfactometer

A four-arm airflow olfactometer, originally developed to study aphid sex pheromones (Pettersson 1970), has been used to measure the orientational responses of hymenoperous parasites to semiochemicals (Vet et al. 1983, Eller et al. 1988a, Turlings et al. 1989). Compared with a Y-tube olfactometer, a

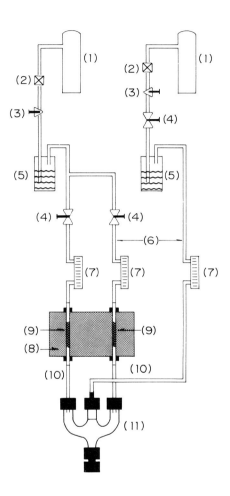

FIGURE 4.11. Schematic representation of a trident olfactometer. 1, gas cylinder with air; 2, shutoff valve; 3, reducing valve; 4, needle valve; 5, gas bubbler filled with water; 6, silicone tubes; 7, gas flowmeter; 8, thermoconstant water bath; 9, tenax adsorption tubes; 10, teflon tubes; 11, trident olfactometer. (From Rembold, H. et al., *J. Appl. Entomol.,* 107, 65, 1989. With permission of Paul Parey.)

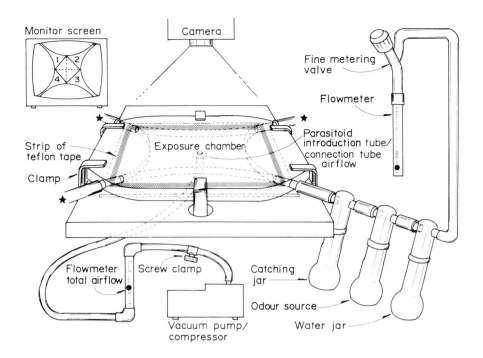

FIGURE 4.12. A perspective view of the four-arm olfactometer (only one arm shown, rest represented by stars). (From Vet, E. M. et al., *Physiol. Entomol.,* 8, 97, 1983. With permission of Blackwell Scientific Publications Ltd.)

four-arm olfactometer has the advantages of allowing more than one odor or different concentrations of one odor to be tested at the same time (Vet et al. 1983). The four-arm olfactometer consists of a central chamber connected to four arms through which air flows into the chamber (Figure 4.12). Air is exhausted through a hole centered in the bottom of the exposure chamber. By balancing the air flows through each arm (300 ml/min), four distinct odor fields can be created in the central chamber. Each arm is connected to a flow meter, a water bubbler (for humidity), an odor chamber, and a catching jar. Test chemicals are placed in the odor chambers and insects that walk up the arms are captured in the catching jars.

4.3.1.1.7. T-Shaped Linear Track Olfactometer

A T-shaped linear track olfactometer with a special wire pathway was built by Sakuma and Fukami (1985) to examine the orientation response of the German cockroach, *Blattella germanica* (L.), toward their aggregation pheromone. Hardie et al. (1990) modified this olfactometer for studying aphid sex pheromones and later Nottingham et al. (1991) used it to show that

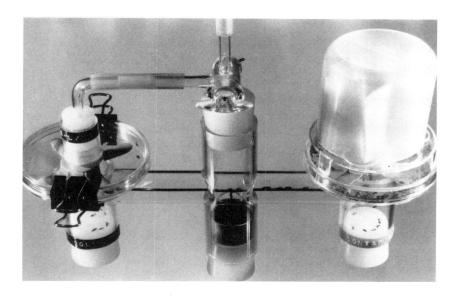

FIGURE 4.13. A linear track olfactometer. (From Sakuma, M. and Fukami, H., *Appl. Entomol. Zool.*, 20, 387, 1985. With permission of Japanese Society of Applied Entomology. Photo courtesy M. Sakuma.)

Aphis fabae Scopoli and *Brevicoryne brassicae* (L.) are attracted by odors from the leaves of their host plants. The olfactometer is constructed from transparent Plexiglass® tubing and steel rods. The steel rods form a T-junction along the axis of crossed olfactometer tubes where the test insects are introduced from the bottom of the T and made to choose a branch at the T-junction (Figure 4.13). Air is drawn through the apparatus at a rate of 1 l/min and meets at the intersection of the vertical and horizontal wires of the T. Thus, the insects could choose to move vertically in only one of the two directions.

4.3.1.1.8. Reaction-Chamber Olfactometer

A reaction-chamber olfactometer was designed by Kamm and Buttery (1983) to permit alfalfa seed chalcids, *B. roddi* (Gussakovsky), to respond to alfalfa volatiles, first by flight towards the odor source and, after landing on the target, by close-range orientation to the odor source. The olfactometer consists of two reaction chambers (22 cm long, 15 cm diameter) made of clear plastic, joined together by a small tubing (10 cm long, 7.5 cm diameter, with screen on each end) with an 8-mm hole in the center (Figure 4.14). Air is drawn into the reaction chamber through the small center holes and out through the screens at a rate of 80 ml/min. A circular pad of white blotter

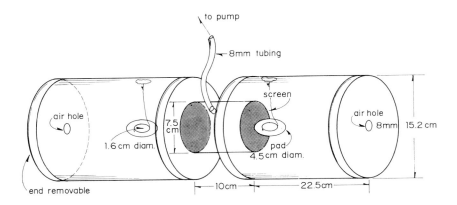

FIGURE 4.14. A reaction-chamber olfactometer. Air drawn into two large reaction chambers and through center tubing, following direction of arrows. (From Kamm, J. A. and Buttery, R. G., *Entomol. Exp. Appl.*, 33, 129, 1983. With permission of Kluwer Academic Publishers.)

paper (4.5 cm diameter) is suspended in each reaction chamber. A small circle (16 mm diameter) is drawn in the center of each pad, and the test material or solvent (control) is applied inside the circle. Similar numbers of test insects are released into each reaction chamber and the number of insects landing on the white blotter paper, the number of insects that entered the circle in the center of the pad, and the number of insects that remain inside the circle 4 sec or longer are recorded.

Using the linear track olfactometer, Kamm and Buttery (1983) demonstrated that *B. roddi* females flew to volatile leaf extracts. At close range, females oriented to extracts of seed pods and leaves but not to flower extracts.

4.3.1.1.9. Multichoice Olfactometer

To test multiple samples, an eight-sample olfactometer was constructed by Huang et al. (1990) by connecting eight sample tubes equidistantly in a circle to the bottom of a petri dish (15 cm diameter) (Figure 4.15), which serves as a bioassay arena. A small air-inlet pore is made on the outside of each tube to eliminate the differences in airflow rate. Small holes are drilled in clusters in the center of the dish to which an exhaust tube is attached and connected to a vacuum pump. All air-inlet pores are enclosed in an air-inlet chamber formed by gluing another dish (15 cm diameter), with holes for the corresponding tubes and exit tube, to the bottom of the bioassay arena.

Air (100 ml/min) passed into the inlet tubes is distributed in the air-inlet chamber, enters each tube at the pores, flows above the samples in each tube, passes into the arena, and is exhausted at the base of the bioassay arena. A

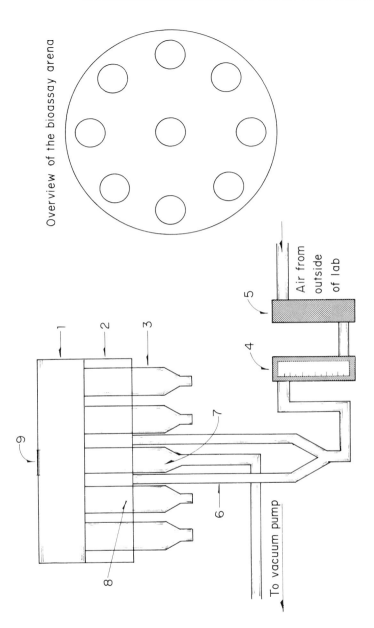

FIGURE 4.15. A multichoice olfactometer. 1, bioassay arena; 2, air-inlet chamber; 3, sample tube; 4, airflow meter; 5, filter filled with activated carbon; 6, air-inlet tube; 7, exit tube; 8, air-inlet pore; 9, insect releasing hole. (From Huang, X. P. et al., *Environ. Entomol.*, 19, 1289, 1990. With permission of the Entomological Society of America.)

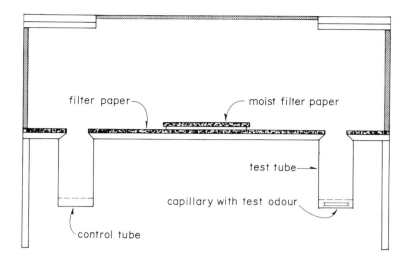

filter paper

moist filter paper

test tube

capillary with test odour

control tube

FIGURE 4.16. A vertical section of a multiple choice pitfall olfactometer. (From Mustaparta, H., *J. Comp. Physiol.,* 102, 57, 1975. With permission of Springer-Verlag.)

loose cotton ball is placed above each test sample. Test insects are placed in the bioassay arena and the number of test insects crawling into each tube is recorded after 30 min or at 5-min intervals for 30 min.

Using this multichoice olfactometer, Huang et al. (1990) demonstrated that significantly more larvae of the lesser cornstalk borer, *E. lignosellus,* oriented toward volatiles eminating from peanut roots, pegs, and young pods, as compared to volatiles from peanut leaves, stems, and older pods.

4.3.1.1.10. Multiple-Choice Pitfall Olfactometer

A multiple choice olfactometer was devised by Mustaparta (1975) to study the orientational response of the pine weevil, *Hylobius abietis* L., to volatile chemicals. The insects are allowed to choose and orient towards six tubes (2.5 cm diameter × 6 cm long) containing test chemicals. The tubes are fastened to the floor of the test chamber (30 cm diameter) and spaced equidistantly 12 cm from the center (Figure 4.16). Test chemicals or solvent controls are placed in the bottom of the test tubes and covered by a cotton plug or nylon mesh to prevent the insect from coming in direct contact with the test chemical. Test insects are released in the middle of the test chamber on a moist filter paper and the number of insects entering control and treated test tubes is recorded.

4.3.1.1.11. Automated Olfactometer

An automated olfactometer, incorporating a photoelectric insect detection and recording system, was developed by Wearing et al. (1973) to study the orientational response of the codling moth, *Laspeyresia pomonella* L., to apple volatiles. This olfactometer helps minimize the amount of time spent on the observation, recording, and analysis of insect activity. To monitor the activity of test insects, two photoelectric detectors emit output pulses each time insects pass beneath them. Pulses are totaled and printed out by an electromechanical counter-printer together with a record of the time lapsed since the beginning of the experiment.

4.3.1.2. Wind Tunnels

Responses of insects to plant volatiles have also been studied in wind tunnels designed for both walking insects (Haskell et al. 1962, Visser 1976, Visser and Ave 1978, Jones et al. 1981, Phelan and Lin 1991) as well as for flying insects (Kennedy and Moorhouse 1969, Hawkes and Coaker 1979, Miller and Roelofs 1978, Eller et al. 1988b, Phelan et al. 1991). Wind tunnels are generally a physical model of the environment, allowing experimental manipulation of variables such as temperature, humidity, and wind velocity. Another key advantage of wind tunnels is that experiments can be performed throughout the year.

Wind tunnels are commonly divided into "high speed" and "low speed" types (Vogel 1969). Only low speed wind tunnels are used to study the olfactory orientation of insects to plant volatiles (Haskell et al. 1962, Kennedy and Moorhouse 1969, de Wilde et al. 1969, Visser 1976, Hawkes and Coaker 1979, Phelan et al. 1991).

Basically a wind tunnel consists of three parts (Figure 4.17): (1) an effuser or entrance zone in which the air is accelerated and the flow is smoothed; (2) a working section; and (3) a diffuser or exhaust zone where the air is decelerated (Vogel 1969). A fan is placed either in the effuser (Figure 4.17) or in the diffuser. Often there is a settling chamber joined to the working section by a contraction. Wind tunnels can be either of the "closed circuit" type (if the air in the tunnel is recirculated) or of an "open circuit" type (if the air is taken from outside and exhausted back outside the tunnels) (Figure 4.17). Wind tunnels can be either "closed throat" or "open jet" types if the working section is open to the outside atmosphere (Figure 4.17).

To use a wind tunnel to study insect responses to plant volatiles, test chemicals are deposited on a muslin cloth or on a filter paper dispenser. After the solvent is evaporated, the dispenser is placed at the upwind end of the tunnel. The location and form of the test chemical plume can be verified by introducing smoke into the dispenser system and observing the smoke trail. A test insect is released into the downwind end of the flight tunnel and is immediately observed for 2 to 5 min. Behavioral responses such as random and oriented movement or flight, contacts, landings and oviposition on extract-treated substrates are recorded.

Some successes have occurred in the evaluation of plant volatiles for insect orientation studies using wind tunnels. Visser and Ave (1978) demonstrated that the odor of fully grown potato plants elicits a positive anemotactic response in the Colorado potato beetle, *Leptinotarsa decemlineata* (Say).

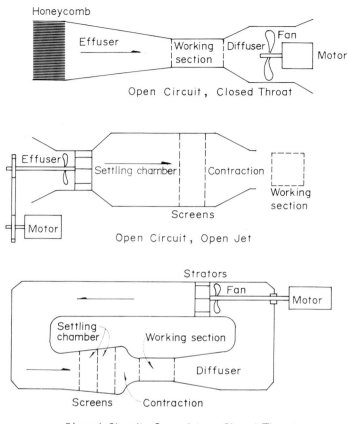

FIGURE 4.17. Examples of several different types of wind tunnels. (From Vogel, S., *Experiments in Physiology and Biochemistry,* Vol. 2, Academic Press, 1969. With permission of Academic Press.)

However, none of the individual component volatiles alone was attractive. In a large wind tunnel, gravid females of the cabbage root fly, *Delia brassicae* (Hoffm. in Wiedeman), showed upwind responses to host plant odor (Hawkes and Coaker 1979). Similarly, in wind tunnel bioassays, apple volatiles were found to be attractive to sexually mature apple maggot flies, *Rhagoletis pomonella* (Walsh) (Fein et al. 1982). Females of *Heliothis virescens* (F.) also responded positively via upwind flight in wind tunnel assays to volatiles emitted from methylene chloride washes of fresh leaves of cotton, tobacco, and a weed species, *Desmodium tortuosum* (Swartz) (Mitchell et al. 1991).

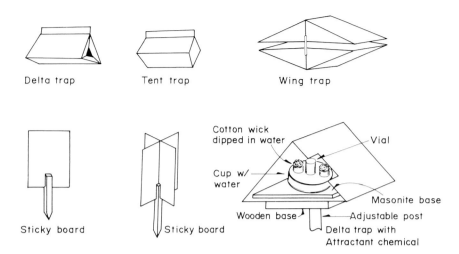

Delta trap   Tent trap   Wing trap

Cotton wick
dipped in water   Vial

Cup w/
water

Masonite base
Wooden base   Adjustable post
Sticky board   Sticky board   Delta trap with
                                              Attractant chemical

Sticky board

FIGURE 4.18. Examples of sticky traps. (Delta trap with attractant chemical from Roseland, C. R. et al., *Entomol. Exp. Appl.,* 62, 99, 1992. With permission.)

4.3.1.3. Field Trials

Field trials of insect responses to traps baited with attractant plant volatiles suggest that such plant volatiles could be successfully evaluated in the field. In many cases, mixtures of volatiles are more attractive than any single chemical (Finch 1980, Miller and Strickler 1984, Finch 1986, Roseland et al. 1992).

Sticky traps (Figure 4.18) and water traps (Figure 4.19) are used most commonly for evaluating the attraction of insects to plant volatiles in the field. Sticky traps are generally made by covering objects such as wooden boards with a thin coating of an adhesive so that the insects settling on them are retained (Figure 4.18). Three of the most common sticky traps commercially available are the Delta trap, the tent trap, and the wing trap (Figure 4.18). In several cases, water traps are preferred to sticky traps if the presence of water vapor itself is also important in attracting insects to volatiles (Finch and Skinner 1982).

The shape, size, alignment, height, background, and position of the trap should be considered during the development of an effective trap. It is often difficult to optimize a trap to suit a range of circumstances or insect species. In addition, the most difficult part of designing an efficient trap is to determine how best to release the attractive chemicals (Finch 1986). McGovern and Beroza (1970) and Beroza (1970) used a wick dipping into a reservoir of volatile chemicals to attract Japanese beetles, *Popillia japonica* Newman. Cotton dental wicks, previously impregnated with mineral oil to prolong

FIGURE 4.19. A water trap with attractant chemical. The attractant chemical placed in a tube is shown by an arrow. (Photograph provided by Z. R. Khan.)

volatilization, were treated with volatile chemicals for field trapping *Diabrotica* spp. (Lampman et al. 1987, Lampman and Metcalf 1987, 1988, Lance 1988). Volz (1988) used a 3 × 5 cm sponge pad enclosed in a polyethylene dispenser bag (0.05 mm low-density PE) to evaluate the attractancy of monoterpenes to bark beetles, *Hylurgops palliatus* Gyllenhal, and *Tomicus piniperda* L. Plant attractants have also been encapsulated into a starch borate matrix to study the behavioral responses of *Diabrotica* sp. adults to plant-derived semiochemicals (Weissling et al. 1989).

4.3.1.4. Small Arenas

Small arenas consisting of petri dishes have been used successfully to evaluate volatile chemicals against insect larvae (Munakata et al. 1959, Matsumoto and Thorsteinson 1968, Matsumoto 1970, Kamm and Buttery 1984, Hibbard and Bjostad 1988). These arenas require no expensive equipment, are capable of detecting if the test chemical is an attractant or a repellent, require little space, and are easily replicated.

Munakata et al. (1959) devised a technique to evaluate the attractancy of the isoflavone oryzanone to rice stem borer, *Chilo suppressalis* (Walker). Small pieces of absorbent cotton wool containing appropriate amounts of the test chemical are placed in two test tubes. The other two test tubes, without test chemical, serve as controls. The four test tubes are placed in a petri dish equidistantly in such a way that their open ends face the center of the petri

dish. Treated and control test tubes are placed alternately. Test insects are released in the center of the petri dish. After 12 or 24 hr, the number of insects
found inside the treated and control tubes and outside the tubes are recorded
as positive (+), negative (−), and indifferent (±), respectively.

Matsumoto and Thorsteinson (1968) and Matsumoto (1970) followed the
method of Munakato et al. (1959) to study olfactory responses of onion maggot larvae, *Hylemya antiqua* Meigen, to volatile organic sulfur compounds.

Kamm and Buttery (1984) used a bioassay arena consisting of a plastic
cylinder (4 cm high × 15 cm diameter) placed on blotting paper, to evaluate
root volatile components of red clover for attraction to the clover root borer,
Hylastinus obscurus (Marsham). Two opposing circles (2 cm diameter) are
drawn on the blotting paper 2 cm from the center of the arena. An appropriate
amount of the test chemical is applied on the blotting paper inside one circle,
and solvent only inside the other circle serves as a control. After the solvent
is evaporated, test insects are introduced into the arena and the number entering each circle during the observation period is recorded.

Hibbard and Bjostad (1988) used a single tube bioassay technique to evaluate behavioral responses of western corn rootworm larvae, *Diabrotica virgifera virgifera* LeConte, to maize seedling volatiles. Plastic petri dishes
(10 cm diameter) are used as bioassay arenas. The top of a sample tube containing the test chemical or control solvent is connected with Teflon tubing to
a 12-mm hole cut in the bottom of the petri dish (Figure 4.20). Larvae are
placed equidistantly in a ring near the wall of the petri dish, the cover is replaced, and the number of larvae entering treated and control tubes is
recorded.

4.3.2. Electroantennogram (EAG)

Identifying volatile compounds which are biologically active toward insects requires the rapid screening of a large number of compounds. An electroantennogram (EAG) technique has been developed that is based on the
principle that biologically active volatile chemicals pass through the micro-
tubules on insect antennae and bind with receptor proteins, initiating reactions in the dendrites of the olfactory cells as the receptor membrane is depo-
larized. Electroantennograms measure the sum of many olfactory receptor
potentials recorded more or less simultaneously by an electrode connected to
the sensory epithelium of an insect.

The EAG technique involves connecting the tip and the base of an excised
(Visser 1979, Tichenor et al. 1979, Anderson et al. 1987, Blust and Hopkins,
1987b, Lampman et al. 1987, Hansson et al. 1989, Valterova et al. 1990) or
an intact (Wadhams et al. 1982, Light et al. 1988, Ramachandran et al. 1990,
Ramachandran and Khan 1990, Ramachandran and Norris 1991, Budenberg
et al. 1993) antenna to a recording and a reference electrode (Figure 4.21). To
measure the response from an intact antenna, the test insect is temporarily
anesthetized with CO_2 and is immobilized in a polypropylene pipette with
paraffin wax. The insect's head and a part of the prothorax protrudes out of

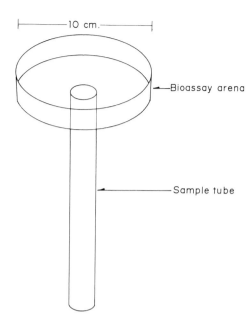

|—————10 cm.—————|

Bioassay arena

Sample tube

FIGURE 4.20. Single-tube bioassay arena to evaluate behavioral responses of *Diabrotica virgifera virgifera*. (From Hibbard, B. E. and Bjostad, L. B., *J. Chem. Ecol.*, 14, 1523, 1988. With permission of Plenum Publishing Corporation.)

the narrow end of the pipette. One antenna of the insect is removed to permit an easy access of the indifferent electrode to the base of the recording antenna. A moist wad of cotton is inserted into the wider end of the pipette to prevent rapid desiccation of the insects. Insects are allowed to recover for 30 min before starting to record responses.

The antenna is inserted into the tip of a glass pipette (20 to 30 μm diameter) with a chlorinated silver wire (Ag-AgCl) electrode (different electrode) filled with saline solution (NaCl, 3.75g; $CaCl_2$, 0.105g; KCl, 0.175g; $NaHCO_3$, 0.1g; H_2O, 500 ml) (Roelofs 1976) containing 10% (V/V) polyrinylpyrolidme. The base of the antenna is connected to a similar saline filled indifferent electrode. The Ag-AgCl electrode is connected to the input of an amplifier, the output of which is displayed on an oscilloscope screen. The EAG peak heights (–mv) are either measured directly from the oscilloscope or are imported into a microcomputer spreadsheet program such as Lotus 1-2-3 (Lotus Development Corporation, MA), using Labtech Notebook (Laboratory Technologies Incorporation, MA) or other suitable software.

The experimental antenna is continuously bathed in a stream of activated charcoal-filtered air (40 ml/sec) delivered by a stainless steel tube positioned 2 cm apart from the antennal preparation. The stimulus from a test chemical

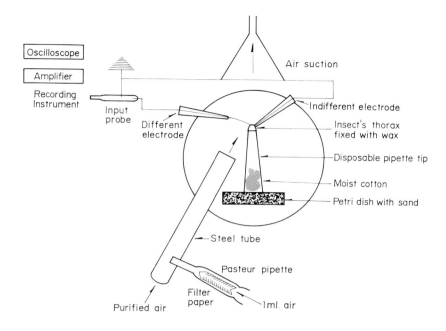

FIGURE 4.21. Line diagram of a set up for measuring electroantennogram response of insects to volatile chemicals. (From Ramachandran, R. and Khan, Z. R., *Int. Rice Res. Newsl.,* 15(5), 22, 1990. Courtesy IRRI.)

is applied from a pasteur pipette. The tip of the pipette is introduced into a small hole located on the side of the steel tube. The other end of the pipette is placed in an injection device which delivers a known quantity of air. The duration of the stimulus is generally 1 sec. Antennal preparations normally remain in good condition for several hours, but their responsiveness decreases gradually. EAG responses to test chemicals are evaluated by measuring the maximum amplitude of negative deflection (–mV) elicited by a given stimulus and by subtracting the amplitude of the response to the accompanying solvent control. Since antennal response diminishes throughout an experiment, the amplitude of response to test compounds is often expressed as a percentage of the EAG response to the standard cis-3-hexen-1-ol (Visser 1979, Guerin and Visser 1980, Dickens 1984, Dickens and Boldt 1985, Light and Jang 1987, Ramachandran et al. 1990, Ramachandran and Norris 1991).

The amplitude of the response, which correlates with the frequency of generated nerve impulses, increases with increasing concentrations of the chemical stimulus until a saturation level is reached. The EAG response is usually a negative potential (Figure 4.22a), but positive potentials may also be obtained depending on the chemistry and biological activity of the compound assayed (Figure 4.22b).

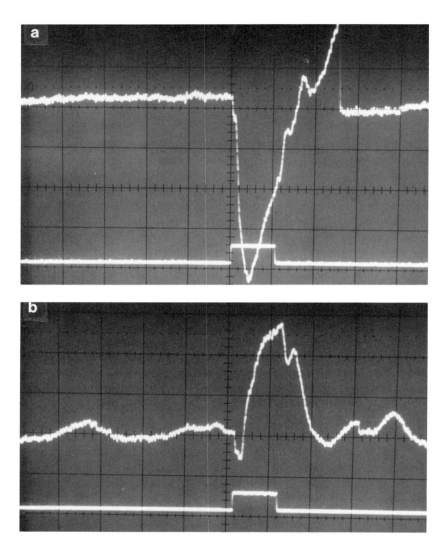

FIGURE 4.22. *Cnaphalocrocis medinalis* female responses to two volatile chemicals as measured by electroantennogram. (a) A typical negative potential elicited in response to stimulation with 1-hexanol, an attractant; and (b) a positive potential elicited in response to thymol, a repellent. (Photo courtesy R. Ramachandran.)

In some cases, EAG results have permitted conclusions about the behavioral significance of volatile allelochemicals. Some EAG results suggest that compounds eliciting very high negative potentials may be attractants at low concentrations, while compounds eliciting positive potentials may be repellents (Contreras et al. 1989, Ramachandran et al. 1990) (Figure 4.22b). However, research should also be conducted with intact insects to determine whether an EAG-active chemical elicits attraction or repellency in the same manner as the antennal preparation.

EAG research has permitted the study of the specificity of the antennal olfactory receptors of pest lepidopterans (Anderson et al. 1987, Hansson et al. 1989, Ramachandran et al. 1990, Valterova et al. 1990, Ramachandran and Norris 1991, van Loon et al. 1992), dipterans (Light and Jang 1987, Light et al. 1988), coleopterans (Visser 1979, Wadhams et al. 1982, Lampman et al. 1987), orthopterans (Blust and Hopkins 1987b), and hymenopterans (Baehrecke et al. 1989, Ramachandran and Norris 1991) to plant volatiles.

4.3.3. Combined Gas Chromatography–Electroantennography

Direct coupling of gas chromatography (GC) and electroantennography (EAG) has added a new dimension to electrophysiological studies of insect pheromones and plant volatiles (Arn et al. 1975, Struble and Arn 1984, Baehrecke et al. 1989, Cork et al. 1992). Combined GC-EAG involves splitting the outlet from a gas chromatograph column between the chromatograph detector and the test insect antenna. After the antenna is prepared, the volatile constituents of the test material are separated on a chromatographic column and the effluent is split between a flame ionization or other detector and the antenna. The effects of individual components of plant aroma on insect olfactory perception are noted by simultaneous recording of the GC signal and the insect's responses, and by comparing the retention time of plant volatiles separated chromographically with the insects EAG response. Moorhouse et al. (1969) reported the first coupled GC-EAG. The GC packed column effluent was split so that 75% went to the flame ionization detector (FID) and 25% was flushed with nitrogen for 3 sec over a standard EAG preparation. This system has identified several insect pheromones (Beevor et al. 1975, Nesbitt et al. 1977, 1980, Rothschild et al. 1982, Kuenen et al. 1990). Arn et al. (1975) reported the first coupling of GC-EAG with the GC effluent passing continuously over the insect antenna. This system uses a high-resolution capillary GC column where peak widths are only a few seconds (<15 sec) and the components reach the antenna with a sharp rise in concentration. Arn et al. (1975) suggested the term "electroantennographic detector" for this system.

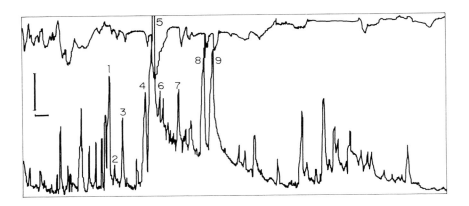

FIGURE 4.23. GC-EAG recordings of *Lobesia botrana* in response to tansy extract. Upper trace, EAG; lower trace, GC signals. 1, ρ-cymene; 2, δ-limonene; 3, α-thujene; 4, α-thujone; 5, β-thujone; 6, thujyl alcohol; 7, terpinen-4-ol; 8, (Z)-verbenol; 9, piperitone. Horizontal bar = 100 sec; vertical bar = 1 mv. (From Gabel, B. et al., *J. Chem. Ecol.,* 18, 693, 1992. With permission of Plenum Publishing Corporation.)

Using the method of Arn et al. (1975), Baehrecke et al. (1989) recorded responses of an ichneumonid parasitoid, *Campoletis sonorensis* (Cameron), to mono- and sesquiterpene volatiles of the cotton plant.

Gabel et al. (1992) used the GC-EAG technique to screen the steam distillate extract constituents of tansy flowers (*Tanacetum vulgare* L.) for responses by female European grapevine moths, *Lobesia botrana* Den. et Schiff. From more than 200 GC peaks, nine monoterpenoid peaks elicited an EAG response in female moths (Figure 4.23).

4.3.4. Single-Cell Recording (SCR)

Although the EAG provides information on the specificity of insect olfactory receptors, a more detailed understanding of the perception of odor molecules by insects can be obtained by recording the responses of individual olfactory cells. There are two methods of single-cell recording.

4.3.4.1. Sensillum Base Recording

In this technique, a fine recording electrode (micropipettete, tungsten, or platinum iridium) is implanted in or near the base of a selected sensillum with the aid of a micromanipulator. The reference electrode is placed in the hemolymph space close to the sensillum under study, usually in the same antennal segment (Figure 4.24) (Boeckh 1967, 1969, Boeckh et al. 1965, Ma and Schoonhoven 1973, Mustaparta 1975, Ma and Visser 1978, Selzer 1981, Wadhams et al. 1982, Bromley and Anderson 1982, Seelinger 1983, Dawson et al. 1987, Nottingham et al. 1991). An AC amplification and filtering sys-

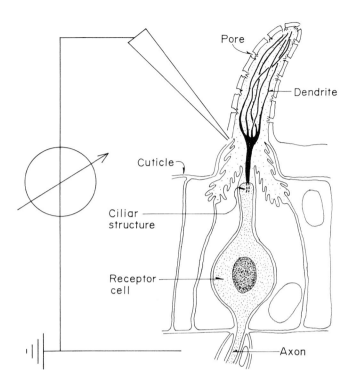

FIGURE 4.24. An antennal olfactory hair (sensillum basiconicum) of a *Necrophorus* sp. beetle. (Reproduced from Boeckh, J., in *Olfaction and Taste* III, page 37. By copyright permission of the Rockefeller University Press, 1969, 648.)

tem is used to produce an extracellular recording which can be viewed on an oscilloscope. A continuous flow of purified air is maintained over the preparation. For antennal stimulation the air flow is passed over a pasteur pipettete containing the odor sample under test.

There are, however, difficulties with this method. Penetrating the insect antennal cuticle may be difficult and positioning the electrode close enough to reasonably record responses of large amplitude may also be difficult (Den Otter 1977). Single-cell preparations are extremely sensitive to mechanical vibration and any movement of the manipulators or vibration through the table may cause loss of contact (Frazier and Hanson 1986).

4.3.4.2. Sensillum Tip Recording

The tip recording technique was first employed by Kaissling (1974) and then used by Den Otter (1977), Den Otter et al. (1978), Van Der Pers and Den Otter (1978), and Hansson et al. (1989). These preparations are less affected by mechanical disturbances (Den Otter 1977, Frazier and Hanson 1986). In this technique, the entire antenna is excised and the base is placed into the tip of an indifferent electrode consisting of a glass pipette filled with Ringer's solution. This electrode contains an Ag/AgCl wire connected to the ground. The recording electrode is prepared from a second glass pipette with a tip diameter of 5 to 10 μm. The pipette is filled with Ringer's solution on an Ag/AgCl wire and the wire is connected to the input probe of a high impedance AC amplifier. The tip of the sensory hair is cut off using two small razor knives to obtain electrical contact with receptor cells associated with the sensillum (Den Otter 1977) or two glass knives (Van Der Pers and Den Otter 1978). The pipette electrode tip is subsequently placed over the end of the remaining part of the sensillum. The electrical activity of the receptor cells is displayed on an oscilloscope.

Recordings from single sensilla trichodea of *Adoxophyes orana* (F.V.R.) males and females revealed a lack of sensitivity to apple tree odors (Den Otter et al. 1978). Comparison of the *Agrotis segetum* Schiff male and female single sensillum response to twelve plant volatiles showed that male antennae produced significantly higher responses than those of females in most of the recordings (Hansson et al. 1989).

4.3.5. Combined Gas Chromatography–Single-Cell Recording

A more recent recording system combines gas chromatography with single-cell recording (GC-SCR) to quantify each stimulus (Frazier and Heitz 1975, Wadhams 1984). Components eluted from the GC column are detected simultaneously by the flame ionization detector (FID) of the gas chromatograph and the olfactory cell of the insect. Both signals are continuously monitored and a determination of sensory activity is made for each chromatographic peak as it is eluted (Figure 4.25).

4.3.6. Feeding Assays

Insect feeding responses to plant volatiles can be assayed in choice or in no-choice assays. Different techniques have also been used for chewing and sucking types of insects.

4.3.6.1. Chewing Insects

The feeding response of chewing insects to plant volatiles may be measured following any of several methods. Velusamy (1990a) evaluated the effects of steam distillate extracts of resistant and susceptible rice varieties on the feeding of the rice leaffolder, *C. medinalis* (Guenée), in a no-choice assay. Leaves cut from a susceptible plant are dipped in the steam distillate extract solution of a susceptible or a resistant plant and placed inside a petri dish (10 cm diameter) lined with moistened filter paper. Leaves cut and

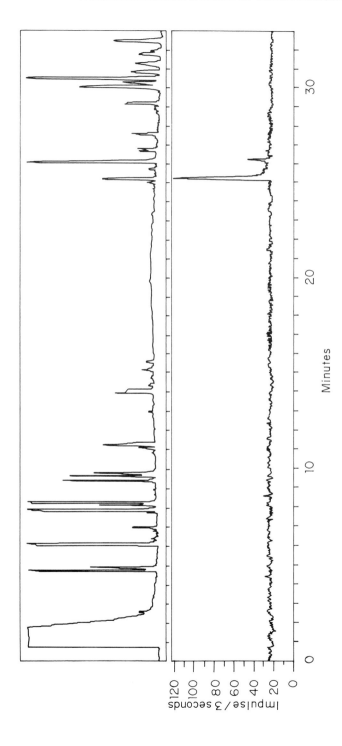

FIGURE 4.25. A coupled GC-single cell recording. (Top) gas chromatography of an air entrainment extract of the volatiles associated with *Scolytus scolytus* female; and (bottom) response of an olfactory cell specialized to α-cubebene, to stimulation with air entrainment extract. (From Wadhams, L. J., in *Techniques in Pheromone Research*, Springer-Verlag, New York, 1984. With permission.)

treated with solvent only are used as a control. Leaves are infested with test insects and after 24 or 48 hr, the area consumed is measured using an automatic leaf area meter. Leaf area consumption by *C. medinalis* larvae was significantly reduced by steam distillate extracts of the highly resistant *Oryza officinalis* and *O. punctata* plants, compared with controls and resistant "Ptb33," "Rathu Heenati," and "Swarnalata" rice plants. Zalkow et al. (1979) used a similar no-choice bioassay to evaluate the volatile oils from rayless goldenrod, *Isocoma wrightii* (Gray), against feeding of the fall armyworm, *Spodoptera frugiperda* (J. E. Smith).

Alfaro et al. (1981) evaluated the effects of volatile cedar leaf oil on the feeding of the white pine weevil, *Pissodes strobi* (Peck), in a two-choice assay. Two Sitka spruce twigs (2 cm long), are connected end to end with a headless entomological pin inserted through the pith. One twig is dipped in the test volatile solution and the second (control) twig is dipped in solvent only. Each pair of connected twigs is placed over a filter paper floor under an inverted jar. Test insects are introduced into each jar and are allowed to feed for 24 hr. *P. strobi* feeding is evaluated by counting the number of feeding punctures on the treated and control twigs. Cedar leaf oil significantly depresses the feeding of *P. strobi* (Alfaro et al. 1981).

Khan et al. (1987) evaluated cabbage looper, *T. ni,* feeding on 18-mm-diameter leaf disks of susceptible or resistant soybean varieties treated with susceptible and resistant soybean steam distillate extracts. Disks are presented in a multichoice arrangement in a 9-cm-diameter petri dish arena. Five leaf disks from a susceptible or resistant plant are positioned equidistantly in a circle with their abaxial sides up. Four of the five disks are treated on their abaxial surfaces with 20 μl of one of the four concentrations (e.g., 500, 1000, 2000, or 4000 ppm) of steam distillate extracts from a susceptible or a resistant plant. The fifth (control) leaf disk is treated only with 20 μl solvent. Before testing, the area of each leaf disk is measured and insects are then placed in the assay arena. After the end of the experiment, the leaf area is again measured and the amount of feeding is determined. *T. ni* larval feeding is significantly less on susceptible leaves treated with resistant plant volatiles as compared to solvent treated leaves. However, volatiles from susceptible plants applied on resistant leaves do not increase larval feeding (Khan et al. 1987).

4.3.6.2. Sucking Insects
The feeding responses of sucking insects to plant volatiles have been measured using several methods. The following descriptions give detailed accounts of each method.

4.3.6.2.1. Choice Test
The feeding choices of sucking insects to volatile chemicals have been measured following the methods of Obata et al. (1981), Saxena and Okech (1985), and Velusamy et al. (1990b). Thirty minutes before exposure to

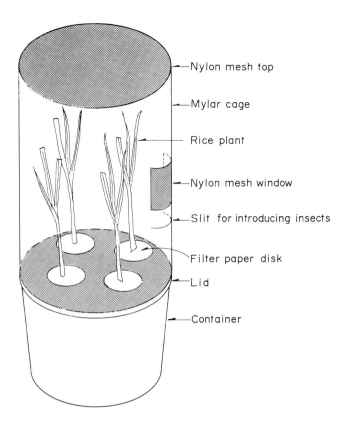

Nylon mesh top

Mylar cage

Rice plant

Nylon mesh window

Slit for introducing insects

Filter paper disk

Lid

Container

FIGURE 4.26. A mylar cage enclosing rice plants for observing influence of rice plant volatiles on feeding response of *Nilaparvata lugens*. (From Saxena, R. C. and Okech, S. H., *J. Chem. Ecol.*, 11, 1601, 1985. With permission of Plenum Publishing Corporation.)

leafhoppers or planthoppers, single tillers of 30-day-old susceptible rice plants, grown in pots 6 cm in diameter, are sprayed individually with 1 ml of susceptible or resistant plant extract. Control plants are sprayed with solvent only. Treated and control plants are arranged equidistantly in a circle in a plastic container partly filled with water. One tiller of each test plant is inserted through holes 1.5 cm in diameter bored equidistantly near the periphery of the lid covering the container (Figure 4.26) and a filter paper disk is fixed at the base of each tiller. Fifty newly emerged, CO_2-anesthetized female leafhoppers or planthoppers are released at the center of the lid which is then covered with a snug-fitting cylindrical Mylar® cage (55 cm high × 30.2 cm diameter). As females feed on their preferred plants and excrete, their honeydew droplets are absorbed on the respective filter paper disks. At the end of the experiment, the filter paper disks are treated with 0.2% ninhydrin-acetone solution and dried at 40°C for 5 min. The relative size of the bluish amino acid spots on treated filter paper disks gives an estimate of the honeydew excretion by females. A visual comparison of honeydew spots on filter paper disks showed that brown planthopper, *Nilaparvata lugens* (Stål), females feed more and consequently excrete more on control and TN1 (susceptible) extract-treated plants than on plants treated with the extract of Ptb33 or ARC6650 (resistant) (Saxena and Okech 1985).

4.3.6.2.2. Gravimetric Method

Ingestion and assimilation of food on extract-treated plants may be measured following the method of Khan and Saxena (1986) developed with *S. furcifera*. Using a camel's-hair brush, 5-cm leaf or leaf sheath portions of intact tillers of potted 40-day-old susceptible rice plants are painted with 0.25 ml of acetone solutions of extracts of susceptible or resistant rice cultivars at a rate of 100, 1000, 2000, or 4000 ppm of extract per tiller. Control plants are untreated or painted with solvent only. One hour after treatment, newly emerged female hoppers (starved for 2 hr but water satiated) are weighed individually on a microbalance and enclosed singly in air-tight Parafilm® sachets (5 cm × 5 cm) sealed around the leaf or leaf sheath of treated or control plants (Figure 2.18). After 24 hr, the weights of each female and its excreta are recorded separately. To assess the loss in insect body weight due to catabolism, a control is similarly established in which the insect has access to a moist cotton swab to prevent desiccation. The amount of food ingested and assimilated by the insect is calculated using the method of Khan and Saxena (1984): Food assimilated = $W1 \times (C1 - C2/C1) + W2 - W1$; where W1 represents the intial weight of insect, W2 the final weight of insect, C1 the initial weight of control insect, and C2 the final weight of control insect. Food ingested = food assimilated + weight of excreta (see Chapter 2.2.3.2). The experiment is replicated 10 to 15 times.

In bioassays, the application of extracts of resistant IR2035-117-3 or moderately resistant "Podiwi A8" foilage to susceptible TN1 plants significantly reduces the ingestion and assimilation of food by *S. furcifera* females compared with the application of acetone or TN1 extracts (Khan and Saxena 1986). However, the application of TN1 extract to resistant IR2035-117-3 plants does not increase ingestion and assimilation of food.

Khan et al. (1988) reported the responses of rice-infesting and *Leersia hexandra* grass-infesting *N. lugens* populations to steam distillate extracts of TN1 rice plants and *L. hexandra* grass. The application of TN1 extract to *L. hexandra* plants significantly reduces the ingestion and assimilation of food by grass-infesting *N. lugens* females compared with the application of acetone or *L. hexandra* extract. Similarly, the application of *L. hexandra* extract to TN1 plants also significantly reduces feeding and assimilation by rice-infesting *N. lugens* females.

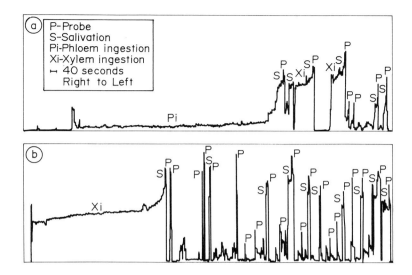

FIGURE 4.27. Electronically recorded waveforms during *Nephotettix virescens* feeding on (a) susceptible TN1 rice seedling (sprayed with acetone/water mixture, and (b) TN1 seedlings sprayed with 4000 ppm steam distillate extract of resistant ASD7 rice plants. (From Khan, Z. R. and Saxena, R. C., *J. Econ. Entomol.*, 78, 562, 1985b. With permission of the Entomological Society of America.)

4.3.6.2.3. Electronic Recording of Feeding Behavior

The feeding behavior of newly emerged leafhopper or planthopper adults can be recorded using an electronic monitor (Khan and Saxena 1984) first developed by McLean and Kinsey (1964). Insect feeding recorded by this method is described in Chapter 2.2.2.1.5. Leaf or leaf sheath portions of single tillers of potted 40-day-old susceptible and resistant rice plants are painted with 0.25 ml of a known concentration of extract of susceptible or resistant rice cultivar using a camel-hair brush. Control plants are treated with acetone. Khan and Saxena (1985b) reported that green leafhoppers, *Nephottetix virescens* Distant shifted from phloem feeding to xylem drinking when allowed to feed on susceptible TN1 seedlings sprayed with resistant ASD7 steam distillate extract (Figure 4.27). The decrease in phloem feeding was accompanied by a corresponding increase in the frequency of probing, salivation period, and xylem drinking on treated plants.

Khan and Saxena (1986) reported that the waveforms associated with *S. furcifera* feeding on susceptible TN1 plants treated with resistant IR2035-117-3 extract and on control TN1 plants treated with acetone also differ. The ingestion period is significantly shorter on IR2035-117-3 extract-treated plants than on control plants. The decrease in feeding duration corresponds to an increase in probing frequency and salivation period. However, the application of TN1 extract to IR2035-117-3 plants does not increase *S. furcifera* feeding.

4.3.6.2.4. Safranine Method

For those insects which feed both on xylem and phloem vessels, it is important to understand their feeding behavior on plants treated with volatile chemicals. The color of leafhopper or planthopper excreta on rice seedlings treated with 0.2% aqueous safranine solution indicates phloem feeding or xylem drinking (Khan and Saxena 1985b). Safranine is highly selective of lignin and the translocated dye colors the xylem vessels red throughout the entire length of rice seedlings. The feeding of leafhoppers and planthoppers on steam distillate extract-treated or control rice plants has been studied using this method (Khan and Saxena 1985b).

Ten-day-old seedlings are removed without damaging their roots and washed thoroughly to remove soil particles. Roots of seedlings are then immersed in an aqueous solution of 0.2% safranine for about 6 hr and the xylem vessels are dyed red. The treated seedlings are then removed, treated with a volatile chemical, and enclosed in the assembly described in Chapter 2.2.2.1.3.

Meanwhile, newly emerged females of leafhoppers or planthoppers are collected from the stock cultures and kept starved but water satiated for 4 hr. Ten hoppers are then released in Mylar® cages on the extract-treated or control seedlings. The honeydew excreted by the hoppers drops onto the filter paper disks and is readily absorbed. Red honeydew spots on the filter paper disks indicate xylem drinking by hopper females on the safranine-treated seedlings. Bluish amino acid spots, however, indicate phloem feeding by the insects when filter paper disks from control seedlings are treated with a 0.1% ninhydrin-acetone solution. Feeding is quantified on the basis of the total area of bluish spots or number of red spots on the filter paper disks. Each treatment, including controls, is replicated several times. Khan and Saxena (1985b) reported that the area of bluish amino acid spots on susceptible TN1 plants sprayed with resistant ASD7 steam distillate extract is significantly smaller at all concentrations tested than on TN1 control plants. The decrease in phloem feeding is accompanied by a significant increase in xylem drinking as indicated by an increase in the number of red honeydew spots (Figure 4.28).

4.3.6.2.5. Phagostimulation

The effect of volatiles from susceptible and resistant plants on phagostimulation may be tested in either choice tests (Saxena and Okech 1985, Khan and Saxena 1985a) or no-choice tests (Khan and Saxena 1986). In both tests, a known amount of the plant extract and 1 ml of 10% sucrose solution is offered to newly emerged test insects in Parafilm® sachets mounted singly on circular plastic cups (1.5 cm high × 3 cm diameter). The feeding chamber is made from a plastic petri dish painted black and has several peripheral holes

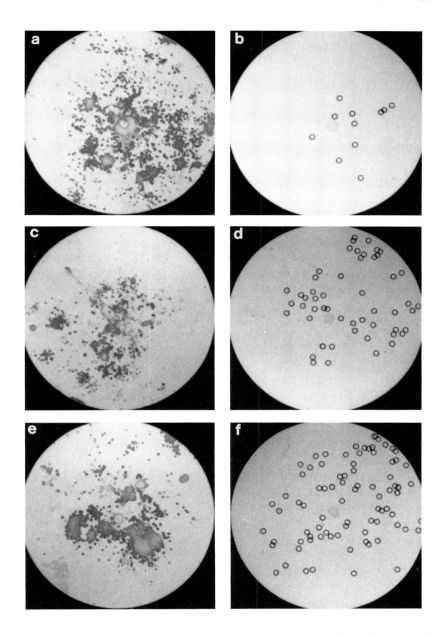

FIGURE 4.28. Filter paper disks on which *Nephotettix virescens* honeydew was collected when it fed on susceptible TN1 rice seedlings sprayed with acetone/water mixture (a and b); TN1 seedlings sprayed with 2000 ppm (c and d); and 4000 ppm (e and f) of steam distillate extract of resistant ASD7 rice plants; and ASD7 seedlings sprayed with acetone/water mixture (g and h). Bluish amino acid spots on ninhydrin-treated filter paper disks indicate phloem feeding (a, c, e, and f) and encircled honeydew droplets (b, d, f, and h) indicate xylem drinking. (From Khan, Z. R. and Saxena, R. C., *J. Econ. Entomol.,* 78, 562, 1985b. With permission of the Entomological Society of America.)

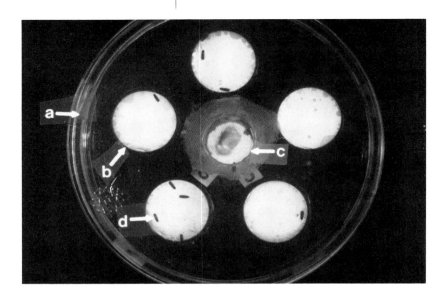

FIGURE 4.29. A feeding chamber for determining phagostimulatory preference of *Nephotettix virescens* in a choice test. (Photograph provided by Z. R. Khan.)

for the choice test (Figure 4.29) or a median hole (3 cm diameter) for the no-choice test (Figure 4.30). One end of the plastic cup is slightly flared to hold a Parafilm® sachet that fits snugly and level with a hole in the feeding chamber. A 2-cm^2 piece of Parafilm® is stretched uniformly over the flared end of the feeding cup. A filter paper disk (2 cm diameter) is then placed on top of the Parafilm® layer and treated with 2 µl of extract dissolved in acetone in a known concentration. Controls are treated with acetone. Treated disks are allowed to stand for 5 min to evaporate the acetone but leave the odorous extracts. A 1-ml aliquot of sucrose solution (1% for cicadellids and 10% for delphacids) is then placed on each filter paper disk with a pipette. Another piece of Parafilm® is stretched over the feeding cup so that the sucrose solution and the extract-treated disk are sandwiched between the Parafilm® membranes (Figure 4.30 A(c)).

The feeding cups and their respective sachets are weighed (W1) and fitted singly or in a multiple-choice arrangement in feeding chambers, each of which are mounted on a clear-bottom, snug-fitting plastic dish with sidewalls painted black. Ten females are introduced into each feeding chamber, which is covered with a black lid. Feeding chambers are arranged in a randomized complete block design on yellow cellophane in an incubator. The yellow light attracts the females to the Parafilm® sachet in each feeding chamber.

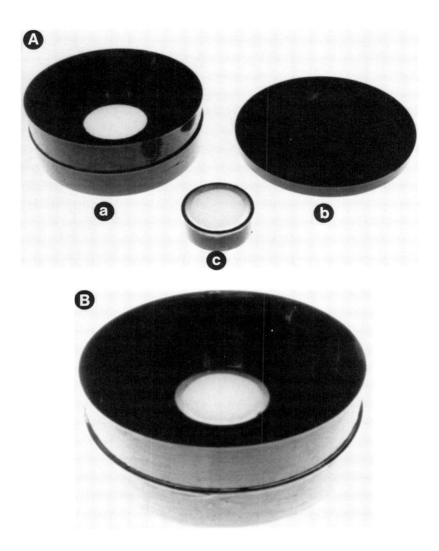

FIGURE 4.30. (A) Parts of a feeding chamber for studying phagostimulatory response of *Sogatella furcifera* in no-choice situation, (a) black painted plastic petri dish with a median hole and snugly fitted with another plastic dish; (b) parafilm sachet mounted on plastic cup; and (c) black painted lid of the feeding chamber; (B) an assembled feeding chamber. (From Khan, Z. R. and Saxena, R. C., *J. Econ. Entomol.,* 79, 928, 1986. With permission of the Entomological Society of America.)

After 24 hr, the females are removed and the feeding cups are reweighed (W2). The difference (W1 – W2) in the weights is the quantity of sucrose solution ingested by the females in 24 hr. The experiment is replicated six to ten times.

To correct the loss in weights of feeding cups due to evaporation, a blank set without females is placed randomly among the other feeding chambers. The mean weight loss in the control is used as a correction factor (CF) for determining the net quantity of sucrose solution ingested by females (i.e., [W1 – W2] + CF).

In the choice test, incorporation of the extract of the susceptible TN1 cultivar into 10% sucrose solution stimulates significantly greater intake by *N. lugens* females than that of sucrose solution alone or intake of the extract of the resistant cultivars Mudgo, Rathu Heenati, Babawee, Ptb33, or ARC6650 (Saxena and Okech 1985). The resistant cultivar ASD7 is an exception, and *N. lugens* females feed as well on sucrose solution with ASD7 extract as with TN1 extract. The extracts of Ptb33 and Rathu Heenati cause maximum inhibition of intake of sucrose solution. In a no-choice test, Khan and Saxena (1986) reported that there was no difference between *S. furcifera* feeding on sucrose solution to which extracts of susceptible or resistant cultivars had been added.

4.3.7. Growth Inhibition

To determine the antibiotic effect of volatile allelochemicals on insect growth, plant extracts or pure compounds can be incorporated in standard insect rearing diets following the method of Zalkow et al. (1979). The volatile extracts or pure chemicals are dissolved in ethanol and added to the weighed hot medium of the insect diet. In the controls, the same quantities of ethanol are added. Neonate larvae of test insects are released separately in treated and control diets, and their weights and development are measured at intervals. A Growth Index [ratio of percent larvae completing the larval period and the duration of the larval period (Saxena et al. 1974)] is calculated for both control and treated diets.

Zalkow et al. (1979) demonstrated that incorporating the volatile oils from goldenrod, *Isocoma wrightii* (Gray), into the diet of the fall armyworm, *S. frugiperda,* at a concentration of 1000 ppm, significantly decreased larval survival and adult emergence, and significantly increases the larval period. To evaluate the effects of steam distillate extracts of susceptible and resistant soybean plants on larval development and adult emergence of the cabbage looper, *T. ni,* Khan et al. (1987) topically treated third-instar larvae and pupae with 5 μg of extract on the dorsum of their first thoracic segment. Control insects are treated similarly with solvent (acetone). Test larvae are maintained on plants or on standard artificial diets. Adults emerging from pupae that received 5 μg extract from resistant PI227687 showed varying degrees of wing deformities.

4.3.8. Toxicity

Plant volatiles from susceptible and resistant plants differ in their toxicity to insects. Toxicity to insect larvae and adults have been measured using several different methods described by Saxena and Okech (1985), Khan and Saxena (1986), Khan et al. (1987, 1988), and Velusamy et al. (1990a).

To determine the toxicity of steam distillate extracts of susceptible and resistant rice varieties to first-instar nymphs of leafhoppers and planthoppers, 10-day-old test plant seedlings are dipped separately in solutions of extracts of susceptible and resistant cultivars and placed individually in test tubes (Saxena and Okech 1985, Khan and Saxena 1986). Control seedlings are untreated or treated with solvent. After the solvent has evaporated, first-instar nymphs of leafhoppers or planthoppers are placed on each of the treated or control seedlings. Nymphal mortality is recorded after 24 hr. Each treatment, including the control, is replicated several times.

Saxena and Okech (1985) reported that significantly more first-instar nymphs of *N. lugens* die on susceptible TN1 plants treated with extracts of the resistant cultivars than on control plants and plants treated with the extract of susceptible plants. Khan and Saxena (1986) reported that the steam distillate extract of a resistant rice cultivar was significantly more toxic to first-instar nymphs of *S. furcifera* than to extracts of highly susceptible, moderately susceptible, and moderately resistant cultivars.

Khan et al. (1988) also demonstrated that steam distillate extracts of the grass *Leersia hexandra,* a rice field weed, applied to susceptible rice plants was significantly more toxic to first-instar nymphs of rice-infesting *N. lugens* than to grass-infesting *N. lugens* nymphs. Similarly, the extract of susceptible rice applied at a similar concentration was significantly more toxic to grass-infesting *N. lugens* nymphs than to rice-infesting nymphs.

Velusamy et al. (1990a) evaluated the toxicity of volatile steam distillate extracts of resistant and susceptible rice varieties to the rice leaffolder, *C. medinalis.* A 6-cm leaf piece of a susceptible rice plant is dipped in a 2000-ppm solution of extract of a susceptible or a resistant plant. The treated and control leaf pieces are placed separately in petri dishes lined with moistened filter paper. After the solvent has evaporated, first-instar larvae are released on each of the treated or control leaf pieces and larval mortality is recorded after 72 hr. Velusamy et al. (1990a) reported that extracts of the wild rices *Oryza officinalis* and *O. punctata* are more toxic to first-instar larvae of *C. medinalis* than extracts of cultivated rice varieties.

Khan et al. (1987) evaluated the toxicity of steam distillate extracts from resistant and susceptible soybean varieties to *T. ni* by incorporating these extracts into a standardized artificial diet. After the solvent is evaporated, newly emerged larvae are released in each diet and larval mortality is observed at 24-hr intervals for 7 days. Steam distillate extracts from resistant PI227687 plants are more toxic to *T. ni* larvae than such extracts from susceptible Davis plants (Khan et al. 1987).

FIGURE 4.31. *Sogatella furcifera* females topically treated with 1 μg of steam distillate extract of (a) susceptible TN1 and (b) resistant IR2035-117-3 resistant rice plants. (From Khan, Z. R. and Saxena, R. C., *J. Econ. Entomol.,* 79, 928, 1986. With permission of the Entomological Society of America.)

The toxicity of plant volatiles to adult insects can be evaluated following the methods of Saxena and Okech (1985) and Khan and Saxena (1986). A known concentration of volatile plant extracts from susceptible or resistant plants is applied topically to the dorsum of newly emerged leafhopper or planthopper females using a microapplicator. Controls are treated with solvent only. Treated insects are maintained on susceptible plants and insect mortality is recorded 24 hr after emergence. Topical application of steam distillate extracts from resistant rice plants reduces the survival of *N. lugens* females (Saxena and Okech 1985) and *S. furcifera* females (Figure 4.31) (Khan and Saxena 1986).

4.3.9. Mating Disruption

Volatile allelochemicals may cause mating disruption in insects. Effects of volatile plant extracts on the mating behavior of leafhoppers and planthoppers, which produce acoustic mating signals, may be recorded following the method of Khan and Saxena (1986).

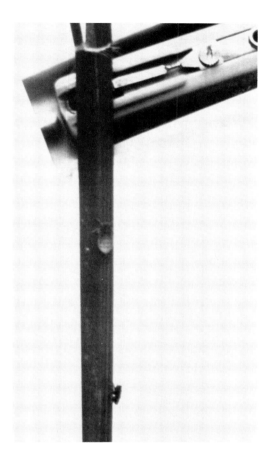

FIGURE 4.32. *Sogatella furcifera* male and female on a rice plant from where their mating signals are recorded through a ceramic cartridge. (Photograph provided by Z. R. Khan.)

Newly emerged unmated leafhopper or planthopper females are anesthetized and individually treated by topically applying the extract on the dorsum with a microapplicator. Control females are treated with solvent only. Treated and control females are then caged on a susceptible plant for 3 to 5 days, depending on the time required for sexual maturation. Mature females are paired singly on a susceptible plant together with a normal, responsive male. Mating signals emitted by males and females are detected as substrate vibrations by means of a high-fidelity, single-needle type ceramic cartridge, amplified and recorded on a high-fidelity sound recording tape using an amplifier (Figure 4.32) (Ichikawa and Ishii 1974). Male and female mating signals are fed to a DC chart recorder through a series of codes and recorded on chart paper (Figure 4.33).

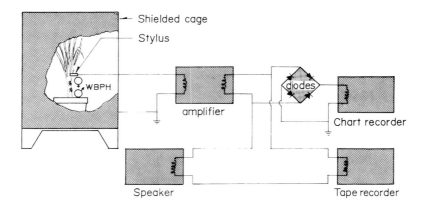

FIGURE 4.33. Schematic diagram of equipment and circuit for recording *Sogatella fur-cifera* mating signals. (From Khan, Z. R. and Saxena, R. C., *J. Econ. Entomol.,* 79, 928, 1986. With permission of the Entomological Society of America.)

Khan and Saxena (1986) reported that *S. furcifera* females treated with extract of a resistant rice stopped emitting normal "drumming" signals and failed to mate with males (Figure 4.34). In contrast, females treated with acetone or extracts of susceptible plants emitted normal signals and ovated successfully.

4.3.10. Ovipositional Responses

Volatile plant chemicals can attract or repel ovipositing females. Such chemicals can be bioassayed by application on the host and nonhost plants or on artificial surfaces, or by exposing ovipositing females to the vapors of volatile chemicals.

Volatile steam distillate extracts from susceptible and resistant rice culti-vars can be bioassayed to determine the ovipositional response they elicit from lepidopterans following the methods of Saxena (1978), Dhaliwal et al. (1988), and Velusamy et al. (1990a). Responses of female moths to these extracts and their fractions are tested by uniformly applying odorous solutions to each of two 2.5 cm × 23 cm paper towel strips that are supported on 18 cm × 23 cm wax paper, that serves as a suitable substrate for moth resting and oviposi-tion. A blank, similarly prepared but treated with solvent only, serves as the control. The treated and the blank strips are then suspended vertically on op-posite sides in a 45 cm × 50 cm cylindrical test chamber made of fine nylon mesh. When the solvent has evaporated, moths are released inside the test chamber. Ovipositional responses are based on the total number of eggs in egg masses laid overnight on the test and blank substrates.

To determine if the moth's ovipositional response can be reversed on resis-tant and susceptible cultivars, the odorous extract of a susceptible rice cultivar, diluted to a known concentration, is sprayed uniformly on a potted resistant

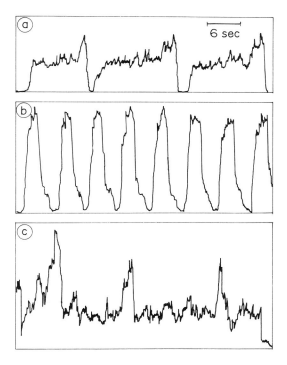

FIGURE 4.34. *Sogatella furcifera* signals emitted before mating: (a) "drumming" signals produced by a sexually mature, acetone-treated (control) female, and (b) "croaking" signals emitted by a responsive male; (c) females treated with extract of resistant IR2036-117-3 rice plant at a rate of 1 µg extract per female stopped emitting normal signals and failed to mate with responsive males. (From Khan, Z. R. and Saxena, R. C., *J. Econ. Entomol.,* 79, 928, 1986. With permission of the Entomological Society of America.)

plant, and vice versa. The treated plants are then exposed to moths in separate cages. Untreated resistant and susceptible plants caged similarly with moths serve as controls. Eggs laid overnight are counted in each treatment.

Saxena (1978) and Dhaliwal et al. (1988) tested striped stem borer, *C. suppressalis,* moths for their response to untreated paper strips and strips treated with the extracts of susceptible and resistant rice varieties. Moths oviposited heavily on strips treated with extracts of susceptible plants, laid no eggs on the resistant plant extract-treated strips, and laid a few eggs on untreated strips. However, almost equal numbers of moths arrived on Rexoro- and TKM6-treated strips, indicating a nearly equal attraction. Only a few moths arrived on the blank strips. Saxena (1978) sprayed the steam distillate extract of resistant plants on susceptible plants and demonstrated a 90% reduction in *C. suppressalis* oviposition over that on untreated susceptible plants. Resistant plants sprayed with susceptible extract received nearly four times more eggs than untreated resistant plants.

Velusamy et al. (1990a) tested the effects of steam distillate extracts of wild and cultivated rice varieties on the oviposition behavior of *C. medinalis*. The moths laid significantly fewer eggs on susceptible plants treated with volatile extracts of highly resistant wild rices, *Oryza officinalis* and *O. punctata*, than on susceptible plants treated with extracts of resistant cultivated rices.

Wearing and Hutchins (1973) described a petri dish bioassay technique to evaluate the effects of volatile chemicals on oviposition by the codling moth, *L. pomonella*. Volatiles in a suitable solvent are pipetted into a 3-cm-diameter ball of cotton wool. A control consists of the same quantity of solvent in cotton wool. After the solvent is evaporated, the cotton wool ball is transferred to petri dishes and offered in choice or in no-choice situations for oviposition by the moth. Gravid females of *L. pomonella* are attracted to oviposit on cotton wool balls impregnated with L-farnesene, which plays an important role in fruit location by *L. pomonella* in the field.

Saxena and Basit (1982) developed a technique to evaluate oviposition by the leafhopper, *Amrasca devastans* (Distant), in the presence of vapors of volatile plant chemicals. A leaf of a host plant is excised with its petiole and the petiole is immersed in a vial of water. The leaf is placed in the upper compartment of the test chamber and test volatiles are placed in the lower compartment and allowed to diffuse into the upper compartment through its nylon net bottom. Gravid females are released in the upper compartment for 24 hr. Eggs laid by leafhoppers are counted following the methods described in Chapter 2.2.6.3. Using this bioassay technique, Saxena and Basit (1982) demonstrated that oviposition by *A. devastans* on its susceptible host plant, cotton, was inhibited by the volatiles of certain plants when these chemicals were presented at a distance from the ovipositional substrate.

Another technique using vapors of volatile plant chemicals was devised by Reed et al. (1988). In this method, air from the exhaust of an air compressor is passed through flasks containing test vapors or distilled water. The volatile-laden air flows into a 13.5 × 10.5 × 3.5-cm plastic box containing an exhaust port on the opposite end. The boxes are lined with cotton-saturated distilled water. Gravid females are introduced in each box and the number of eggs laid in each box is counted 24 and 48 hr after release. Using this method, Reed et al. (1988) reported a strong ovipositional response of the lesser peachtree borer, *S. pictipes*, to volatiles of peach wood.

4.3.11. Egg Hatchability

Volatile allelochemicals from resistant varieties adversely affect the embryonic development of insects and reduce egg hatchability. Freshly laid eggs on susceptible plants are sprayed with or dipped in a known concentration of volatile test chemical. Eggs sprayed with or dipped in the solvent serve as controls. The excess liquid is drained and the eggs are incubated in petri dishes lined with a moistened filter paper disk. The number of hatched and unhatched eggs are recorded after 72 hr.

Medina and Tryon (1986) and Velusamy et al. (1990a) reported that steam distillate extracts of resistant rice plants significantly reduced the hatching of *C. medinalis* eggs as compared with susceptible extracts or solvent. Similarly, Velusamy et al. (1990b) reported that application of steam distillate extract of the resistant wild rice, *Oryza officinalis,* on a susceptible plant adversely affected the hatchability of *N. lugens* eggs.

4.4. EVALUATION OF NONVOLATILE CHEMICALS

This section summarizes the methods of assay of nonvolatile chemicals that elicit or inhibit insect responses. Some techniques described here were developed for testing chemicals not of plant origin, but they can also be utilized for the evaluation of plant chemicals.

4.4.1. Feeding Assays

Several methods have been used for testing the phagodeterrent or phago-stimulatory activity of allelochemicals or plant extracts to both piercing and sucking and chewing insects. These methods involve presentation of the test material on an inert substrate on plant tissue or in an artificial diet. Leaf disks are commonly used for chewing insects but are not recommended for work with sucking insects because of the physiological changes that occur in phloem sap (Lewis and van Emden 1986). The principal artificial substrate that has been used for sucking insects is a chemically defined liquid presented between natural (Hamilton 1930) or artificial membranes (Mittler and Dadd 1962, Yoshihara et al. 1980, Kim et al. 1985, Smith et al. 1991).

Inert artificial substrates such as styropor, filter paper, or glass fibers, while being botanically inert, have the disadvantage of being less acceptable to the test insect than the natural host-plant tissues. In whatever substrate is chosen, the application of allelochemical should not result in textural or color differences between control and treated test substrates. Most frequently, a treated substrate and a control substrate are introduced into a test arena. In larger arenas, more numbers of treated and control substrates are used to ensure that the test insect comes into contact with them. Generally, control and treated substrates are positioned alternately at the periphery of the arena. In multichoice experiments, several test substrates with different treatments are placed in each arena. In no-choice experiments, only one treated or one control substrate is used in each arena. Usually, one insect is introduced into each arena, but some researchers have used groups of insects, especially if small insects are being tested, to obtain an easily measurable feeding response.

Several different methods have been used to assess the effects of test chemicals on insect feeding. These include weight loss of the substrate and substrate area loss measured using graph paper, planimeter, or photometer (see following sections on bioassays). Substrate area loss is generally used only if the insects feed completely through the substrate.

Several workers have used a subjective visual assessment of feeding (Hsiao and Fraenkel 1968, Mehrotra and Rao 1972, Kon et al. 1978, Landis and Gould 1989). Kon et al. (1978) devised a 0 to 6 rating scale for cereal leaf beetle, *Oulema melanopus* (L.), feeding on agar gel strips, where 0 = not damaged; 1 = less than 1/4 of one strip side with feeding damage; 2 = approximately 1/4 of one strip side with feeding damage; 3 = approximately 1/2 of one strip side with feeding damage; 4 = most of one strip side well damaged; 5 = most of one side plus 1/2 second side damaged; 6 = 80 to 100% of entire strip damaged. The average rating value of the experiment provided the activity index of test chemicals. Feeding activity of the test insect can also be assessed by counting the number of feeding marks, ridges, or punctures (LaPidus et al. 1963).

Phagostimulation or deterrence of a test chemical can also be determined by measuring the distribution of a group of insects feeding on treated and control substrates at different time intervals during the experiment (Loschiavo 1965, Robinson et al. 1982, Gershenzon et al. 1985, Chapman et al. 1988, Bowers and Puttick 1988, Isman et al. 1989). Fecal production has also been used to assess the effects of allelochemicals on feeding activity (Dadd 1960, Hsiao and Fraenkel 1968, Ma 1972, Ascher and Meisner 1973).

As an indicator of the relative activity of phytochemicals, several workers have developed formulae to calculate insect feeding indices. Bentley et al. (1984), Alford et al. (1987), and Benz et al. (1989) calculated the percentage feeding reduction or antifeedant index (AFI) as follows:

$$\text{AFI} = \left(1 - \frac{\text{treatment consumption}}{\text{control consumption}}\right) \times 100$$

Dimock et al. (1991) used a feeding deterrent index (FDI) as follows:

$$\text{FDI} = 100 \left(\frac{\text{Control consumption} - \text{Treatment consumption}}{\text{Control consumption} + \text{Treatment consumption}}\right)$$

4.4.1.1. Chewing Insects

4.4.1.1.1. Use of Plant Tissues

Surface Application

Phytochemicals can be presented on leaf disks cut with a cork borer from the insect's host plant. Compounds in a suitable solvent (e.g., acetone, ethanol, methanol) are spread evenly over the surfaces of leaf disks with a

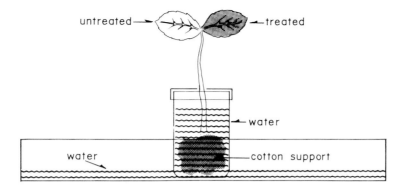

FIGURE 4.35. Feeding bioassay by surface application of antifeedant. Larvae are placed at the branching point of the leaves, and the treated and untreated leaves are scored at appropriate time intervals. (From Nakanishi, K., in *Insect Biology in the Future,* Academic Press, New York, 1980. With permission.)

polyethylene-tipped micropipette. Control disks are treated with equal volumes of solvent (Leskinen et al. 1984, Khan et al. 1986, Alford et al. 1987, Lindroth et al. 1988, Usher et al. 1988, Pelletier and Smilowitz 1990, Benz et al. 1989, Dimock et al. 1991). Leaf disks may also be dipped for 5 to 10 sec in treatment solution or solvent alone (Landis and Gould 1989).

Wolfson and Murdock (1987) applied phytochemicals to leaf surfaces by incorporating them into gelatin solutions. These solutions are diluted with an appropriate amount of phytochemical in water or a buffered chemical solution. Leaves of a host plant are excised from greenhouse grown plants and their petioles immediately placed into water. With leaves at full turgor, warm gelatin solution is painted onto the leaf surface with a paint brush and allowed to cool. [^{3}H] methemoglobin is used to calibrate the application rates of the gelatin solutions. [^{3}H] methemoglobin is diluted with gelatin to give off 1 ml gelatin with 800,000 counts per minute (cpm) at the appropriate gelatin concentration for the leaf type being calibrated. [^{3}H] methemoglobin-containing gelatin at different concentrations (5, 10, 15, and 20 µl) are applied to the leaf surfaces. After the gelatin has cooled, areas containing the gelatin samples are then cut from the leaf and placed into separate 1 ml volumes of buffer at 50° C for 1 hr. The test samples are then vigorously mixed, then 200 µl of the buffer wash is added to scintillation fluid and is counted in the liquid scintillation counter. A linear regression of volume of [^{3}H] methemoglobin and counts per minute is obtained which is used to calibrate the gelatin application rate. Differences between treated substrates are based on weight or area loss of the test substrate, or a visual estimation of feeding damage.

Feeding bioassays can also be carried out using two intact plant leaves of the same size on the plant itself (Nakanishi 1980, Leskinen et al. 1984). Two top leaves of the same size from the same plant are chosen for each replication (Figure 4.35). One leaf is coated with test compound and the other coated with solvent alone. The base of the stem is coated with vaseline to prevent the insect from crawling off. All plants are kept inside a cage for the test period to prevent test insect escape. One or two larvae of the test insect are then placed on each leaf and the leaf area eaten is determined from 2 hr (Ritter 1967) to 7 days (Benz et al. 1989) after the test begins.

To evaluate antifeedant allelochemicals against termites, wood slices have been used as a feeding bioassay substrate (Scheffrahn et al. 1988). In accordance to the weight of each wood slice, gravimetrically determined concentrations of test solutions are deposited by pipette to yield a desired concentration (weight solute/weight wood) after drying. Wood slices are offered to termites for feeding along with solvent-treated control wood slices. These experiments are normally of several weeks duration. Differences between treatments are based on wood block weight losses.

Passive Transpiration

Test phytochemicals may be applied through passive transpiration either by standing plants or shoots in solutions of chemicals (Erikson and Feeny 1974, Bodnaryk 1991) or by soaking the roots in the compound (Meisner et al. 1978). Harrison and Mitchell (1988) used a passive transpiration method to treat potato leaves with plant alkaloids. Leaf petioles are cut with a sharp razor blade, weighed, and immersed in a test tube containing plant allelochemical solution in water for 4 hr. The chemicals accumulate in leaves as transpiration occurs. From a knowledge of percent water content of the test plant and the concentration of test solution, the approximate amount of phytochemical accumulating in a leaf is calculated by weight loss of test chemical solution. Control leaves are placed in deionized water for the same length of time as treated leaves before being tested. Differences between the effects of test chemicals are determined by comparing final weights of test and control substrates.

Seed Treatment

Treated seeds have also been used as substrates to bioassay feeding deterrents against soil pests (Villani and Gould 1985, Villani et al. 1985). In these experiments, maize seeds were used to bioassay plant extracts as antifeedants against the wireworm, *Melanotus communis* Gyllenhal, and the southern corn rootworm, *Diabrotica undecimpunctata howardi* Barber. Seeds were soaked in an aqueous solution of the test chemical or distilled water (control) for 2 hr and air dried. Feeding assays were performed in plexiglass chambers filled with moist sand or soil.

Vacuum Infiltration

Kuhn and Low (1955) vacuum infiltrated the leaves of cultivated potato with glycosides from *Solanum* species resistant to the Colorado potato beetle, *L. decemlineata,* to determine feeding deterrence. The plant material is immersed in water or in a solution of host plant extract and is subjected to a vacuum with a faucet filter pump until the air no longer bubbles from the intercellular spaces in the leaves. The vacuum is then released, allowing the intercellular spaces to fill up with the surrounding fluid. Extracts containing organic solvents are evaporated and plant residue emulsified with water using an ultrasonic disintegrator before infiltration into the plant. Harris and Mohyuddin (1965) used vacuum infiltrated lettuce leaves in bioassays with three species of Lepidoptera and two of Coleoptera to test acceptability of host plant extracts. Akeson et al. (1967) used vacuum infiltrated root disks from sweet clover to bioassay for feeding deterrents to the sweet clover weevil, *Sitona cylindricolis* (F.). Differences in the effects of phytochemicals tested is determined by comparing the area loss of treated and control root disks. Much of the success of this method depends on choosing a plant that does not contain antifeedants.

4.4.1.1.2. Use of Artificial Media

Paper
Filter papers and paper towels have been widely used for the bioassay of allelochemicals (Figure 4.36) (Yamamoto and Fraenkel 1960, Thorsteinson and Nayar 1963, McMillian et al. 1970, Rust and Reierson 1977). Plant extracts are pipetted onto sections of filter paper or paper towel at a desired rate. After they are air dried, the treated papers are placed in each of two sections of a quadrant petri dish, and the solvent-treated ones are placed in the two remaining quadrants. Dishes are infested and differences in the effects of plant extracts are determined by area loss comparisons of treatment and control substrates. However, matted filter paper fibers hinder all but the strongest insects from biting off pieces to swallow (Harris and Mohyuddin 1965, Cook 1976). An exception is the scraping rasping feeding of some Coccinellid beetles, such as the Mexican bean beetle, *Epilachna varivestis* (Mulsant). To overcome this difficulty, Niimura and Ito (1964) charred filter paper at 200° C for 15 hr, treated it with 2% sucrose solution and then air dried it. The host plant extract was then applied and the solvent evaporated.

Serit et al. (1991) used an edible coloring matter, Brilliant Blue FCF, to detect feeding deterrency in a plant chemical against the termite, *Reticulitermes speratus* Kolbe. Each test sample is dissolved in an appropriate solvent and applied to a colored paper disk, whereas the control disk receives the solvent only. When termite larvae feed on paper disks containing coloring matter, their abdomens become colored.

Pith
Elderberry pith has been used by several workers (Harris 1963, Heron 1965, Losehiavo 1965, Norris and Baker 1967, Ritter 1967) to bioassay phytochemicals. Pith disks sliced from pitch sections cut with a cork borer

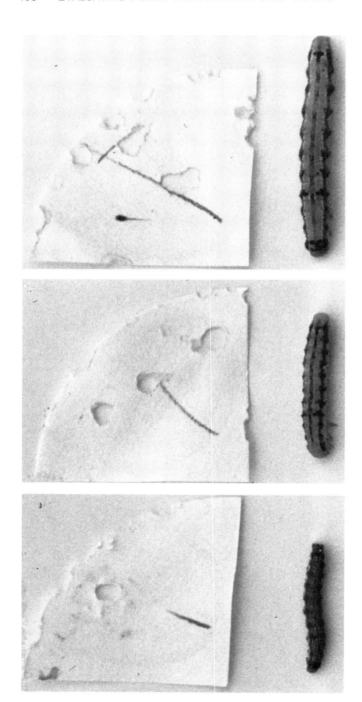

FIGURE 4.36. Consumption of filter paper disks impregnated with water extract of a susceptible maize silks by different instars of *Heliothis zea* larvae. (From McMillian, W. W. et al., *Ann. Entomol. Soc. Am.*, 63, 371, 1970. With permission of the Entomological Society of America. Photo courtesy B. R. Wiseman.)

FIGURE 4.37. Bioassay apparatus showing *Pissodes strobi* feeding through lens paper into agar disk containing 2% Sitka spruce bark. (From Alfaro, R. I. et al., *J. Chem. Ecol.,* 5, 663, 1979. With permission of Plenum Publishing Corporation. Photo Courtesy H. D. Pierce, Jr.)

are soaked in a 2% sucrose solution and dried. The host plant extract is then applied to disks and the solvent is evaporated. Control disks are treated with solvent only. Treatment differences are determined gravimetrically or photometrically.

Agar/Agar Cellulose

Agar and agar cellulose, presented in petri dishes or as disks or plugs, have been widely used for evaluating plant chemicals (Hsiao and Fraenkel 1968, Ma 1972, Yamamoto and Fraenkel 1960, Kon et al. 1978, Huang and Mack 1989). Test chemicals are applied to the agar surface or mixed into a hot solution of 3% agar or into a mixture of 4% agar and 4% cellulose dissolved in water. Alfaro et al. (1979) bioassayed antifeedant chemicals against the white pine weevil, *P. strobi,* by covering the top surface of the agar gel disk with a lens paper and immersing it in paraffin (Figure 4.37). In this

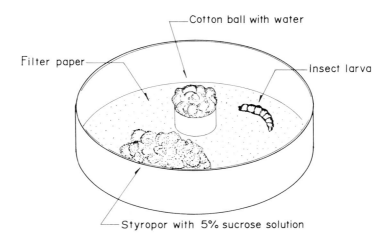

FIGURE 4.38. The 'stryopor method' used for testing phagostimulation with *Spodoptera littoralis* larvae. (From Ascher, K. R. S. and Meisner, J., *Entomol. Exp. Appl.*, 16, 101, 1973. With permission of Kluwer Academic Publishers.)

method, feeding stimulants or deterrents are applied to the lens paper covering of one of the disks while the other is kept as a control. The feeding activity of candidate antifeedants is assessed as the number of feeding punctures and is correlated with the amount of agar ingested from treated and control disks.

Gelatin

Bodnaryk (1991) reported a feeding assay against the bertha armyworm, *Manestria configurata* Walker, and a flea beetle, *Phyllotreta cruciferae* Goeze, using a gelatin substrate. Plant chemicals are incorporated into gelatin for analysis of feeding inhibition or stimulation. Test insects are allowed to feed directly on chemical-incorporated or control gelatin disks and the amount of feeding is estimated gravimetrically or visually.

Styropor

The styropor bioassay method developed by Meisner and Ascher (1968) has been used successfully to test the effects of allelochemicals on the feeding of several insects (Figure 4.38) (Meisner et al. 1972, Ascher and Meisner 1973, Meisner et al. 1974, Meisner and Skatulla 1975, Meisner et al. 1977a, Ascher and Nemmy 1978). This method consists of using the lamellae of foamed polystyrene (styropor) as a carrier for phagostimulatory or phagodeterrent compounds, as styropor itself is not fed upon by larvae of several insects. Ma and Kubo (1977) used styrofoam instead of styropor, because nonpolar solvents cannot be applied to styropor. Ascher and Nemmy (1978) used polyurethane in place of styropor for the same reason. Differences between treatments are determined by comparing differential weight loss of the substrate.

Cellulose Acetate/Glass Fiber/Nitrocellulose Membrane Disks

The feeding responses of insects to plant extracts have also been determined in experiments where phytochemicals are presented to insects on glass fiber disks (Adams and Bernays 1978, Woodhead and Bernays 1978, Woodhead 1983, Chapman et al. 1988), cellulose acetate disks (Albert and Parisella 1985, Torto et al. 1991) or nitrocellulose membrane disks (Bristow et al. 1979, Blust and Hopkins 1987a). Test samples in solvents are applied topically to both sides of the disks, which are dried in a stream of warm air. Each test disk is then treated with a 0.1-M sucrose solution to serve as a feeding stimulant. Where the test chemicals are water soluble, they are dissolved in a 1-M sucrose solution. After feeding, differences between allelochemicals or extracts are determined by gravimetric or photometric comparison.

4.4.1.1.3. Use of Artificial Diet

Artificial diets have been very widely used to bioassay the activity of allelochemicals against various insect pests such as the gypsy moth, *Malalenca leucadendron* (Trial and Dimond 1979, Doskotch et al. 1980); the cotton bollworm, *H. zea* (Klocke and Chan 1982); the armyworm *Spodoptera littoralis* (Boisduval) (Meisner et al. 1977b, 1982); the European corn borer, *Ostrinia nubilalis* (Hübner) (Robinson et al. 1982); the rice leaffolder, *C. medinalis* (Ramachandran and Khan 1991); the southern armyworm, *Spodoptera eridania* (Cramer); the migratory grasshopper, *M. sanguinipes*; and the sunflower moth, *Homoeosoma electellum* (Hulst) (Gershenzon et al. 1985).

In these experiments, test chemicals were either applied to the surface of the diet or mixed into it. However, Freedman et al. (1979) reported that larvae penetrating into such diets received insufficient exposure to the test chemicals applied to the surface only. It is, therefore, desirable to mix the allelochemicals throughout the diet. Water extracts of the host plant can be added directly to the diet solution, whereas phytochemicals soluble in organic solvents should be coated onto alphacel, removed under vacuum, and added to the diet as a portion of the alphacel component (Chan et al. 1978) (see Chapter 7.2.2).

To test the antifeedant activity of limonin against the Colorado potato beetle, *L. decemlineata,* and the fall armyworm, *S. frugiperda,* Mendel et al. (1991) presented the test chemical in artificial diet in choice and no-choice situations. Weight loss of the diet, presented as circular disks, represented insect feeding on treated and control plants. In both choice and no-choice situations, insect feeding was significantly more on untreated control diet as compared to limonin-treated diets.

4.4.1.2. Sucking Insects

It is difficult to visually assess when and if sucking insects are actually feeding. However, sucking insects produce stylet sheaths when they are allowed feed on an aqueous solution of sucrose or other probing media (Sogawa 1971, 1974, 1976, Sekido and Sogawa 1976, Miles 1972, Kim et al. 1985). The stylet sheaths deposited on a Parafilm® membrane can then be used to quantitate the effects of phytochemicals and the feeding sheaths can be clearly observed with a microscope after dying with a suitable dye (Sekido and Sogawa 1976, Kim et al. 1985).

In some past research, allelochemical solutions have been commonly dipped or sprayed onto plant surfaces (Saxena and Khan 1985b). This procedure is sometimes aided by mixing a surfactant such as Tween®, Triton®, or Teepol® (Saxena et al. 1981, Saxena and Khan 1985b) to the allelochemical to evaluate the feeding responses of sucking insects. Artificial media has also been used to evaluate the effects of phytochemicals on the feeding of sucking insects. Test chemicals or plant extracts are dissolved in distilled water, artificial diet, agar, or sucrose solutions at known concentrations and offered to insects between Parafilm® membranes. Test chemicals can also be painted on one half of the Parafilm® membrane containing an artificial diet, and the number of insects feeding on treated and untreated halves are counted. The following discussion of the responses of sucking insects to allelochemicals serves to illustrate the general procedures described above.

4.4.1.2.1. Probing Response

Test chemicals or plant extracts are dissolved in a sucrose solution and the solutions are added to a watch glass. Test insects are confined in a glass tube (2 cm in diameter × 3 cm high), one end of which is closed with a sheet of stretched Parafilm® (Kim et al. 1985). The tube is then placed on the watch glass (7.5 cm in diameter) containing the test solution enclosing the medium between the Parafilm® membrane and the watch glass (Figure 4.39). The open end of the tube is closed with a plastic cup. Insects are allowed to probe the test medium through the Parafilm® membrane for 24 hr at 27° C. Distilled water control feeding chambers are prepared in a similar manner. At the end of the experiment, the stylet sheaths deposited through the Parafilm® membrane are washed with water, stained with 0.1% rhodamine B aqueous solution, and observed under a microscope. The average length and percentages of elongated and branched sheaths are recorded as parameters for evaluation of the probing response. The average length of the stylet sheaths is calculated by sampling 30 to 50 single tubular sheaths from each feeding apparatus at random. Significant differences between them are analyzed using the t test after square root transformation ($\sqrt{100X}$; where X is the length of the stylet sheath). Elongated stylet sheaths are also classified according to their forking patterns as nonbranched (coefficient 1), two-branched (coefficient 2), three-branched (coefficient 3), and those with four or more branches (coefficient 4). The probing activity is expressed by a response value that is the summation of the number of the observed sheaths multiplied by their coefficient.

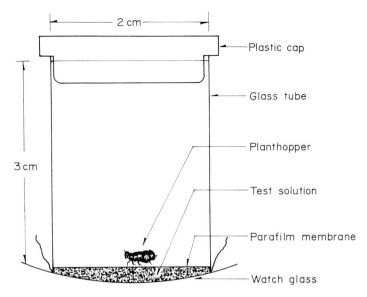

FIGURE 4.39. A bioassay apparatus for measuring probing response of planthoppers to various test media. (From Kim, M. et al., *J. Chem. Ecol.*, 11, 441, 1985. With permission of Plenum Publishing Corporation.)

Using this technique, Sogawa (1974) bioassayed the probing response of the brown planthopper, *N. lugens,* to plant saps from rice, and the nonhost plants *Echinochloa crus-galli* var. *hispidula, Cyperus microiria, Brassica oleracea* var. *capitata* and *Ipomoea batatas.* Planthoppers deposit longer and more branched stylet sheaths in rice plant sap and nonhost plant saps than in distilled water. However, branched sheaths are less frequently produced in nonhost plant saps than in rice plant saps. Sogawa (1976) determined that the rice flavonoids tricin-5-glucoside, glucotricin, orizatin, and homoinetin stimulate the probing of *N. lugens.* Of the 13 other flavonoids evaluated, luteorin also stimulated the probing of *N. lugens.* Later Kim et al. (1985) isolated eight C-glycosylflavones from rice plants that stimulated *N. lugens* probing to the same degree as the whole rice plant.

4.4.1.2.2. Sucking Response
The effect of plant extracts and other known chemicals on the feeding behavior of sucking insects can be evaluated using both choice and no-choice methods. In no-choice tests, sucking responses can be determined from the amount of honeydew excreted by insects in feeding chambers composed of translucent plastic cups (6 cm in diameter × 2.5 cm high) and watch glasses (Yoshihara et al. 1979). The open end of a cup is covered with a Parafilm® membrane. Five test insects are placed in the cup through a small hole at the bottom, which is then closed with a small piece of cotton or sponge (Figure 4.40). The plastic cup is inverted on a watch glass containing about 5 ml of test solution and the solution is enclosed between the Parafilm® membrane

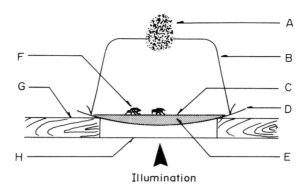

Illumination

FIGURE 4.40. A bioassay apparatus for measuring *Nilaparvata lugens* sucking on rice plant extracts. A, sponge; B, plastic cup; C, parafilm membrane; D, watch glass; E, test solution; F, *N. lugens* adult females; G, wooden plate; H, yellow cellophane paper. (From Yoshihara, T. et al., *Entomol. Exp. Appl.*, 26, 314, 1979. With permission of Kluwer Academic Publishers.)

and watch glass. The entire apparatus is then mounted in a hole (1 cm in diameter) in the bottom of a dark chamber (painted black) kept at about 27° C and 70% RH for 19 to 20 hr and illuminated from the underside by fluorescent lamps throughout the test. A sheet of yellow cellophane is used to cover the bottom of the holes and the test insects are attracted by the yellow light to the Parafilm® membrane over the test solution.

To stimulate insect sucking, sucrose (5 to 10%) is added to 10 ml of each aqueous solution of plant extract or test chemical (Sakai and Sogawa 1976). An equivalent sucrose solution is used as a control. The test solution is adjusted to pH 6 to 7 with diluted hydrochloric acid or sodium hydroxide solutions. The amount of honeydew excreted during the period is scored visually as – = no excretion; ± = trace amounts of excretion; + = excretion noticeable but less than on control; and ++ = excretion as much as or more than on control. The scores – and ± indicate an inhibitory response, while ++ indicates a stimulatory response.

In another method developed by Sogawa (1974), the sucking response of *N. lugens* was measured by incorporating $^{32}PH_3PO_4$ into the probing media. After ingesting labeled media, insects are digested in concentrated nitric acid by heating under an infrared lamp, and their radioactivity is measured with an Aloka® SC-IC automatic 2π thin window glass flow G.M. counter. Insects that feed on the labeled media can also be frozen, and the radioactivity of each insect's body measured directly in the same manner.

Kurata and Sogawa (1976) described a different bioassay to evaluate the inhibitory effects of aromatic amines on *N. lugens* sucking. Aromatic amines are dissolved in 5% sucrose solution at known concentrations, and naphthol yellow S (2, 4-dinitro-1-naphthol-7-sulfuric acid) is added to the test solutions at 0.2% as an indicator to chlorimetrically measure the amount of insect sucking. A sucrose solution containing only naphthol yellow S is used as a control.

Bioassays are conducted using an apparatus made of a glass ring (25 mm I.D. × 20 mm high). The one open end of the ring is covered by tetoron gauze, leaving a small opening at the center that is closed with a cotton plug. A certain volume of each test solution is encapsulated in a Parafilm® sachet (two sheets of stretched Parafilm®) covering the opening of the apparatus.

Five newly emerged female leafhoppers or planthoppers are introduced into the apparatus through the small opening of tetoron gauze and allowed to ingest the test solution for 22 to 24 hr at 27° C and 70% RH under continuous light. At the end of the test period, the insects are removed from the apparatus, and the test solution remaining in the Parafilm® sachet is carefully transferred into a glass tube with a capillary pipette. The solution is diluted with distilled water and the absorbance of the collected solution is measured at 430 nm with a spectrophotometer. The amount of insect imbibition (I) is calculated:

$$I = Y(Y + Z) \ \frac{A - B}{A(Y + Z) - BY}$$

where Y represents the amount of test solution enclosed in the Parafilm® sachet; Z the amount of distilled water used for dilution; A the absorbance in the control; and B the absorbance after the insect sucking. The amount of imbibition can be calculated from a standard calibration curve. The bioassays for each chemical are repeated 5 to 10 times and the results averaged.

Sogawa (1974) reported that salicylic acid adversely affects the intake of 20% sucrose solutions by *N. lugens*. With the presence of salicylic acid above 0.001 *M,* the intake of sucrose solution decreases in proportion to the increase in salicylic acid concentration. Insect sucking is almost completely inhibited at a concentration of 0.008 *M.*

The effects of several nutrient compounds on the sucking response of *N. lugens* were determined by Sakai and Sogawa (1976) using a radioactive tracer. Fluid intake by *N. lugens* is greatly enhanced in 20% sucrose solution, and the acceptability of the sucrose solution is further improved when incorporated with 0.5 to 1.0% amino acids, 0.007 to 0.014% vitamins, 0.07% mineral salts, or 0.0035% metal sequestrenes. Transaconitic acid, an antifeedant from barnyard grass, *Echinochloa crus-galli* (L.) Beauv. var *oryzicola* (Vasing.) Ohwi., is a potent inhibitor of *N. lugens* feeding at concentrations of 0.25 and 0.5% in a 15% sucrose solution.

Kurata and Sogawa (1976) determined that the aromatic amines phenethylamine hydrochloride, tyramine hydrochloride, and hordenine sulfate reduce the sucking of rice leafhoppers and planthoppers on 5% sucrose by 80 to 90% at 100 ppm and about 50% at 10 to 50 ppm.

Using a modified method of Sakai and Sogawa (1976), Shigematsu et al. (1982) identified asparagine as a sucking stimulator and β-sitosterol as a sucking inhibitor of *N. lugens*. The addition of asparagine to a 15% sucrose solution increased honeydew production 150% with 0.2% asparagine, and 200% with 1.0% asparagine. Conversely, the sucrose solutions containing 50 ppm of β-sitosterol caused a complete inhibition of *N. lugens* sucking.

In a no-choice feeding deterrent test, Dreyer et al. (1981, 1987a) visually counted the number of greenbug, *Schizaphis graminum* (Rondani), individuals feeding and wandering 2 to 3 hr after the start of the experiment and identified ρ-hydroxybenzaldehyde, dhurrin, and procyanidine (all isolated from sorghum) as major *S. graminum* feeding deterrents.

Multichoice arenas involving areas of Parafilm® which contain artificial solution have often been used (Cartier and Auclair 1964, Khan and Saxena 1985a) to assess the feeding deterrency of allelochemicals to sucking insects. The feeding chamber consists of a plastic petri dish (2.5 cm high, 14 cm diameter) painted black with 5 holes (2.5 cm diameter) equidistantly arranged in a circle (Figure 4.29). Plant extracts are placed in Parafilm® sachets that are mounted singly on circular plastic cups (3 cm long, 2.45 cm diameter) and arranged equidistantly inside the feeding chamber. Each cup has one end slightly flared to hold a Parafilm® sachet and fits snugly and level with any of the five holes in the feeding chamber. The feeding cups, along with their respective sachets, are weighed (W_1) and randomly arranged in five holes in the feeding chamber, allowing the insects a choice of five feeding sites (four treated and one control). After 24 hr the insects are removed and the feeding cups are reweighed (W_2). The difference in the weights ($W_1 - W_2$) of the feeding cups indicate the quantity of plant extract or control solution ingested by insects in 24 hr. Phagostimulation by different plant extracts is calculated as the amount of test solution ingested divided by the amount of control solution ingested. Insect phagostimulation or deterrence by a test chemical can also be determined by measuring the distribution of test insects on treated and control feeding cups at the end of the bioassay (Dreyer et al. 1981).

4.4.1.2.3. Settling Response

The antifeedant properties of phytochemicals towards sucking insects can also be tested by treating one half of the leaves of a host plant with a test compound and comparing the numbers of settled insects on treated and untreated leaf surfaces during the experiment (Dawson et al. 1982, Asakawa et al. 1988). Similar antifeedant tests can also be conducted using a Parafilm® membrane stretched on a plastic ring. Test compounds are painted onto one half of the lower surface of the membrane and insects placed in the ring are allowed to feed on the artificial diet held on the upper surface of the Parafilm® membrane. The numbers of insects settling on treated and untreated surfaces are counted during the experiment (Asakawa et al. 1988).

4.4.2. Electrophysiological Assays

4.4.2.1. Chewing Insects

In addition to feeding bioassays, phytochemicals can also be evaluated using electrophysiological assays, that are based on the sensory codes by which phytophagous insects distinguish palatable from unpalatable foods. The chemosensory systems of different insects differ markedly from each other. For example, the maxillary styloconic sensilla of lepidopteran larvae and the locust, *Locusta migratoria* (L.), are important in the recognition of food (Schoonhoven 1969, Blaney 1974, Städler and Hanson 1976, Blaney et al. 1987, Blaney and Simmonds 1988, Fu-Shun et al. 1990, Waladde et al. 1990), whereas contact chemosensilla on tibias and tarsi of the American grasshopper, *Schistocerca americana* (Drury), also play important roles in food selection (White and Chapman 1990, Chapman et al. 1991).

Antifeedants may be perceived by either stimulation of specialized deterrent receptors or by their capacity to distort the normal function of neurons which perceive phagostimulating chemicals (Schoonhoven 1982, 1987). Many antifeedants may influence feeding activity through a combination of those two modes of action.

Electrophysiological recordings are obtained using the classical sensillum tip recording method of Hodgson et al. (1955) and Blaney (1974). Briefly, a 15 mm long (1 mm outside diameter) glass tube is drawn to a tip diameter of about 0.05 mm and filled with the test phytochemical solution in 0.05 M sodium chloride. The tube containing the solution serves both as stimulator and a recording electrode.

The test insect is anesthetized briefly (<2 min) with CO_2 and immobilized using sticky wax to expose the chemosensory sensillum being tested. The tube containing stimulating solution is connected to a cathode followed by a Ag/AgCl wire inserted into the large end of the tube. The glass capillary with the test solution and the recording electrode is applied to the sensillum (Figure 4.41). An indifferent electrode of Ag/AgCl is inserted into the haemolymph adjacent to the chemosensory sensillum. The recording electrode is maneuvered by a micromanipulator until the tip of a single chemosensory hair just penetrates the surface film of the electrode solution. The indifferent electrode is then connected via a back-off DC source and a calibration source to earth. The calibration source provides a square wave of 50 mV (peak to peak) with a frequency of 1 kHz.

Electrical signals from the sensilla are fed from the recording electrode through an amplifier to a cathode follower (Figure 4.42). Amplified signals can be displayed on a cathode ray oscilloscope (CRO) and recorded on a tape recorder. For analysis, these signals can also be printed on a strip chart recorder or photographed on oscillograph paper.

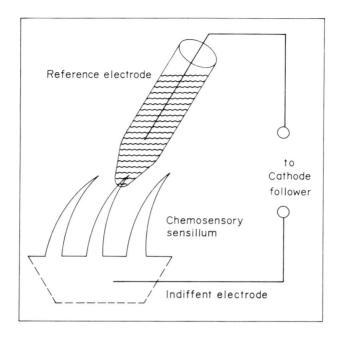

FIGURE 4.41. Diagram of experimental preparation for electrophysiological recordings from a sensillum tip. (From Hodgson, E. S. et al., *Science,* 122, 417, Copyright 1955 by the American Association for the Advancement of Science. With permission.)

Alterations of chemoreceptor function by plant-derived extracts or pure chemicals have been shown in *L. migratoria* with electrolytes and nonelectrolytes (Blaney 1974); in the nutgrass armyworm, *Spodoptera exempta* (Walker), larvae with a sesquiterpenoid (Ma 1977); in *S. littoralis, H. virescens* and *H. armigera* with a range of sugars, amino acids, and alcohols (Blaney and Simmonds 1988); in *S. exempta, S. littoralis,* the fall armyworm, *S. frugiperda, H. armigera* (Hubner), and *H. virescens* (F.) with natural and synthetic clerodane diterpenoids (Blaney et al. 1988); in the sugarcane stem borer, *Eldana saccharina* (Walker), with limonids (Waladde et al. 1989); and in *S. americana* with nicotine hydrogen tartrate, quinine, hordenine, and salicin (Chapman et al. 1991).

The tarsal receptors of adult lepidopterans have also been shown to respond to salts and other plant chemicals (Ma and Schoonhoven 1973, Waladde 1983). Waladde et al. (1989) compared the neural responses from the maxillary styloconic sensilla of *C. partellus* to aqueous extracts from susceptible and resistant maize varieties, and demonstrated that the extracts stimulated different functional forms of impulse frequencies, suggesting differences in the quality of gustatory stimuli in the two maize varieties.

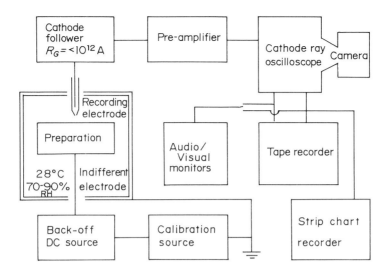

FIGURE 4.42. A schematic diagram of recording arrangements to study electrophysiological responses of the terminal sensilla on the maxillary palps of *Locusta migratoria* to some electrolytes and nonelectrolytes. (From Blaney, W. M., *J. Exp. Biol.*, 60, 275, 1974. With permission of the Company of Biologists Ltd.)

4.4.2.2. Sucking Insects

An electronic method of recording insect feeding on plants (McLean and Kinsey 1964) (see Chapter 2.2.2.1.5) can be used successfully to evaluate plant chemicals by recording subtle changes in insect feeding activity through time. Using a DC variant of the electronic monitor, Saxena and Khan (1985a) monitored the feeding behavior of *N. virescens* on rice plants sprayed with neem (*Azadirachta indica* A. Juss) oil. Using a quick-spray atomizer, different concentrations of neem oil were sprayed on rice plants and the feeding activity of *N. virescens* was monitored using the method described earlier (see Section 4.3.6.2.3). The application of neem oil to rice plants disrupted the normal feeding behavior of *N. virescens* on neem-treated plants, significantly reducing phloem feeding, increasing the frequency of probing, and the duration of salivation and xylem feeding.

The effects of phytochemicals, incorporated in an artificial diet at different concentrations and encapsulated in a Parafilm® membrane, can also be evaluated with the electronic feeding monitor. Raman et al. (1979) evaluated total glycoalkoid fractions extracted from 10 resistant *Solanum* sp. for their effects on potato leafhopper, *Empoasca fabae* (Harris), feeding. Waveform patterns associated with *E. fabae* artificial diet containing total glycoalkaloid fractions demonstrated that the mean salivation-ingestion period was significantly shorter on diets containing glycoalkaloid fractions from resistant *Solanum* sp. The reduction in feeding time resulted from repeated withdrawal of stylets from the artificial diet.

4.4.3. Growth and Survival

Phytochemicals that interfere with the digestion and assimilation of insect food are often associated with antibiotic resistance and are of major importance in regulating insect growth and survival. The antibiotic activity of these chemicals can be evaluated by incorporating them in an insect's standard diet and comparing insect growth and survival on the treated diet with a control diet (see Chapters 2.2.4.1 and 7.2.3). Plant extracts or pure allelochemicals (equal to those naturally occurring in plants) are weighed and mixed with the diet. Water soluble compounds are dissolved in water and added to the diet in place of the prescribed amount of water. Chemicals or extracts soluble in organic solvents are added in the diet by either (a) diluting them in the solvent with which they were extracted, adding them to the appropriate amount of dry diet mix and evaporating them to dryness in a rotary evaporator under reduced pressure; or by (b) coating them onto an inert carrier such as alphacel and drying off the solvent with a stream of nitrogen in a vacuum dessicator. The phytochemical-coated alphacel is then mixed with a dry portion of an artificial diet. Controls consist of artificial diets amended with untreated alphacel.

By placing eggs or newly emerged larvae on treated and control diets, antibiosis can be determined by recording larval weight, pupal weight, larval life, pupal life, larval mortality, pupal mortality, and adult mortality.

Using such a technique, Meisner et al. (1977b) demonstrated that as little as 0.25% gossypol acetate in a semisynthetic diet significantly increased larval mortality and significantly reduced the pupation of *S. littoralis* larvae, compared to a control diet. Pinatol and maysin, larval growth inhibitors from maize, inhibit the growth of corn earworm, *H. zea,* larvae when added to an artificial diet (Dreyer et al. 1979, Reese et al. 1982). Chavicol, another larval growth inhibitor extracted from leaves of *Viburnum japonicum* Spreng., kills larvae of the fruit fly, *Drosophila melanogaster,* at a concentration of 1700 ppm (Ohigashi and Koshimizu 1976). Sendanin, a limonoid, also inhibits growth of the pink bollworm, *Pectinophora gosypiella* Saunders, *H. zea, H. virescens* and *S. frugiperda* (Kubo and Klocke 1982).

In order to understand the mechanisms of resistance of two lucerne varieties to a French biotype of the pea aphid, *Acyrthosiphon pisum* Harris, growth and survival of the aphid were investigated on holidic diets differing either in their amino acid or sucrose concentrations, or by the relative proportions of some amino acids (Rahbe et al. 1988). Groups of 20 newly emerged larvae were transferred directly onto artificial diet sachets. The mortality during the developmental period, the weights of adults, and the developmental time were noted. Aphid growth and survival on artificial diet only partially confirmed the assays of resistance on plants because of the lack of understanding of allelochemical factors present in phloem sap.

Todd et al. (1971) evaluated effects of phenolic and flavonoid compounds on the growth of the greenbug, *S. graminum,* by incorporating these chemicals into an artificial diet. Weight gain and survival of *S. graminum* was significantly reduced on diets containing catechol, gentisic acid, quercetin, tannic acid, and chlorogenic acid. Inclusion of ascorbic acid, farneson, syringadehyde, and quinic acid in the artificial diet allowed the greenbugs to gain more weight. A similar technique was used by Cress and Chada (1971) to evaluate effects of zinc, iron, manganese, and copper on the development of *S. graminum.*

4.4.4. Ovipositional Responses

For many phytophagous insects, the selection of a host plant by mobile adults is crucial in determining the fitness of their less mobile offspring. Egg-laying females identify suitable plants by chemical, visual and tactile cues that characterize such plants. Plant allelochemicals serve in the final stages of oviposition behavior as the cues permitting females to discriminate and to recognize their hosts (David and Gardiner 1962, Nishida 1977, Saxena and Goyal 1978, Rodman and Chew 1980). Oviposition by a particular insect may also be deterred by chemicals that occur in nonhost plants, but that do not occur in host plants (Mitchell and Heath 1985). Oviposition deterrency by plant compounds could play a significant role in future efforts to breed insect resistant-plants.

Various bioassay techniques have been developed to assess the role of phytochemicals on insect oviposition. Depending upon the oviposition behavior of the test insect, phytochemicals can be applied to the host plant (Tingle and Mitchell 1984, Tabashnik 1985, 1987, Mitchell and Heath 1987, Renwick et al. 1989, Unithan and Saxena 1990, Severson et al. 1991, Dimock and Renwick 1991, Jackson et al. 1991), filter paper (Nishida and Fukami 1989, Reed et al. 1989, Harris and Rose 1990), muslin or broadcloth (Saxena and Goyal 1978, Mitchell and Heath 1987, Mitchell et al. 1990), fiberglass (Elsey and McFadden 1981), or to an artificial surface or media (Sekido and Sogawa 1976, Fatzinger and Merkel 1985, Eisemann and Rice 1985, Honda 1990, Städler and Schoni 1990). Generally, an ovipositional preference test is conducted with a control in a two-choice or a multichoice arrangement. The ovipositional response of an insect to a chemical is measured either by observing the insect's response to the substrate, such as curling of the abdomen of butterflies (Nishida and Fukami 1989), or by counting the numbers of eggs laid on treated and control surfaces. An index of oviposition discrimination has been calculated as $(E - C/E + C) \times 100$; where E is the sum of all eggs laid on all test substrates and C is the sum of all eggs laid on the corresponding control substrates (Städler and Schöni 1990).

Tingle and Mitchell (1984) calculated the percent reduction in oviposition due to deterrents as:

$$\frac{\% \text{ eggs in control} - \% \text{ eggs in treatment}}{\% \text{ eggs in control}} \times 100.$$

4.4.4.1. Using Plants or Plant Parts

4.4.4.1.1. Laboratory Bioassays

In choice tests, one plant treated with control solution and another treated with a phytochemical are often placed in diagonally opposite corners of a cage to assess insect oviposition. Eggs are counted after each trial, and plant positions in the cage are alternated in successive trials. Several workers have used tests consisting of four or five treated plants that are placed alternately with four or five untreated (control) plants in a circular pattern inside a cage. Each plant is rotated clockwise or counterclockwise daily for the duration of the experiment. In no-choice tests, females are placed on either a control plant or a treated plant, placed in the center of a cage, and the numbers of eggs are recorded after each trial.

To obtain further information, female behavior is also observed during the assays to determine how phytochemicals influence their pre- and postalighting behavior. General contact is often recorded when females touch a plant with one or more legs.

To determine if an insect's oviposition response can be reversed on susceptible and resistant plants, the extract of a susceptible plant, diluted at a known concentration in a suitable solvent is sprayed uniformly on potted resistant plants and vice versa. The treated plants are then exposed to gravid females in separate cages. Untreated resistant and susceptible plants caged similarly with females serve as controls. The number of eggs laid in each treatment are then counted.

Unithan and Saxena (1990) demonstrated that seedling maize, a nonhost for the sorghum shoot fly, *Atherigora soccata* Rondani, when sprayed with an acetone extract of seedling sorghum, was equivalent to sorghum seedlings for *A. soccata* oviposition. In a laboratory experiment, Khan and Hatchett (1993) demonstrated that oat plants, a nonhost, when sprayed with chloroform extract of wheat plants, received significantly more Hessian fly, *Mayetiola destructor* (Say), eggs than control oat plants. Conversely, wheat plants sprayed with oat extract received significantly fewer eggs than control wheat plants.

Mitchell and Heath (1985) used treated and control leaves of the same intact plant to determine the effects of allelochemicals of *Amaranthus hybridus* L. on the oviposition behavior of the armyworms, *Spodoptera exigua* (Hübner) and *S. eridania* (Cramer). Test plants were stripped to six leaves or three leaves per plant and sprayed with leaf extract. Plants were exposed to ovipositing females in a cage and the number of eggs laid on control and treated leaves on each plant was recorded.

Wilson et al. (1988) and Nottingham et al. (1989a) developed a technique to bioassay oviposition stimulants and deterrents for the sweet potato weevil, *C. formicarius elegantulus,* using sweet potato storage root cores. Sweet potato cores, each cut with no. 11 corkborer through the periderm, were pushed into wells of a 24-well Falcon tissue culture plate so that the periderm was exposed for oviposition. The sweet potato cores, each treated with a test chemical or control solvent, were arranged randomly. Thirty to sixty adult females, held for 3 to 6 hr without food, were then allowed to oviposit on all sweet potato cores and the number of eggs laid in each core was counted after 48 hr.

4.4.4.1.2. Field Tests

Extract and solvent-sprayed plants may be exposed to oviposition by either natural insect populations in open fields (Tingle and Mitchell 1984, Dimock and Renwick 1991) or laboratory-reared gravid females used to supplement natural field populations (Unithan and Saxena 1990). Dimock and Renwick (1991) demonstrated that cabbage plants sprayed in the field with a butanol extract of *Erysimum cheiranthoides,* a wild nonhost crucifer, received significantly fewer eggs than solvent-treated control plants. In a similar field-grown experiment, maize seedlings sprayed with sorghum seedling extract received numbers of *A. soccata* eggs similar to that laid on sorghum plants (Unithan and Saxena 1990).

The percent change in oviposition in field experiments has been calculated by Tingle and Mitchell (1984) as

$$\frac{\% \text{ of eggs in control} - \% \text{ of eggs in treatment}}{\% \text{ eggs in control}} \times 100.$$

4.4.4.2. Using Surrogate Leaves

Surrogate or artificial green plastic leaves, shaped like a host plant leaf, can be used to evaluate the ovipositional response of an insect to a phytochemical (Honda 1986, 1990, Städler and Schöni 1990). The surrogate leaf is dipped into an aqueous solution of a test chemical of known concentration and air dried at room temperature. If the chemical is dissolved into an organic solvent, the surrogate leaf is coated with a thin layer of paraffin wax by a short immersion into warm water (54 to 56° C) with a floating layer (2 to 6 mm) of the melted wax on top. The test solutions or control solvents are sprayed evenly onto the whole surface of the surrogate leaves using a sprayer. Treated and control surrogate leaves are presented to ovipositing females in a cage in a choice test and the numbers of eggs laid on treated and control leaves are then recorded.

Städler and Schöni (1990) coated surrogate plastic leaves with a thin layer of paraffin wax and treated them with 0.1 g equivalent of a raw ethanolic cabbage extract. When exposed to these surrogate leaves, cabbage root fly, *Delia radicum* (L.), females displayed the same sequence of behavioral patterns as on cabbage leaves and oviposited similarly on the two surfaces.

4.4.4.3. Using Filter Paper

Filter paper can be displayed in numerous shapes and sizes to serve as an oviposition substrate. Test samples have been applied either to intact filter paper disks (Reed et al. 1989), strips of cut filter paper (Harris and Rose 1990), or fan-shaped pieces of filter paper (Nishida and Fukami 1989).

Reed et al. (1989) identified glucosinolate oviposition stimulants for the diamondback moth, *Plutella xylostella* (L.), from three plant species of Brassicacaeae, using whole filter paper discs. Bioassays are conducted in clear plastic chambers (30 × 30 × 20 cm) with screened tops. A drawer in the side of the chamber is attached to three parallel horizontal metal rods allowing for the insertion of a 18.5 cm diameter Whatman no. 1 filter paper disk containing plant extracts. The filter paper is marked off into 6 to 12 equal-sized radial sectors, and test materials or solvent control are placed on every sector randomly. Disks are allowed to dry and are then exposed to females for oviposition.

Harris and Rose (1990) and Khan and Hatchett (1993) used filter paper strips (10 × 0.5 cm) to evaluate the ovipositional behavior of *M. destructor* to extracts of host and nonhost plants (Figure 4.43). Appropriate amounts of extracts are applied to filter paper strips and the solvent is allowed to evaporate for 60 min before the treated paper strips are placed in cages. Chloroform controls are prepared similarly. One end of the filter paper strip is placed (long axis vertical) in a metal clip standing in moist sand in a pot (15 cm diameter). Cylindrical plastic cages (15 cm diameter and 20 cm high) with mesh screen tops are then placed over the sand. Within each pot, two treated and two control filter paper strips are placed alternately and equidistantly from the center. Ten or fifteen female *M. destructor* are released in the center of each cage and the number of eggs deposited on each filter paper strip is counted. Both studies demonstrate that *M. destructor* females oviposit significantly more on filter paper strips treated with chloroform extract from wheat seedlings than control strips and those treated with oat or barley extracts.

Nishida and Fukami (1989) applied allelochemicals to a piece of filter paper by cutting a filter paper disk (9 cm diameter) into six equal fan-shaped pieces to study the ovipositional response of a Japanese pipevine swallowtail butterfly, *Atrophaneura alcinous* Klug (Figure 4.44). An artificial blend of arisolochic acid and sequoyitol applied to a filter paper "fan" stimulated an ovipositional response identical to that elicited by intact leaves of *Aristolochia debilis* Sieb et Zucc.

FIGURE 4.43. An oviposition chamber with two treated (applied with epicuticular lipid extracts from wheat plants) and two control filter paper strips to test ovipositional response of Hessian flies. (Photograph provided by Z. R. Khan.)

4.4.4.4. Using Fiber Glass

Elsey and McFadden (1981) used pieces of fiber glass building insulation 56 cm^2 × 1 cm thick as an artificial surface for pickleworm, *Diaphania nitidalis* (Stoll), oviposition assays. Plant extract or solvent is sprayed on a fiberglass sheet, the solvents are allowed to evaporate and the sheet is then hung in a cage of moths. The ethanol extract of squash leaves applied to the fiberglass sheet is equivalent to fresh squash leaves as an oviposition site for *D. nitidalis.*

4.4.4.5. Using Gelatin-Alphacel/Agar

Nair and McEwen (1976) and Nair et al. (1976) developed bioassays to evaluate the effects of glucosinolates and common nutrients on the oviposition behavior of the cabbage maggot, *Hylemya brassicae* (Bouche). The adult flies lay eggs in the soil close to the host plant chemicals, and little oviposition occurs when access to such allelochemicals is denied. Oviposition stimulant/deterrent bioassays were conducted in glass petri dishes 9 cm in diameter filled with soil surrounding a gelatin-alphacel block (3 cm diameter) containing test allelochemicals. To prepare the block, 4 ml of a warm (37° C) 15% aqueous solution of gelatin and 1 ml of water are added to 250 mg alphacel. The mixing is continued until solidification begins, and test materials are then

FIGURE 4.44. Ovipositional response of *Atrophaneura alcinous* female to a piece of filter paper treated with methanolic extract of *Aristolochia debilis*. (From Nishida, R. and Fukami, H., *J. Chem. Ecol.*, 15, 2565, 1989. With permission of Plenum Publishing Corporation.)

incorporated into the mixture. Water-soluble chemicals are dissolved in water and an appropriate volume of the solution replaces the equivalent volume of water used in preparation of the block. Chemicals soluble in organic solvent are applied on the surface of the block and the solvent removed by vacuum evaporation. Control blocks are prepared similarly. Blocks are placed in separate petri dishes and the soil is lightly packed around them. *H. brassicae* females are then given a choice of treated or control gelatin-alphacel blocks and the number of eggs laid in the soil of each dish is recorded. These bioassays indicate that glucosinolates are major oviposition-inducing substances present in cruciferous plants (Nair and McEwen 1976, Nair et al. 1976).

Eiseman and Rice (1985) developed an agar substrate bioassay to determine the effects of some salts and sugars on the oviposition behavior of the fruit fly, *Dacus tryoni* (Frogg.), in specially adapted 9-cm-diameter petri dishes with their bases partitioned into quadrants (Figure 4.45). The quadrants are filled with 0.75% agar gel, either plain or amended with a test

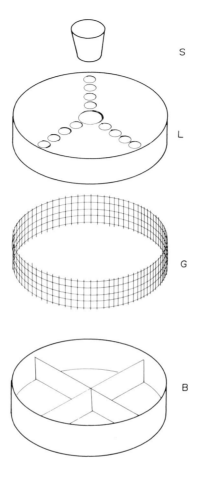

FIGURE 4.45. An ovipositional chamber for use with agar substrates for studying oviposition behavior of *Dacus tryoni*. S, rubber stopper; I, petri dish lid with ventilation holes; G, wire gauze cylinder; B, modified partitioned petri dish base. (From Eisemann, C. H. and Rice, M. J., *Entomol. Exp. Appl.*, 39, 61, 1985. With permission of Kluwer Academic Publishers.)

chemical. In choice experiments, opposite quadrants have the same substrate mixtures, adjacent quadrants have different mixtures, and comparisons are made within each dish. In no-choice experiments, all quadrants contain the same substrate and comparisons are made between treatment dishes. Eight to ten gravid female flies are added to each of the test chambers. Flies are either allowed to walk on the surface of the agar and lay eggs, or the surface is covered with a layer of Parafilm® membrane through which the flies readily oviposit. Petri dishes are examined under a dissecting microscope for eggs. β-D(–) fructose stimulated oviposition by *D. tryoni* while calcium chloride deters flies from ovipositing on agar (Eiseman and Rice 1985).

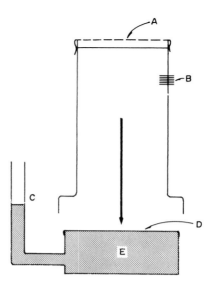

FIGURE 4.46. Schematic representation of an oviposition cage to collect rice leafhopper and planthopper eggs. A, screen mesh; B, hole for introducing insects; C, a side arm; D, stretched parafilm; E, liquid medium. (From Mitsuhashi, J., *Appl. Entomol. Zool.*, 5, 47, 1970. With permission of the Japanese Society of Applied Entomology and Zoology.)

4.4.4.6. Using Liquid Media

A method using a liquid medium for measuring the ovipositional response of rice leafhoppers and planthoppers to different chemicals was developed by Mitsuhashi (1970) and later modified by Mitsuhashi and Koyama (1975), Koyama and Mitsuhashi (1976), and Sekido and Sogawa (1976).

The oviposition cage (Figure 4.46) described by Mitsuhashi (1970) was a modification of the feeding cage used for leafhoppers designed by Fulton and Chamberlin (1934). In this cage, a stretched Parafilm® membrane is used to separate the test insects from a liquid medium (5 to 10% sucrose solution) or plant extract. The liquid medium is introduced to the medium container through the side arm after fixing the stretched Parafilm® membrane to the top of the container. Gravid female test insects are placed into the assembled cage kept at 27° C in a 16L:8D photoperiod. The insects suck the medium through the Parafilm® and also oviposit through it. Some eggs sink to the bottom of the container, but most are attached to the Parafilm® membrane. The numbers of eggs laid on each test medium are counted and expressed as the ovipositional response.

In the modified method described by Mitsuhashi and Koyama (1975), insects are given a choice of two test solutions for oviposition. Gravid females are individually caged in a glass cylinder (33 mm in diameter × 45 mm long), both openings of which are covered with a stretched Parafilm® membrane. The test solutions are placed on both sides of the stretched membranes, and enclosed with another sheet of Parafilm®. The experiment is conducted at 27° C, 16L:8D photoperiod, and the cages are placed horizontally. The insects are allowed to oviposit for 24 hr, and the number of eggs laid on both sides are recorded.

Mitsuhashi and Koyama (1975) evaluated various carbohydrates for the ovipositional responses of the smaller brown planthopper, *Laodelphax striatellus* (Fallén). Their results determined the maximum oviposition response of *L. striatellus* on a 5% sucrose solution. Koyama and Mitsuhashi (1976) reported that among 23 amino acids tested individually, arginine, glutamic acid, tryosine, and valine showed inhibitory effects upon oviposition of *L. striatellus,* while the influence of cystine was stimulative. Sekido and Sogawa (1976) reported that *N. lugens* oviposited preferably in a salicylic acid media, while only a few eggs were laid in a salicylic acid-free control medium. However, salicylic acid media was not preferred for oviposition by *L. striatellus.*

4.4.5. Egg Hatchability

The hatchability of insect eggs can differ between varieties of plants (Saxena and Pathak 1977). Several plant chemicals are ovicidal in nature and cause a decrease in the rate of egg hatch. The bioassays of Dreyer et al. (1987b) or Saxena and Puma (1979) serve well to illustrate the effects of allelochemicals on egg hatchability.

Dreyer et al. (1987b) evaluated the effects of alfalfa allelochemicals on egg hatchability of the Egyptian alfalfa weevil, *Hypera brunneipennis* (Boheman). A filter paper is impregnated with an aliquot of plant chemical solution. The solvent is allowed to evaporate, and the filter paper is dampened with sterile water, placed in a petri dish with eggs on the paper, and is incubated at 25° C and 80% humidity. The numbers of eggs hatched are counted after 5 to 7 days and compared with oviposition on the appropriate controls kept under the same conditions. This bioassay indicated that 3-acyl-4-hydroxylcoumarins with short chain alkyl groups reduce hatchability of *H. brunneipennis* eggs.

Saxena and Puma (1979) determined the effects of trans-aconitic acid on the egg hatchability of *N. lugens*. Freshly laid eggs of *N. lugens* were immersed into different concentrations of trans-aconitic acid solution for different periods. Saxena and Puma (1979) reported that immersion of *N. lugens* eggs in 0.5% trans-aconitic acid solution for 3 days or longer significantly reduced their hatchability.

4.5. CONCLUSIONS

There are tremendous advantages to the full elucidation of the allelochemical bases of insect resistance in plants. The identification of the chemicals that confer insect resistance or susceptibility has greatly improved the ability to breed for resistant varieties, but much greater use of this information could occur. Such studies will also open new avenues for the manipulation of insect behavior for use in insect pest management programs.

The research techniques described in this chapter have greatly improved the accuracy of allelochemical/insect pest research and aided in the understanding of the chemical bases of insect resistance. Advances in allelochemical extraction, collection, identification, and bioassay have added to our understanding of the effects of both volatile and nonvolatile plant allelochemicals on insect plant selection, acceptance, feeding, and reproduction. However, there is a continued need for improved rapid, accurate, economical microanalytical techniques to determine the allelochemical factors mediating insect resistance in plants.

REFERENCES

Adams, C. M. and Bernays, E. A., The effect of combinations of deterrents on the feeding behavior of *Locusta migratoria, Entomol. Exp. Appl.,* 23, 101, 1978.

Akeson, W. R., Manglitz, G. R., Gorz, H. J. and Haskins, F. A., A bioassay for detecting compounds which stimulate or deter feeding by the sweet clover weevil, *J. Econ. Entomol.,* 60, 1082, 1967.

Albert, P. J. and Parisella, S., Tests for induction of feeding preferences in larvae of eastern spruce budworm using extracts from three host plants, *J. Chem. Ecol.,* 11, 809, 1985.

Alfaro, R. I., Pierce, Jr., H. D., Borden, J. H. and Oehlschlager, A. C., A quantitative feeding bioassay for *Pissodes strobi* Peck (Coleoptera: Curculionidae), *J. Chem. Ecol.,* 5, 663, 1979.

Alfaro, R. I., Pierce, Jr., H. D., Borden, J. H. and Oehlschlager, A. C., Insect feeding and oviposition deterrents from western red cedar foliage, *J. Chem. Ecol.,* 7, 39, 1981.

Alford, A. R., Cullen, J. A., Storch, R. H. and Bentley, M. D., Antifeedant activity of limonin against the Colorado potato beetle (Coleoptera: Chrysomelidae), *J. Econ. Entomol.,* 80, 575, 1987.

Altman, D. W., Stipanovic, R. D. and Benedict, J. H., Terpenoid aldehydes in upland cottons. II. Genotype-environment interactions, *Crop Sci.,* 29, 1451, 1989.

Andersen, J. F., Mikolajczak, K. L. and Reed, D. K., Analysis of peach bark volatiles and their electroantennogram activity with lesser peachtree borer, *Synanthedon pictipes* (Grote and Robinson), *J. Chem. Ecol.,* 13, 2103, 1987.

Arn, H., Stadler, E. and Rauscher, S., The electroantennographic detector—a sensitive tool in the gas chromatographic analysis of insect pheromones, *Z. Naturforsch.,* 30C, 722, 1975.

Asakawa, Y., Dawson, G. W., Griffiths, D. C., Lallemand, J.-Y., Ley, S. V., Mori, K., Mudd, A., Pezechk-Leclaire, M., Pickette, J. A., Watanabe, H., Woodcock, C. M. and Zhong-Ning, Z., Activity of drimane antifeedants and related compounds against aphids, and comparative biological effects and chemical reactivity of (–)- and (+)-polygodial, *J. Chem. Ecol.,* 14, 1845, 1988.

Ascher, K. R. S. and Meisner, J., Evaluation of a method for assay of phagostimulants with *Spodoptera littoralis* larvae under various conditions, *Entomol. Exp. Appl.,* 16, 101, 1973.

Ascher, K. R. S. and Nemny, N. E., Use of foamed polyurethane as a carrier for phagostimulant assays with *Spodoptera littoralis* larvae, *Entomol. Exp. Appl.,* 24, 346, 1978.

Ascoli, A. and Albert, P. J., Orientation behavior of second-instar larvae of eastern budworm, *Choristoneura fumiferana* (Clem.) (Lepidoptera: Tortricidae) in a Y-tube olfactometer, *J. Chem. Ecol.,* 11, 837, 1985.

Baehrecke, E. H., Williams, H. J. and Vinson, S. B., Electroantennogram responses of *Campoletis sonorensis* (Hymenoptera: Ichneumonidae) to chemicals in cotton (*Gossypium hirsutum* L.), *J. Chem. Ecol.,* 15, 37, 1989.

Barrows, W. M., The reactions of the pomace fly, *Drosophila ampelophila* Lowe., to odorous substances, *J. Exp. Zool.,* 4, 515, 1907.

Beevor, P. S, Hall, D. R., Lester, R., Poppi, R. G., Read, J. S. and Nesbitt, B. F., Sex pheromones of the armyworm moth, *Spodoptera exempta* (Wlk.), *Experientia,* 31, 22, 1975.

Benincasa, M., Buiarelli, F., Cartoni, G. P. and Coccioli, F., Analysis of lemon and bergamot essential oils by HPLC with microbore columns, *Chromatographia,* 30, 271, 1990.

Bentley, M. D., Leonard, D. E., Stoddard, W. F. and Zalkow, L. H., Pyrrolizidine alkaloids as larval feeding deterrents for spruce budworm, *Choristoneura fumiferana* (Lepidoptera: Tortricidae), *Ann. Entomol. Soc. Am.,* 77, 393, 1984.

Benz, G., Abivardi, C. and Muckensturm, B., Antifeedant activity of bisabolangelone and its analogs against larvae of *Pieris brassicae, Entomol. Exp. Appl.,* 53, 257, 1989.

Berezkin, V. G., Alishoyev, V. R. and Nemirovskaya, I. B., *Gas Chromatography of Polymers,* Elsevier, New York, 1977, 233.

Beroza, M., Current usage and some recent developments with insects attractants in the U.S.D.A., in *Chemicals Controlling Insect Behavior,* M. Beroza, ed., Academic Press, New York, 1970, 170.

Besson, E., Dellamonica, G., Chopin, J., Markham, K. R., Kim, M., Koh, H. and Fukami, H., C-glycosylflavones from *Oryza sativa, Phytochemistry,* 24, 1061, 1985.

Binder, R. G. and Waiss, Jr., A. C., Effects of soybean leaf extracts on growth and mortality of bollworm (Lepidoptera: Noctuidae) larvae, *J. Econ. Entomol.,* 77, 1585, 1984.

Blaney, W. M., Electrophysiological responses of the terminal sensilla on the maxillary palps of *Locusta migratoria* (L.) to some electrolytes and non-electrolytes, *J. Exp. Biol.,* 60, 275, 1974.

Blaney, W. M. and Simmonds, S. J., Food selection in adults and larvae of three species of Lepidoptera: a behavioural and electrophysiological study, *Entomol. Exp. Appl.,* 49, 111, 1988.

Blaney, W. M., Simmonds, M. S. J., Ley, S. V. and Katz, R. B., An electro-physiological and behavioural study of insect antifeedant properties of natural synthetic drimane-related compounds, *Physiol. Entomol.,* 12, 281, 1987.

Blaney, W. M., Simmonds, M. S. J., Ley, S. V. and Jones, P. S., Insect antifeedants: a behavioural and electrophysiological investigation of natural and synthetically derived clerodane diterpenoids, *Entomol. Exp. Appl.,* 46, 267, 1988.

Blust, M. H. and Hopkins, T. L., Gustatory responses of a specialist and a generalist grasshopper to terpenoids of *Artemisia ludoviciana, Entomol. Exp. Appl.,* 45, 37, 1987a.

Blust, M. H. and Hopkins, T. L., Olfactory responses of a specialist and a generalist grasshopper to volatiles of *Artemisia ludoviciana* Nutt. (Asteraceae), *J. Chem. Ecol.,* 13, 1893, 1987b.

Bodnaryk, R. P., Developmental profile of sinalbin (p-hydroxybenzyl glucosinolate) in mustard seedlings, *Sinapis alba* L., and its relationship to insect resistance, *J. Chem. Ecol.,* 17, 1543, 1991.

Boeckh, J., Inhibition and exitation of single insect olfactory receptor cells and their role as a primary sensory code, in *Olfaction and Taste II,* Vol. VIII, Pergamon Press, London, 1967, 835.

Boeckh, J., Electrical activity in olfactory receptor cells, in *Olfaction and Taste III,* C. Pfaffmann, ed., Rockefeller University Press, New York, 1969, 648.

Boeckh, J., Kaissling, K. E. and Schneider, D., Insect olfactory receptors, *Cold Spring Harbor Symp. Quant. Biol.,* 30, 263, 1965.

Bowers, M. D. and Puttick, G. M., Response of generalist and specialist insects to quantitative allelochemical variation, *J. Chem. Ecol.,* 14, 319, 1988.

Brewer, G. J., Sorensen, E. L. and Horber, E. K., Attractiveness of glandular and simple-aired *Medicago* clones with different degrees of resistance to the alfalfa seed chalcid (Hymenoptera: Eurytomidae) tested in an olfactometer, *Environ. Entomol.,* 12, 1504, 1983.

Bristow, P. R., Doss, R. P. and Campbell, R. L., A membrane filter bioassay for studying phagostimulatory material in leaf extracts, *Ann. Entomol. Soc. Am.,* 72, 16, 1979.

Bromley, A. K. and Anderson, M., An electrophysiological study of olfaction in the aphid *Nasonovia biris-nigri, Entomol. Exp. Appl.,* 32, 101, 1982.

Budenberg, W. J., Ndiege, I. O., Karago, F. W. and Hansson, B. S., Behavioral and electrophysiological responses of the banana weevil, *Cosmopolites sordidus* to host plant volatiles, *J. Chem. Ecol.,* 19, 267, 1993.

Busch, K. L. and Cooks, R. G., Mass spectrometry of large, fragile, and involatile molecules, *Science,* 218, 247, 1982.

Cannon, Jr., W. N., Olfactory response of eastern spruce budworm larvae to spruce needles exposed to acid rain and elevated levels of ozone, *J. Chem. Ecol.,* 16, 3255, 1990.

Cartier, J. J. and Auclair, J. L., Pea aphid behaviour: colour preference on a chemical diet, *Can. Entomol.,* 96, 1240, 1964.

Chan, B. G., Waiss, Jr., A. C., Stanley, W. L. and Goodban, A. E., A rapid diet preparation method for antibiotic phytochemical bioassay, *J. Econ. Entomol.,* 71, 366, 1978.

Chapman, R. F., Bernays, E. A. and Wyatt, T., Chemical aspects of host-plant specificity in three *Larrea*-feeding grasshoppers, *J. Chem. Ecol.,* 14, 561, 1988.

Chapman, R. F., Ascoli-Christensen, A. and White, P. R., Sensory coding for feeding deterrence in the grasshopper *Schistocerca americana, J. Exp. Biol.,* 158, 241, 1991.

Charpentier, B. A., Sevenants, M. R. and Sanders, R. A. Comparison of the effects of extraction methods on the flavor volatile composition of Shitake mushrooms (*Lentinus edodes*) via GC/MS and GC/FTIR, *Dev. Food Sci.,* 12, 413, 1986.

Chyau, C. C. and Wu, C. M., Differences in volatile constituents between inner and outer flesh-peel of guava (*Psidium guajava* L.) fruit, *Lebensm. Wiss. Technol.,* 22, 104, 1989.

Contreras, M. L., Perez, D. and Rozab, R., Empirical correlations between electroantennograms and bioassays for *Periplaneta americana, J. Chem. Ecol.,* 15, 2539, 1989.

Cook, A. G., A critical review of the methodology and interpretation of experiments designed to assay the phagostimulatory activity of chemicals to phytophagous insects, *Symp. Biol. Hung.,* 16, 47, 1976.

Cork, A., Boo, K. S., Dunkelblum, E., Hall, D. R., Ja-Rajunga, K., Kehat, M., Kong Jie, E., Park, K. C., Tepgidagarn, P. and Xun, L., Female sex pheromone of oriental tobacco budworm, *Helicoverpa assulta* (Guenée) (Lepidoptera: Noctuidae): Identification and field testing, *J. Chem. Ecol.,* 18, 403, 1992.

Cress, D. C. and Chada, H., Development of a synthetic diet for the greenbug, *Schizaphis graminum.* II. Greenbug development as affected by zinc, iron, manganese, and cooper, *Ann. Entomol. Soc. Am.,* 64, 1245, 1971.

Crippen, R. C., *Identification of Organic Compounds with the Aid of Gas Chromatography,* McGraw-Hill, New York, 1973, 331.

Dadd, R. H., Observations on the palatability and utilization of food by locusts, with particular reference to the interpretation of performance in growth trials using synthetic diets, *Entomol. Exp. Appl.,* 3, 283, 1960.

David, W. A. L. and Gardiner, B. O. C., Oviposition and the hatching of the eggs of *Pieris brassicae* (L.) in a laboratory culture, *Bull. Entomol. Res.,* 43, 91, 1962.

Dawson, G. W., Gibson, R. W., Griffiths, D. C., Pickett, J. A., Rice, A. D. and Woodcock, C. M., Aphid alarm pheromone derivatives affecting settling and transmission of plant viruses, *J. Chem. Ecol.,* 8, 1377, 1982.

Dawson, G. W., Griffiths, D. C., Pickett, J. A., Wadhams, L. J. and Woodcock, C. M., Plant derived synergists of alarm pheromone from turnip aphid, *Lipaphis (Hyadaphis) erysimi* (Homoptera, Aphididae), *J. Chem. Ecol.,* 13, 1663, 1987.

Demchenko, A. P., *Ultraviolet Spectroscopy of Proteins,* Springer-Verlag, New York, 1986, 312.

Den Otter, C. J., Single sensillum responses in male moth *Adoxophyes orana* (F.V.R.) to female sex pheromone components and their geometrical isomers, *J. Comp. Physiol.,* 121, 205, 1977.

Den Otter, C. J., Schuil, H. A. and Sander-van Oosten, A., Reception of host-plant odours and female sex pheromone in *Adoxophyes orana* (Lepidoptera: Tortricidae): electrophysiology and morphology, *Entomol. Exp. Appl.,* 24, 370, 1978.

de Wilde, J., HilleRisLambers-Suverkropp, K. and Tol, A. V., Responses to air flow and airborne plant odour in Colorado beetle, *Neth. J. Plant Pathol.,* 75, 53, 1969.

Dhaliwal, G. S., Pathak, M. D. and Vega, C. R., Effect of plant age on resistance in rice varieties to *Chilo suppressalis* (Walker)—allelochemical interactions, *J. Insect Sci.,* 1, 143, 1988.

Dickens, J. C., Olfaction in the boll weevil, *Anthonomus grandis* Boh. (Coleoptera: Curculionidae): Electroantennogram studies, *J. Chem. Ecol.,* 10, 175a, 1984.

Dickens, J. C., Orientation of boll weevil, *Anthonomus grandis* Boh. (Coleoptera: Curculionidae), to pheromone and volatile host compound in the laboratory, *J. Chem. Ecol.,* 12, 91, 1986.

Dickens, J. C. and Boldt, P. E., Electroantennogram responses of *Trirhabda bacharides* (Weber) (Coleoptera: Chrysosomelidae) to plant volatiles, *J. Chem. Ecol.,* 11, 767, 1985.

Dimock, M. B. and Renwick, J. A. A., Oviposition by field populations of *Pieris rapae* (Lepidoptera: Pieridae) deterred by an extract of a wild crucifer, *Environ. Entomol.,* 20, 802, 1991.

Dimock, M. B., Renwick, J. A. A., Radke, C. D. and Sachdev-Gupta, K., Chemical constituents of an unacceptable crucifer, *Erysimum cheiranthoides,* deter feeding by *Pieris rapae, J. Chem. Ecol.,* 17, 525, 1991.

Doskotch, R. W., Cheng, H-Y., Odell, T. M. and Girard, L., Nerolidol: an antifeeding sesquiterpene alcohol for gypsy moth larvae from *Malalenca leucadendron, J. Chem. Ecol.,* 6, 845, 1980.

Dreyer, D. L., Binder, R. G., Chan, B. G., Waiss, Jr., A. C., Hartwig, E. E. and Beland, G. L., Pinitol, a larval growth inhibitor for *Heliothis zea* in soybeans, *Experientia,* 35, 1182, 1979.

Dreyer, D. L., Reese, J. C. and Jones, K. C., Aphid feeding deterrents in sorghum: bioassay, isolation and characterization, *J. Chem. Ecol.,* 7, 272, 1981.

Dreyer, D. L., Jones, K. C. and Jurd, L., Lack of chemical factors in host-plant resistance of alfalfa toward alfalfa weevil: ovicidal activity of some coumarin derivatives, *J. Chem. Ecol.,* 13, 917, 1987a.

Dreyer, D. L., Jones, K. C., Jurd, L. and Campbell, B. C., Feeding deterrency of some 4-hydroxycoumarins and related compounds: relationship to host-plant resistance of alfalfa towards pea aphid (*Acyrthosiphon pisum*), *J. Chem. Ecol.,* 13, 925, 1987b.

Duniz, J. D., *X-ray Analysis and Structure of Organic Molecules,* Cornell University Press, Ithaca, NY, 1979, 514.

Eisemann, C. H. and Rice, J. J., Oviposition behaviour of *Dacus tryone*: the effects of some sugars and salts, *Entomol. Exp. Appl.,* 39, 61, 1985.

Eller, F. J., Tumlinson, J. H. and Lewis, W. J., Beneficial arthropod behavior mediated by airborne semiochemicals. II. Olfactometric studies of host location by the parasitoid *Microplitis croceipes* (Cresson) (Hymenoptera: Braconidae), *J. Chem. Ecol.,* 14, 425, 1988a.

Eller, F. J., Tumlinson, J. H. and Lewis, W. J., Beneficial arthropod behavior mediated by airborne semiochemicals: sources of volatiles mediating the host location flight behavior of *Microplitis croceipes* (Cresson) (Hymenoptera: Braconidae), a parasitoid of *Heliothis zea* (Boddie) (Lepidoptera: Noctuidae), *Environ. Entomol.,* 17, 745, 1988b.

Elliger, C. A., Chan, B. G. and Waiss, Jr., A. C., Growth inhibitors in tomato (*Lycopersicon*) to tomato fruitworm (*Heliothis zea*), *J. Chem. Ecol.,* 7, 753, 1981.

Elsey, K. D. and McFadden, T. L., Pickleworm: oviposition on an artificial surface treated with cucurbit extract, *J. Econ. Entomol.,* 74, 473, 1981.

Erikson, J. M. and Feeny, P., Sinigrin: a chemical barrier to the black swallowtail butterfly, *Papilio polyxenes, Ecology,* 55, 103, 1974.

Fatzinger, C. W. and Merkel, E. P., Oviposition and feeding preferences of the southern pine coneworm (Lepidoptera: Pyralidae) for different host-plant materials and observations on monoterpenes as an oviposition stimulant, *J. Chem. Ecol.,* 11, 689, 1985.

Fein, B. L., Reissig, W. H. and Roelofs, W. L., Identification of apple volatile attractive to the apple maggot, *Rhagoletis pomonella, J. Chem. Ecol.,* 12, 1473, 1982.

Finch, S., Chemical attraction of plant-feeding insects to plants, *Appl. Biol.,* 5, 67, 1980.

Finch, S., Assessing host-plant finding by insects, in *Insect-Plant Interactions,* J. R. Miller and T. A. Miller, eds., Springer-Verlag, New York, 1986, 342.

Finch, S. and Skinner, G., Trapping cabbage root flies in traps baited with plant extracts and with natural and synthetic isothiocyanates, *Entomol. Exp. Appl.,* 31, 133, 1982.

Folsom, J. W., A chemotropometer, *J. Econ. Entomol.,* 24, 827, 1931.

Frazier, J. L. and Hanson, F. E., Electrophysiological recording and analysis of insect chemosensory responses, in *Insect-Plant Interactions,* J. R. Miller and T. A. Miller, eds., Springer-Verlag, New York, 1986, 342.

Frazier, J. L. and Heitz, J. R., Inhibition of olfaction in the moth, *Heliothis virescens,* by the sulfhydryl reagent fluorescein mercuric acetate, *Chem. Senses Flavor,* 1, 271, 1975.

Freedman, B., Nowak, L. J., Kwolek, W. F., Berry, E. C. and Guthrie, W. D., A bioassay for plant-derived control agents using the European cornborer, *J. Econ. Entomol.,* 72, 541, 1979.

Fulton, R. A. and Chamberlin, J. C., An improved technique for the artificial feeding of the best leafhopper with notes on its ability to synthesize glycerides, *Science,* 79, 346, 1934.

Fu-Shun, Y., Evans, K. A., Stevens, L. H., Van Beek, T. A. and Schoonhoven, L. M., Deterrents extracted from the leaves of *Ginkgo biloba*: effects on feeding and contact chemoreceptors, *Entomol. Exp. Appl.,* 54, 57, 1990.

Gabel, B., Thiery, D., Suchy, V., Marion-Poll, F., Hradsky, P. and Farkas, P., Floral volatiles of *Tanacetum vulgare* L. attractive to *Lobesia botrana* Den. et Schiff. females, *J. Chem. Ecol.,* 18, 693, 1992.

Gershenzon, J., Rossiter, M., Mabry, T. J., Rogers, C. E., Blust, M. H. and Hopkins, T. L., Insect antifeedant terpenoids in wild sunflower, a possible source of resistance to the sunflower moth, in *Bioregulators for Pest Control,* P. A. Hedin, ed., American Chemical Society, Washington, DC, 1985, 540.

Gudzinowicz, B. J., Gudzinowicz, M. J. and Martin, H. F., *Fundamentals of Integrated GC-MS,* Part I, *Gas Chromatography,* Part II, *Mass Spectrometry,* Marcel Dekker, New York, 1976, 381.

Gudzinowicz, B. J., Gudzinowicz, M. J. and Martin, H. F., *Fundamentals of GC-MS,* Part II, *Mass Spectrometry,* Marcel Dekker, New York, 1977, 326.

Guerin, P. M. and Visser, J. H., Electroantennogram responses of the carrot fly, *Psila rosae* to volatile plant components, *Physiol. Entomol.,* 5, 111, 1980.

Hamilton, M. A., Notes on the culturing of insects for virus work, *Ann. Appl. Biol.,* 17, 487, 1930.

Hancock, W. S., *High Performance Liquid Chromatography in Biotechnology,* John Wiley & Sons, New York, 1990, 564.

Hansson, B. S., vander Pers, J. N. C. and Lofqvist, J., Comparison of male and female olfactory cell response to pheromone compounds and plant volatile in turnip moth, *Agrotis segetum, Physiol. Entomol.,* 14, 147, 1989.

Hardie, J., Holyoak, M., Nicholas, J., Nottingham, S. F., Pickett, J. A., Wadhams, L. J. and Woodcock, C. M., Aphid sex pheromone components: age-dependent release by females and species-specific male response, *Chemoecology,* 1, 63, 1990.

Harris, M. O. and Rose, S., Chemical, color, and tactile cues influencing oviposition behavior of Hessian fly (Diptera: Cecidomyiidae), *Environ. Entomol.,* 19, 303, 1990.

Harris, P., Host specificity of *Calophasia lunula* (Hufn.) (Lepidoptera: Noctuidae), *Can. Entomol.,* 95, 101, 1963.

Harris, P. and Mohyuddin, A. I., The bioassay of insect feeding tokens, *Can. Entomol.,* 97, 830, 1965.

Harrison, G. D. and Mitchell, B. K., Host-plant acceptance by geographic populations of the Colorado potato beetle, *Leptinotarsa decemlineata,* role of solanaceous alkaloids as sensory deterrents, *J. Chem. Ecol.,* 14, 777, 1988.

Haskell, P. T., Paskin, M. W. J. and Moorhouse, J. E., Laboratory observations on factors affecting the movements of hoppers of the desert locust, *J. Insect Physiol.,* 8, 53, 1962.

Hawkes, C. and Coaker, T. H., Factors affecting the behavioural responses of the adult cabbage root fly, *Delia brassicae,* to host plant odour, *Entomol. Exp. Appl.,* 25, 45, 1979.

Haynes, K. F., Zhao, J. Z. and Latif, A., Identification of floral compounds from *Abelia grandiflora* that stimulate upwind flight in cabbage looper moths, *J. Chem. Ecol.,* 17, 637, 1991.

Henschen, A., *High Performance Liquid Chromatography in Biochemistry,* VCH Publishers, Deerfield, FL, 1985, 638.

Heath, R. R. and Tumlinson, H., Techniques for purifying, analyzing, and identifying pheromones, in: *Techniques in Pheromone Research,* H. E. Hummel and T. A. Miller, eds., Springer-Verlag, New York, 1984, 464.

Hernandez, H. P., Hsieh, T. C. Y., Smith, C. M. and Fisher, N. H., Foliage volatiles of two rice cultivars, *Phytochemistry,* 28, 2959, 1989.

Heron, R. J., The role of chemotactic stimuli in the feeding behaviour of spruce budworm larvae on white spruce, *Can. J. Zool.,* 43, 247, 1965.

Hibbard, B. E. and Bjostad, L. B., Behavioral responses of western corn rootworm larvae to volatile semiochemicals from corn seedlings, *J. Chem. Ecol.,* 14, 1523, 1988.

Hibbard, B. E. and Bjostad, L. B., Isolation of corn semiochemicals attractive and repellent to western corn rootworm larvae, *J. Chem. Ecol.,* 16, 3425, 1990.

Hodgson, E. S., Lettvin, J. Y. and Roeder, K. D., Physiology of a primary chemoreceptor unit, *Science,* 122, 417, 1955.

Holyoke, C. W. and Reese, J. C., Acute insect toxicants from plants, in *Handbook of Natural Pesticides, Vol. III, Part B,* E. D. Morgan and N. B. Mandava, eds., CRC Press, Boca Raton, FL., 1987.

Honda, K., Flavanone glycosides as oviposition stimulants in a papilionid butterfly, *Papilio protenor, J. Chem. Ecol.,* 12, 1999, 1986.

Honda, K., Identification of host-plant chemicals stimulating oviposition by swallowtail butterfly, *Papilio protenor, J. Chem. Ecol.,* 16, 325, 1990.

Hopkins, T. L. and Young, H., Attraction of the grasshopper, *Melanoplus sanguinipes,* to host plant odors and volatile components, *Entomol. Exp. Appl.,* 56, 249, 1990.

Howell, D. E. and Goodhue, L. D., A simplified insect olfactometer, *J. Econ. Entomol.,* 58, 1027, 1965.

Hsiao, T. H. and Fraenkel, G., The influence of nutrient chemicals on the feeding behavior of the Colorado potato beetle, *Leptinotarsa decemlineata* (Coleoptera: Chrysomelidae), *Ann. Entomol. Soc. Am.,* 61, 44, 1968.

Huang, X. P. and Mack, T. P., Effects of peanut fractions on lesser cornstalk borer (Lepidoptera: Pyralidae) larval feeding, *Environ. Entomol.,* 18, 763, 1989.

Huang, X. P., Mark, T. P. and Berger, R. S., Olfactory responses of lesser cornstalk borer (Lepidoptera: Pyralidae) larvae to peanut parts, *Environ. Entomol.,* 19, 1289, 1990.

Ichikawa, T. and Ishii, S., Mating signals of the brown planthopper, *Nilaparvata lugens* (Stål) (Homoptera: Delphacidae): vibration of the substrate, *Appl. Entomol. Zool.,* 9, 196, 1974.

Isman, M. B., Brard, N. L., Nawrot, J. and Harmatha, J., Antifeedant and growth inhibitory effects of bakkenolide-A and other sesquiterpene lactones on the variegated cutworm, *Peridroma saucia* Hübner (Lepidoptera: Noctuidae), *J. Appl. Entomol.,* 107, 524, 1989.

Jackson, D. M., Severson, R. F., Sisson, V. A. and Stephenson, M. G., Ovipositional response of tobacco budworm moths (Lepidoptera: Noctuidae) to cuticular labdanes and sucrose esters from the green leaves of *Nicotiana glutinosa* L. (Solanaceae), *J. Chem. Ecol.,* 17, 2489, 1991.

Jaffe, H. H. and Orchin, M., *Theory and Applications of Ultraviolet Spectroscopy,* John Wiley & Sons, New York, 1962, 624.

Jaffery, J. W., *Methods in X-ray Chrystallography,* Academic Press, New York, 1971, 571.

Jones, O. T., Lomer, R. A. and Howse, P. E., Responses of male Mediterranean fruit flies, *Ceratitis capitata,* to *trimedlure* in a wind tunnel of novel design, *Physiol. Entomol.,* 6, 175, 1981.

Kaissling, K. E., Sensory transductions in insect olfactory receptors, in *Biochemistry of Sensory Functions,* L. Jaenicke, ed., Springer-Verlag, New York, 1974, 641.

Kamm, J. A. and Buttery, R. G., Response of the alfalfa seed chalcid *Bruchophagus roddi,* to alfalfa volatiles, *Entomol. Exp. Appl.,* 33, 129, 1983.

Kamm, J. A. and Buttery, R. G., Root volatile components of red clover: identification and bioassay with the clover root borer (Coleoptera: Scolytidae), *Environ. Entomol.,* 13, 1427, 1984.

Katsoyannos, B. I., Boller, E. F. and Remund, U., A simple olfactometer for investigation of sex pheromones and other olfactory attractants in fruit flies and moths, *Z. Angew. Entomol.,* 90, 105, 1980.

Kennedy, J. S., Olfactory responses to distant plants and other sources, in *Chemical Control of Insect Behaviour: Theory and Application,* H. H. Shorey and J. J. McKelvey, Jr., eds., John Wiley & Sons, New York, 1977, 414.

Kennedy, J. S. and Moorhouse, J. E., Laboratory observations on locust responses to wind-borne grass odour, *Entomol. Exp. Appl.,* 12, 487, 1969.

Khan, Z. R. and Hatchett, J. H., Influence of epicuticular lipid extracts from wheat and oat seedlings on oviposition of Hessian fly, *Entomol. Exp. Appl.,* 1993 (in press).

Khan, Z. R. and Saxena, R. C., Electronically recorded waveforms associated with the feeding behavior of *Sogatella furcifera* (Homoptera: Delphacidae) on susceptible and resistant rice plants, *J. Econ. Entomol.,* 77, 1479, 1984.

Khan, Z. R. and Saxena, R. C., Behavior and biology of *Nephotettix virescens* (Homoptera: Cicadellidae) on tungro virus-infected rice plants: epidemiology implications, *Environ. Entomol.,* 14, 297, 1985a.

Khan, Z. R. and Saxena, R. C., Effect of steam distillate extract of a resistant rice variety on feeding behavior of *Nephotettix virescens* (Homoptera: Cicadellidae), *J. Econ. Entomol.,* 78, 562, 1985b.

Khan, Z. R. and Saxena, R. C., Effect of steam distillate extracts of resistant and susceptible rice cultivars on behavior of *Sogatella furcifera* (Homoptera: Delphacidae), *J. Econ. Entomol.,* 79, 928, 1986.

Khan, Z. R., Norris, D. M., Chiang, H. S., Weiss, N. E. and Oosterwyk, A. S., Light-induced susceptibility in soybean to cabbage looper, *Trichoplusia ni* (Lepidoptera: Noctuidae), *Environ. Entomol.,* 15, 803, 1986.

Khan, Z. R., Ciepiela, A. and Norris, D. M., Behavioral and physiological responses of cabbage looper, *Trichoplusia ni* (Hübner), to steam distillate extracts from resistant versus susceptible soybean plants, *J. Chem. Ecol.,* 13, 1903, 1987.

Khan, Z. R., Saxena, R. C. and Rueda, B. P., Responses of rice-infesting and grass-infesting populations of brown planthopper to rice plants and *Leersia* grass, *J. Econ. Entomol.,* 81, 1080, 1988.

Kim, M., Koh, H. S. and Fukami, H., Isolation of C-glycosylflavones as probing stimulant of planthoppers in rice plant, *J. Chem. Ecol.,* 11, 441, 1985.

King, R. R., Calhoun, L. A., Singh, R. P. and Boucher, A., Sucrose esters associated with glandular trichomes of wild *Lycopersicon* species, *Phytochemistry,* 29, 2115, 1990.

Klocke, J. A. and Chan, B. G., Effects of cotton condensed tanin on feeding and digestion in the cotton pest, *Heliothis zea, J. Insect Physiol.,* 28, 911, 1982.

Kon, R. T., Zabik, M. J., Webster, J. A. and Leavitt, R. A., Cereal leaf beetle response to biochemicals from barley and pea seedlings. I. Crude extract, hydrophobic and hydrophilic fractions, *J. Chem. Ecol.,* 4, 511, 1978.

Koritsas, V. M., Lewis, J. A. and Fenwick, G. R., Glucosinolate responses of oil seed rape, mustard and kale to mechanical wounding and infestation by cabbage stem flea beetle (*Psylliodes chrysocephala*), *Ann. Appl. Biol.,* 118, 209, 1991.

Koyama, K. and Mitsuhashi, J., Differences in oviposition of the smaller brown planthopper, *Laodelphax striatellus,* in response to various amino acids solutions (Hemiptera: Delphacidae), *Appl. Entomol. Zool.,* 11, 33, 1976.

Kubo, I. and Hanke, F. J., Chemical methods for isolating and identifying phytochemicals biologically active in insects, in *Insect-Plant Interactions,* J. R. Miller and T. A. Miller, eds., Springer-Verlag, New York, 1976, 342.

Kubo, I. and Klocke, J. A., An insect growth inhibitor from *Trichilia roka* (Meliaceae), *Experientia,* 38, 639, 1982.

Kuenen, L. P. S., Peacock, J. W., Silk, P. J. and Wright, S. S., Identification of the primary sex pheromone component of the mimosa webworm (Lepidoptera: Plutellidae), *Environ. Entomol.,* 19, 1, 1990.

Kuhn, R. and Low, I., Resistance factors against *Leptinotarsa decemlineata* Say isolated from the leaves of wild *Solanum* species, in *Resistance to Toxic Agents,* Academic Press, New York, 1955.

Kurata, S. and Sogawa, S. Sucking inhibitory action of aromatic amines for the rice plant- and leafhoppers (Homoptera: Delphacidae, Deltocephalidae), *Appl. Entomol. Zool.,* 11, 89, 1976.

Lampman, R. L. and Metcalf, R. L., Multicomponent kairomonal lines for southern and western corn rootworms (Coleoptera: Chrysomelidae: *Diabrotica* spp.), *J. Econ. Entomol.,* 80, 1137, 1987.

Lampman, R. L. and Metcalf, R. L., The comparative response of *Diabrotica* species (Coleoptera: Chrysomelidae) to volatile attractants, *Environ. Entomol.,* 17, 644, 1988.

Lampman, R. L., Metcalf, R. L. and Andersen, J. F., Semiochemical attractants of *Diabrotica undecimpunctata howardi* Barber, southern corn rootworm, and *Diabrotica virgifera virgifera* LeConte, the western corn rootworm (Coleoptera: Chrysomelidae), *J. Chem. Ecol.,* 13, 959, 1987.

Lance, D. R., Potential of 8-methyl-2-decyl propanoate and plant-derived volatiles for attracting corn rootworm beetles (Coleoptera: Chrysomelidae) to toxic bait, *J. Econ. Entomol.,* 81, 1359, 1988.

Landis, D. A. and Gould, F., Investigating the effectiveness of feeding deterrents against the southern corn rootworm using behavioral bioassays and toxicity testing, *Entomol. Exp. Appl.,* 51, 163, 1989.

LaPidus, J. B., Davidson, R. H., Fisk, F. W. and Augustine, M. G., Chemical factors influencing host selection by the Mexican bean beetle, *Epilachna varivestis* Muls., *J. Agric. Food Chem.,* 11, 462, 1963.

Leskinen, V., Polonsky, J. and Bhatnagar, S., Antifeedant activity of quassinoids, *J. Chem. Ecol.,* 10, 1497, 1984.

Lewis, A. C. and van Emden, H. F., Assays for insect feeding, in *Insect-Plant Interactions,* J. R. Miller and T. A. Miller, eds., Springer-Verlag, New York, 1986, 342.

Light, D. M. and Jang, E. B., Electroantennogram responses of the oriental fruit fly, *Dacus dorsalis,* to a sprectrum of alcohol and aldehyde plant volatiles, *Entomol. Exp. App.,* 45, 55, 1987.

Light, D. M., Jang, E. B. and Dickens, J. C., Electroantennogram responses of the Mediterranean fruit fly, *Ceratitis capitata,* to a spectrum of plant volatiles, *J. Chem. Ecol.,* 14, 159, 1988.

Lin, H. and Phelan, L., Identification of food volatiles attractive to dusky sap beetle, *carpophilus lugubris* (Coleoptera: Nitidulidae), *J. Chem. Ecol.,* 17, 6, 1991.

Lindroth, R. L., Scriber, J. M. and Hsia, M. T. S., Chemical ecology of the tiger swallowtail: mediation of host use by phenolic glycosides, *Ecology,* 69, 814, 1988.

Littlewood, A. B., *Gas Chromatography Principles, Techniques, and Applications,* Academic Press, New York, 1970, 546.

Liu, S. H., Norris, D. M. and Marti, E., Behavioral responses of female adult *Trichoplusia ni* to volatiles from soybean versus a preferred host, lima bean, *Entomol. Exp. Appl.,* 49, 99, 1988.

Loschiavo, S. R., Methods for studying aggregation and feeding behaviour of the confused flour beetle, *Tribolium confusum* (Coleoptera: Tenebrionidae), *Ann. Entomol. Soc. Am.,* 58, 383, 1965.

Lyons, P. C., Hipskind, J. D., Wood, K. V. and Nicholson, R. L., Separation and quantification of cyclic hydroxamic acids and related compounds by high-pressure liquid chromatography, *J. Agric. Food Chem.,* 36, 57, 1988.

Ma, W-C., Dynamics of feeding responses in *Pieris brassicae* Linn. as a function of chemosensory input: a behavioural, ultrastructural and electrophysiological study, *Meded. Landbouwhogesch. Wageningen,* 72, 11, 1972.

Ma, W-C., Alterations of chemoreceptor function in armyworm larvae (*Spodoptera exempta*) by a plant-derived sesquiterpenoid and by sulfhydryl reagents, *Physiol. Entomol.,* 2, 199, 1977.

Ma, W-C. and Kubo, I., Phagostimulants for *Spodoptera exempta*: identification of adenosine from *Zea mays, Entomol. Exp. Appl.,* 22, 107, 1977.

Ma, W-C. and Schoonhoven, L. M., Tarsal contact chemosensory hairs on the large white butterfly *Pieris brassicae* and their possible role in oviposition behaviour, *Entomol. Exp. Appl.,* 16, 343, 1973.

Ma, W-C. and Visser, J. H., Single unit analysis of odor quality coding by the olfactory antennal receptor system of the Colorado beetle, *Entomol. Exp. Appl.,* 24, 320, 1978.

Mackenzie, M. W., ed., *Advances in Applied Fourtier Transform Infrared Spectroscopy,* John Wiley & Sons, New York, 1988, 353.

Maeshima, K., Hayashi, N., Murakami, T., Takahashi, F. and Komae, H., Identification of chemical oviposition stimulants from rice grain for *Sitophilus zeamais* Motschulsky (Coleoptera: Curculionidae), *J. Chem. Ecol.,* 11, 1, 1985.

Masada, Y., *Analysis of Essential Oils by Gas Chromatography and Mass Spectrometry,* John Wiley & Sons, New York, 1976, 334.

Massage, G. M., *Practical Aspects of Gas Chromatography/Mass Spectrometry,* John Wiley & Sons, New York, 1984, 351.

Matsumoto, Y., Volatile organic sulphur compounds as insect attractants with special reference to host selection, in *Control of Insect Behavior by Natural Products,* D. L. Wood, R. M. Silverstein and M. Nakajima, eds., Academic Press, New York, 1970, 343.

Matsumoto, Y. and Thorsteinson, A. J., Olfactory response of larvae of the onion maggot, *Hylemya antiqua* Meigen (Diptera: Anthomyiidae) to organic sulfur compounds, *Appl. Entomol. Zool.,* 3, 107, 1968.

McGovern, T. P. and Beroza, M., Volatility and compositional changes of Japanese beetle attractant mixtures and means of dispensing sufficient vapor having a constant composition, *J. Econ. Entomol.,* 63, 1475, 1970.

McIndoo, N. E., An insect olfactometer, *J. Econ. Entomol.,* 12, 545, 1926.

McLean, D. L. and Kinsey, M. G., A technique for electronically recording aphid feeding and salivation, *Nature (London),* 202, 1358, 1964.

McMillan, W. W., Wiseman, B. R. and Sekul, A. A., Further studies on the responses of corn earworm larvae to extracts of corn silks and kernels, *Ann. Entomol. Soc. Am.,* 63, 371, 1970.

Medina, E. B. and Tryon, E. H., Response of leaffolder *Cnaphalocrocis medinalis* to extracts of resistant *Oryza sativa* and *O. brachyantha, Int. Rice Res. Newsl.,* 15, 23, 1986.

Mehrotra, K. N. and Rao, P. J., Phagostimulants for locusts: studies with edible oils, *Entomol. Exp. Appl.,* 15, 208, 1972.

Meisner, J. and Ascher, K. R. S., A method to assay the phagostimulatory effect towards insects of plant extracts applied to styropor discs, *Riv. Parassit.,* 29, 74, 1968.

Meisner, J., Skatulla, U., Phagostimulation and phagodeterrency in the larvae of the gypsy moth, *Porthetria dispar* L., *Phytoparasitica,* 3, 19, 1975.

Meisner, J., Ascher, K. R. S. and Flowers, H. M., The feeding response of the larvae of Egyptian cotton leafworm, *Spodoptera littoralis* Boisd., to sugars and related compounds. 1. Phagostimulatory and deterrent effects, *Comp. Biochem. Physiol.,* 42, 899, 1972.

Meisner, J., Ascher, K. R. S. and Lavie, D., Phagostimulants for the larvae of the potato tuber moth, *Gnorimoschema operculella* Zell., *Z. Ang. Entomol.,* 77, 77, 1974.

Meisner, J., Ascher, K. R. S. and Zur, M., Phagodeterrency induced by pure gossypol and leaf extracts of a cotton strain with high gossypol content in the larvae of *Spodoptera littoralis, J. Econ. Entomol.,* 70, 149, 1977a.

Meisner, J., Navon, A., Zur, M. and Ascher, K. R. S., The response of *Spodoptera littoralis* larvae to gossypol incorporated in an artificial diet, *Environ. Entomol.,* 6, 243, 1977b.

Meisner, J., Kehat, M., Zur, M. and Eizick, C., Responses of *Earis insulana* Boisd. larvae to neem (*Azadirachta indica* A. Juss) kernel extract, *Phytoparasitica,* 6, 85, 1978.

Meisner, J., Fleischer, A. and Eizick, C., Phagodeterrency induced by (–)-carvone in the larva of *Spodoptera littoralis* (Lepidoptera: Noctuidae), *J. Econ. Entomol.,* 75 462, 1982.

Mendel, M. J., Alford, A. R. and Bentley, M. D., A comparison of the effects of limonin on Colorado potato beetle, *Leptinotarsa decemlineata,* and fall armyworm, *Spodoptera frugiperda,* larval feeding, *Entomol. Exp. Appl.,* 58, 191, 1991.

Messerschmidt, R. G. and Harthcock, M. A., *Infrared Microscopy: Theory and Applications,* Marcel Dekker, New York, 1988, 282.

Miles, P. W., The saliva of Hemiptera, *Adv. Insect Physiol.,* 9, 183, 1972.

Miller, J. R. and Roelofs, W. L., Sustained-flight tunnel for measuring insect responses to wind-borne sex pheromone, *J. Chem. Ecol.,* 4, 187, 1978.

Miller, J. R. and Strickler, K. L., Finding and accepting host plants, in *Chemical Ecology of Insects,* W. J. Bell and R. T. Carde, eds., Chapman and Hall, London, 1984, 524.

Mitchell, E. R. and Heath, R. R., Influence of *Amaranthus hybridus* L. allelochemics on oviposition behavior of *Spodoptera exigua* and *S. eridania* (Lepidoptera: Noctuidae), *J. Chem. Ecol.,* 11, 609, 1985.

Mitchell, E. R. and Heath, R. R., *Heliothis subflexa* (Gn.) (Lepidoptera: Noctuidae): demonstration of oviposition stimulant from ground cherry using novel bioassay, *J. Chem. Ecol.,* 13, 1849, 1987.

Mitchell, E. R., Tingle, F. C. and Heath, R. R., Ovipositional response of three *Heliothis* species (Lepidoptera: Noctuidae) to allelochemicals from cultivated and wild host plants, *J. Chem. Ecol.,* 16, 1817, 1990.

Mitchell, E. R., Tingle, F. C. and Heath, R. R., Flight activity of *Heliothis virescens* (F.) females (Lepidoptera: Noctuidae) with reference to host-plant volatiles, *J. Chem. Ecol.,* 17, 259, 1991.

Mitsuhashi, J., Advice for collecting planthopper and leafhopper eggs (Hemiptera, Delphacidae and Deltocephalidae), *Appl. Entomol. Zool.,* 5, 47, 1970.

Mitsuhashi, J. and Koyama, K., Oviposition of smaller brown planthopper, *Laodelphax striatellus* into various carbohydrate solutions through a parafilm membrane (Hemiptera: Delphacidea), *Appl. Entomol. Zool.,* 10, 123, 1975.

Mittler, T. E. and Dadd, R. H., Artificial feeding and rearing of the aphid, *Myzus persicae* (Sulzer), on a completely defined synthetic diet, *Nature (London),* 195, 404, 1962.

Moorhouse, J. E., Yeadon, R., Beevor, P. S. and Nesbitt, B. F., Method for use in studies of insect chemical communication, *Nature (London),* 223, 1174, 1969.

Mullen, K. and Pregosin, P. S., *Fourier Transform NMR Techniques: A Practical Approach,* Academic Press, New York, 1976, 149.

Munakata, K., Saito, T., Ogawa, S. and Ishii, S., Oryzanone, an attractant of the rice stem borer, *Bull. Agric. Chem. Soc. Jpn.,* 23, 64, 1959.

Mustaparta, H., Behavioral responses of the pine weevil *Hylobius abietis* L. (Col.: Curculionidae) to odours activating different groups of receptor cells, *J. Comp. Physiol.,* 102, 57, 1975.

Muzika, R. M., Campbell, C. L., Hanover, J. W. and Smith, A. L., Comparison of techniques for extracting volatile compounds from conifer needles, *J. Chem. Ecol.,* 16, 2713, 1990.

Nair, K. S. S. and McEwen, F. L., Host selection by the adult cabbage maggot, *Hylemya brassicae* (Diptera: Anthomyiidae): effect of glucosinolates and common nutrients on oviposition, *Can. Entomol.,* 108, 1021, 1976.

Nair, K. S. S., McEwen, F. L. and Snieckus, V., The relationship between glucosinolate content of cruciferous plants and oviposition preferences of *Hylemya brassicae* (Diptera: Anthomyiidae), *Can. Entomol.,* 108, 1031, 1976.

Nakanishi, K., Insect antifeedants from plants, in *Insect Biology in the Future,* M. M. Locke and D. S. Smith, eds., Academic Press, New York, 1980, 977.

Nesbitt, B. F., Beevor, P. S., Hall, D. R., Lester, R., Sternlicht, M. and Goldenberg, S., Identification and synthesis of the female sex pheromone of the citrus flower moth, *Prays citri, Insect Biochem.,* 7, 355, 1977.

Nesbitt, B. S., Beevor, P. S., Hall, D. R., Lester, R. and Williams, J. R., Components of the sex pheromone of the female sugarcane borer, *Chilo sacchariphagus* (Bojer) (Lepidoptera: Pyralidae). Identification and field trials, *J. Chem. Ecol.*, 6, 385, 1980.

Niimura, M. and Ito, T., Feeding on filter paper by larvae of silkworm, *Bombyx mori* L., *J. Insect Physiol.*, 10, 425, 1964.

Nishida, R., Oviposition stimulants of some papilionid butteflies contained in their host plants, *Botyu-Kagaku*, 42, 133, 1977.

Nishida, R. and Fukami, H., Oviposition stimulants of an Aristolochiaceae-feeding swallowtail butterfly, *Atrophaneura alcinous, J. Chem. Ecol.*, 15, 2565, 1989.

Norris, D. M. and Baker, J. E., Feeding responses of the beetle *Scolytus* to chemical stimuli in the bark of *Ulmus, J. Insect Physiol.*, 13, 955, 1967.

Nottingham, S. F., Son, K. C., Wilson, D. D., Severson, R. F. and Kays, S. J., Feeding and oviposition preferences of sweat potato weevil, *Cylas formicarius elegantulus* (Summers), on storage roots of sweet potato cultivars with different surface chemistries, *J. Chem. Ecol.*, 15, 895, 1989a.

Nottingham, S. F., Son, K. C., Severson, R. F., Arrendale, R. F., and Kays, S. J., Attraction of adult sweet potato weevils, *Cylas formicarius elegantulus* (Summers), (Coleoptera: Curculionidae), to sweet potato leaf and root volatiles, *J. Chem. Ecol.*, 15, 1095, 1989b.

Nottingham, S. F., Hardie, J., Dawson, G. W., Hick, A. J., Pickett, J. A., Woodhams, L. J. and Woodcock, C. M., Behavioral and electrophysiological responses of aphids to host and nonhost plant volatiles, *J. Chem. Ecol.*, 17, 1231, 1991.

Obata, T., Kim, M., Koh, H. and Fukami, H., Planthopper attractant(s) in rice plant, *Jpn. J. Appl. Entomol. Zool.*, 25, 47, 1981.

Obata, T., Koh, H., Kim, M. and Fukami, H., Constituents of planthopper attractants in rice, *Appl. Entomol. Zool.*, 18, 161, 1983.

Odham, G., Larsson, L. and Mardh, P. A., *Gas Chromatography/Mass Spectrometry Applications in Microbiology,* Plenum Press, New York, 1984, 444.

Ohigashi, H. and Koshimizu, K., Charicol, a larva-growth inhibitor, from *Viburnum japonicum* Spreng., *Agr. Biol. Chem.*, 40, 2283, 1976.

Painter, R. H., Director's biennial report, *Bienn. Rep. Kans. Agric. Exp. Sta.*, 116, 1930.

Payne, T. L., Hart, E. R., Edson, L. J., McCarty, F. A., Billings, P. M. and Coster, J. E., Olfactometer for assay of behavioral chemicals for the southern pine beetle, *Dendroctonus frontalis* (Coleoptera: Scolytidae), *J. Chem. Ecol.*, 2, 411, 1976.

Pelletier, Y. and Smilowitz, Z., Effects of trichome B exudate of *Solanum berthaultii* Hawkes on consumption by the Colorado potato beetle, *Leptinotarsa decemlineata* (Say), *J. Chem. Ecol.*, 16, 1547, 1990.

Pettersson, J., An aphid sex attractant, I. Biological studies, *Entomol. Scan.* 1, 63, 1970.

Phelan, P. L. and Lin, H., Chemical characterization of fruit and fungal volatiles attractive to dried-fruit beetle, *Carpophilus hemipterus* (L.) (Coleoptera: Nitidulidae), *J. Chem. Ecol.,* 17, 1253, 1991.

Phelan, P. L., Roelofs, C. J., Youngman, R. R. and Baker, T. C., Characterization of chemicals mediating ovipositional host-plant finding by *Amyelois transitella* females, *J. Chem. Ecol.,* 17, 599, 1991.

Pierce, A. M., Borden, J. H. and Oehlschlager, A. C., Olfactory response to beetle-produced volatiles and host-food attractants by *Oryzaephilus surinamensis* and *O. mercator, Can. J. Zool.,* 59, 1980, 1981.

Pierce, A. M., Pierce, H. D., Oehlschlager, A. C. and Borden, J. H., Attraction of *Oryxalphilus Surinamensis* (L.) and *Oryzalphilus mercator* (Fauvel) (Coleoptera: Cucujidae) to some common volatiles of food, *J. Chem. Ecol.,* 16, 465, 1990.

Pierce, A. M., Pierce Jr., H. D., Borden, J. H., and Oehlschlager, A. C., Fungal volatiles: semiochemicals for stored-product beetles (Coleoptera: Cucujidae), *J. Chem. Ecol.,* 17, 581, 1991.

Pierce, Jr., H. D., Vernon, R. S., Borden, J. H. and Oehlschlager, A. C., Host selection by *Hylemya antiqua* (Meigen), identification of three new attractants and oviposition stimulants, *J. Chem. Ecol.,* 4, 65, 1978.

Phillips, J. K. and Burkholder, W. E., Evidence for a male-produced aggregation pheromone in the rice weevil, *J. Econ. Entomol.,* 74, 539, 1981.

Pouzat, J., Host plant chemoscensory influence on oogenesis in the bean weevil, *Acanthoscelides obtectus* (Coleoptera: Bruchidae), *Entomol. Exp. Appl.,* 24, 401, 1978.

Price, G. D., Harris, R. H. and Smith, J., An uncommon system for manual control of olfactometer are relative humidity, *J. Med. Entomol. Honolulu,* 14, 715, 1978.

Pu, F., Zhang, Z. J. and Shi, Y., Chemical composition of *Fissistigma shangtzeense* Tsiang of P.T. Li flower extract, *Flavour and Fragrance Journal,* 3, 171, 1988.

Rahbe, Y., Febvay, G., Delobel, B. and Bournoville, R., *Acyrthosiphon pisum* performance in response to the sugar and animo acid composition of artificial diets, and its relation to lucerne varietal resistance, *Entomol. Exp. Appl.,* 48, 283, 1988.

Ramachandran, R. and Khan, Z. R., Electroantennogram technique for studying olfactory sensitivity of insects to volatile compounds, *Int. Rice Res. Newsl.,* 15, 22, 1990.

Ramachandran, R. and Khan, Z. R., Mechanisms of resistance in wild rice, *Oryza brachyantha* to rice leaffolders *Cnaphalocrocis medinalis* (Guenée) (Lepidoptera: Pyralidae), *J. Chem. Ecol.,* 17, 41, 1991.

Ramachandran, R. and Norris, D. M., Volatiles mediating plant-herbivore-natural enemy interactions: electroantennogram responses of soybean looper, *Pseudoplusia includens* and a parasitoid, *Microplitis demolitor,* to green leaf volatiles, *J. Chem. Ecol.,* 17, 1665, 1991.

Ramachandran, R., Khan, Z. R., Caballero, P. and Juliano, B. O., Olfactory sensitivity of two sympatric species of rice leaffolders (Lepidoptera: Pyralidae) to plant volatiles, *J. Chem. Ecol.,* 16, 2647, 1990.

Raman, K. V., Tingey, W. M. and Gregory, P., Potato glycoalkaloids: effects on survival and feeding behavior of the potato leafhopper, *J. Econ. Entomol.,* 72, 337, 1979.

Reed, D. K., Midolajczak, K. L., Krause, C. R. Ovipositional behavior of lesser peachtree borer in presence of host-plant volatiles, *J. Chem. Ecol.*, 14, 237, 1988.

Reed, D. W., Pivnick, K. A. and Underhill, E. W., Identification of chemical oviposition stimulants for the diamondback moth, *Plutella xylostella*, present in three species of Brasicaceae, *Entomol. Exp. Appl.*, 53, 277, 1989.

Reese, J. C., Chan, B. G. and Waiss, Jr., A. C., Effects of cotton condensed tannin, maysin (corn) and pinital (soybean) on *Heliothis zea* growth and development, *J. Chem. Ecol.*, 8, 1429, 1982.

Reid, C. D. and Lampman, R. L., Olfactory responses of Orius insidious (Hemiptera: Anthocridae) to volatiles of corn silk, *J. Chem. Ecol.*, 15, 1109, 1989.

Rembold, H., Wallner, P. and Singh, A. K., Attractiveness of volatile chickpea (*Cicer arietinum* L.) seed components to *Heliothis armigera* larvae (Lep., Noctuidae), *J. Appl. Entomol.*, 107, 65, 1989.

Renwick, J. A. A., Radke, C. D. and Sachdev-Gupta, K., Chemical constituents of *Erysimum cheiranthoides* deterring oviposition by the cabbage butterfly, *Pieris rapae*, *J. Chem. Ecol.*, 15, 2161, 1989.

Ritter, F. J., Feeding stimulants for the Colorado beetle, *Meded. Fac. Landbouwet. Rijksuniv. Gent.*, 32, 291, 1967.

Robinson, J. F., Klun, J. A., Guthrie, W. D. and Brindley, T. A., European corn borer (Lepidoptera: Pyralidae) leaf feeding resistance: Dimboa bioassays, *J. Kans. Entomol. Soc.*, 55, 357, 1982.

Roboz, J., *Introduction to Mass Spectrometry: Instrumentation and Techniques*, Interscience Publ., New York, 1968, 539.

Rodman, J. E. and Chew, F. S., Phytochemical correlates of herbivory in a community of native and naturalized cruciferae, *Biochem. Syst. Ecol.*, 8, 43, 1980.

Roelofs, W. L., The scope and limitation of electroantennogram technique in identifying the pheromone components, in *The Evaluation of Biological Activities*, N. R. McFourlance, ed., Academic Press, New York, 1976.

Roseland, C. R., Bates, M. B., Carlson, R. B. and Oseto, C. Y., Discrimination of sunflower volatiles by the red sunflower seed weevil, *Entomol. Exp. Appl.*, 62, 99, 1992.

Rothschild, G. H. L., Nesbitt, B. F., Beevor, P. S., Cork, A., Hall, D. R. and Vickers, R. A., Studies of the female sex pheromone of the native budworm, *Heliothis punctiger*, *Entomol. Exp. Appl.*, 31, 3955, 1982.

Rowlands, M. L. J., A fusiform olfactometer chamber, *Entomol. Monthly Mag.*, 121, 163, 1985.

Rust, M. K. and Reierson, D. A., Using wood extracts to determine the feeding preferences of the western drywood termite, *Incisitermes minor* (Hagen), *J. Chem. Ecol.*, 3, 391, 1977.

Sakai, T. and Sogawa, K., Effects of nutrient compounds on sucking response of the brown planthopper, *Nilaparvata lugens* (Homoptera: Delphacidae), *Appl. Entomol. Zool.*, 11, 82, 1976.

Sakuma, M. and Fukami, H., The linear track olfactometer: an assay device for taxes of German cockroach, *Blatella germanica* (L.) (Dictyoptera: Blattedae) toward their aggregation pheromone, *Appl. Entomol. Zool.*, 20, 387, 1985.

Saxena, K. N. and Basit, A., Inhibition of oviposition by volatiles of certain plants and chemicals in the leafhopper, *Amrasca devastans* (Distant), *J. Chem. Ecol.*, 8, 329, 1982.

Saxena, K. N. and Goyal, S., Host-plant relations of the citrus butterfly, *Papilio demoleus* L., orientational and ovipositional responses, *Entomol. Exp. Appl.,* 24, 1, 1978.

Saxena, K. N., Gandhi, J. R. and Saxena, R. C., Patterns of relationship between certain leafhoppers and plants. I. Responses to plants, *Entomol. Exp. Appl.,* 17, 303, 1974.

Saxena, R. C., Biochemical basis of resistance in crop plants, IRRI Saturday seminar, 11 March 1978, *International Rice Research Institute,* Los Baños, Laguna, Philippines, 1978 (mimeo).

Saxena, R. C. and Khan, Z. R., Electronically recorded disturbances in feeding behavior of *Nephotettix virescens* (Homoptera: Cicadellidae) on neem oil-treated rice plants, *J. Econ. Entomol.,* 78, 222, 1985a.

Saxena, R. C. and Khan, Z. R., Effect of neem oil on survival of *Nilaparvata lugens* (Homoptera: Delphacidae) and on grassy stunt and ragged stunt virus transmission, *J. Econ. Entomol.,* 78, 647, 1985b.

Saxena, R. C. and Okech, S. H., Role of plant volatiles in resistance of selected rice varieties to brown planthopper, *Nilaparvata lugens* (Stål) (Homoptera: Delphacidae), *J. Chem. Ecol.,* 11, 1601, 1985.

Saxena, R. C. and Pathak, M. D., Factors affecting resistance of rice varieties to the brown planthopper, *Nilaparvata lugens,* Proc. 8th Ann. Conf. Pest Control Counc. Philippines, 1977, 34.

Saxena, R. C. and Puma, B. C., Effects of trans-aconitic acid, a barnyard grass allelochemic, on hatching of eggs of brown planthopper and green leafhopper, Paper Presented at the 10th Annual Conference of the Pest Control Council of the Philippines, Manila, May 2–5, 1979.

Saxena, R. C., Liquido, N. J. and Justo, Jr., H. D., Neem seed oil, a potential antifeedant for the control of the rice brown planthopper, *Nilaparvata lugens,* in *Natural Pesticides from the Neem Tree* (Azadirchta indica) *A. Juss,* H. Schmutterer and K. R. S. Ascher, K. R. S., German Agency for Technical Cooperation, 1981.

Scheffrahn, R. H., Hsu, R.-C., Su, N-Y., Huffman, J. B., Midland, S. L. and Sims, J. J., Allelochemical resistance of bald cypress, *Taxodium distichum* Heartwood, to the subterranean termite, *Coptotermes formosanus, J. Chem. Ecol.,* 14, 765, 1988.

Schlunegger, U. P., *Advanced Mass Spectrometry: Applications in Organic and Analytical Chemistry,* Pergamon Press, New York, 1980, 143.

Schoonhoven, L. M., Sensitivity changes in some insect chemoreceptors and their effect on food selection behaviour, *Proc. Kon. Ned. Ser. C,* 72, 491, 1969.

Schoonhoven, L. M., Biological aspects of antifeedants, *Entomol. Exp. Appl.,* 31, 57, 1982.

Schoonhoven, L. M. Perception of antifeedants by Lepidopterous larvae, in *Pesticide Science and Biotechnology, Proc. Sixth Int. Congress of Pesticide Chemistry,* R. Greenhalgh and T. R. Roberts, eds., Blackwell Scientific Publ., London, 1987, 604.

Sebedio, J. L., Le-Quere, J. L., Semon, E., Morin, O., Prevost, J. and Grandgirard, A., Heat treatment of vegetable oils, II. GC-MS and GC-FTIR spectres of some isolated cyclic fatty acid monomers, *J. Am. Oil Chem. Soc.,* 64, 1324, 1987.

Seelinger, G., Response characteristics and specificity of chemoreceptors in *Hemilepistus reaumuri* (Crustacea, Isopoda), *J. Comp. Physiol.,* 152, 219, 1983.

Sekido, S. and Sogawa, K., Effects of salicylic acid on probing and oviposition of the rice plant- and leafhoppers (Homoptera: Delphacidae and Deltocephalidae), *Appl. Entomol. Zool.,* 11, 75, 1976.

Selzer, R., The processing of a complex food odor by antennal olfactory receptors of *Periplaneta americana, J. Comp. Physiol.,* 144, 509, 1981.

Serit, M., Ishida, M., Kim, M., Yamamoto, T. and Takahasi, S., Antifeedants from *Citrus natsudaidai* Hayata against termite *Reticulitermes speratus* Kolbe, *Agric. Biol. Chem.,* 55, 2381, 1991.

Severson, R. F., Jackson, D. M., Johnson, A. W., Sisson, V. A. and Stephenson, M. G., Ovipositional behavior of tobacco budworm and tobacco hornworm, in *Naturally Occurring Pest Bioregulators,* P. A. Hedin, ed., ACS Symp. Ser. 449, Americal Chemical Society, Washington, DC, 1991, 456.

Shaw, D., *Fourier Transform N.M.R. Spectroscopy,* Elsevier, Amsterdam, 1984, 344.

Shaw, G. J., Allen, J. M. and Visser, F. R., Volatile flavor components of babaco fruit (*Carica pentagona,* Heilborn), *J. Agric. Food Chem.,* 33, 795, 1985.

Shigematsu, Y., Murofushi, N., Ito, K., Kaneda, C., Kawabe, S. and Takahashi, N., Sterols and asparagine in the rice plant, endogenous factors related to resistance against the brown planthopper (*Nilaparvata lugens*), *Agric. Biol. Chem.,* 46, 2877, 1982.

Smith, C. M., Khan, Z. R. and Caballero, P., Techniques and methods to evaluate the chemical bases of insect resistance in the rice plant, in *Rice Insects: Management Strategies,* E. A. Heinrichs and T. A. Miller, eds., Springer-Verlag, New York, 1991, 347.

Sogawa, K., Preliminary assay of antifeeding chemicals for the brown planthopper, *Nilaparvata lugens* (Stål) (Hemiptera, Delphacidae), *Appl. Entomol. Zool.,* 6, 215, 1971.

Sogawa, K., Studies on the feeding habits of the brown planthopper, *Nilaparvata lugens* (Stål) (Hemiptera: Delphacidae) IV. Probing stimulant, *Appl. Entomol. Zool.,* 9, 204, 1974.

Sogawa, K., Studies on the feeding habits of the brown planthopper, *Nilaparvata lugens* (Stål) (Hemiptera: Delphacidae) V. Probing stimulatory effect of rice flavonoid, *Appl. Entomol. Zool.,* 11, 160, 1976.

Städler, E. and Hanson, F. E., Influence of induction of host preference on chemoreception of *Manduca sexta*: behavioural and electrophysiological studies, *Symp. Biol. Hung.,* 16, 267, 1976.

Städler, E. and Schöni, R., Oviposition behavior of the cabbage root fly, *Delia radicum* (L.), influenced by host plant extracts, *J. Insect Behavior,* 3, 195, 1990.

Stewart, J. E., *Infrared Spectroscopy: Experimental Methods and Techniques,* Marcel Dekker, New York, 1970, 636.

Stipanovic, R. D., Altman, D. W., Begin, D. L., Greenblatt, G. A. and Benedict, J. H., Terpenoid aldehydes in upland cottons: analysis by airline and HPLC methods, *J. Agric. Food Chem.,* 36, 509, 1988.

Struble, D. L. and Arn, H., Combined gas chromatography and electroantennogram recording of insect olfactory responses, in *Techniques in Pheromone Research,* H. E. Hummel and T. A. Miller, eds., Springer-Verlag, New York, 1984, 464.

Stubbs, M. R., Chambers, J., Schofield, S. B. and Wilkins, J. P. G., Attractancy to *Oryzaephilus surinamensis* (L.) of volatile materials isolated from vacuum distillate of heat-treated carolus, *J. Chem. Ecol.,* 11, 565, 1985.

Tabashnik, B. E., Deterrence of diamondback moth (Lepidoptera: Plutellidae) oviposition by plant compounds, *Environ. Entomol.,* 14, 575, 1985.

Tabashnik, B., Plant secondary compounds as oviposition deterrents for cabbage butterfly, *Pieris rapae* (Lepidoptera: Pieridae), *J. Chem. Ecol.,* 13, 309, 1987.

Takeoka, G. R., Flath, R. A., Mon, T. R., Teranishi, R. and Geuntert, M., Volatile constituents of apricot (*Prunus armeniaca*), *J. Agric. Food Chem.,* 38, 471, 1990.

Tatsuka, K., Suekane, S., Sakai, Y. and Sumitani, H., Volatile constituents of kiwi fruit flowers: simultaneous distillation and extraction versus headspace sampling, *J. Agric. Food Chem.,* 38, 2176, 1990.

Thorsteinson, A. J. and Nayar, J. K., Plant phospholipids as feeding stimulants for grasshopper, *Can. J. Zool.,* 41, 931, 1963.

Tichenor, L. H., Seigler, D. S. and Sternburg, J. G., An apparatus for the bioassay of certain plant-chemical-insect interactions, *Bioscience,* 29, 364, 1979.

Tingle, F. C. and Mitchell, E. R., Aqueous extracts from indigenous plants as oviposition deterrents for *Heliothis virescens, J. Chem. Ecol.,* 10, 101, 1984.

Todd, G. W., Getahun, A. and Cress, D. C., Resistance in barley to the greenbug, *Schizaphis graminum.* I. Toxicity of phenolic and flavonoid compounds and related substances, *Ann. Entomol. Soc. Am.,* 64, 718, 1971.

Torto, B., Hassanali, A., Saxena, K. N. and Nokoe, S., Feeding responses of *Chilo partellus* (Swinhoe) (Lepidoptera: Pyralidae) larvae to sorghum plant phenolics and their analogs, *J. Chem. Ecol.,* 17, 67, 1991.

Trial, H. and Dimond, J. B., E modin in buckthorn: a feeding deterrent to phytophagous insects, *Can. Entomol.,* 111, 207, 1979.

Turlings, T. C. J., Tumlinson, J. H., Lewis, W. J. and Vet, L. E. M., Beneficial arthropod behavior mediated by airborne semiochemicals. VIII. Learning of host-related odors induced by a brief contact experience with host by-products in *Cotesia marginiventris* (Cresson), a generalist larval parasite, *J. Insect Behavior,* 2, 217, 1989.

Turner, C. J., Multipulse NMR in liquids, *Prog. NMR Spectroscopy,* 16, 311, 1984.

Unithan, G. C. and Saxena, K. N., Diversion of oviposition by *Atherigona soccata* (Diptera: Muscidae) to nonhost maize with sorghum seedling extract, *Environ. Entomol.,* 19, 1432, 1990.

Usher, B. F., Bernays, E. A. and Barbehenn, R. V., Antifeedant tests with larvae of *Pseudaletia unipuncta*: variability of behavioral response, *Entomol. Exp. Appl.,* 48, 203, 1988.

Valterova, I., Bolgar, T. S., Kalinova, B., Kovalev, B. G. and Verkoc, J., Host plant components from maize tassel and electroantennogramme responses of *Ostrinia nubilalis* to the identified compounds and their analogues, *Acta Entomol. Bohemoslov.,* 87, 435, 1990.

Van Der Pers, J. N. C. and Den Otter, C. J., Single cell responses from olfactory receptors of small ermine moths to sex-attractants, *J. Insect Physiol.,* 24, 337, 1978.

van Loon, J. J. A., Frentz, W. H. and van Eeuwijk, F. A., Electroantennogram resonses to plant volatiles in two species of *Pieris* butterflies, *Entomol. Exp. Appl.,* 62, 253, 1992.

Velusamy, R., Thayumanavan, B. and Sadasivam, S., Effect of steam distillate extracts of selected resistant cultivated and wild rices on behavior of leaffolder *Cnaphalocrocis medinalis* (Guenée) (Lepidoptera: Pyralidae), *J. Chem. Ecol.,* 16, 2291, 1990a.

Velusamy, R., Thayumanavan, B., Sadasivan, S. and Jayaraj, S., Effect of steam distillate extract of resistant wild rice *Oryza officinalis* on behavior of brown planthopper *Nilaparvata lugens* (Stål) (Homoptera: Delphacidae), *J. Chem. Ecol.,* 16, 809, 1990b.

Vet, L. E. M., Van Lenteren, J. C., Heymans, M. and Meelis, E., An airflow olfactometer for measuring olfactory responses of hymneopterous parasitoids and other small insects, *Physiol. Entomol.,* 8, 97, 1983.

Villani, M. and Gould, F., Butterfly milkweed extract as a feeding deterrent of the wireworm, *Melanotus communis, Entomol. Exp. Appl.,* 37, 95, 1985.

Villani, M. G., Meinke, L. J. and Gould, F., Laboratory bioassay of crude extracts as antifeedants for the southern corn rootworm (Coleoptera: Chrysomelidae), *Environ. Entomol.,* 14, 617, 1985.

Visser, J. H., The design of a low-speed wind tunnel as an instrument for the study of olfactory orientation in the Colorado beetle (*Leptinotarsa decemlineata*), *Entomol. Exp. Appl.,* 20, 275, 1976.

Visser, J. H., Electroantennogram responses of the Colorado beetle, *Leptinotarsa decemlineata* to plant volatiles, *Entomol. Exp. Appl.,* 25, 86, 1979.

Visser, J. H., Differential sensory perceptions of plant compounds by insects, in *Plant Resistance to Insects,* P. A. Hedin, ed., ACS Symp. Ser. No. 208, American Chemical Society, Washington, DC, 1983, 375.

Visser, J. H. and Ave, D. A., General green leaf volatiles in the olfactory orientation of the Colorado beetle, *Leptinotarsa decemlineata, Entomol. Exp. Appl.,* 24, 738, 1978.

Visser, J. H., Van Straten, S. and Maarse, H., Isolation and identification of volatiles in the foliage of potato, *Solanum tuberosum,* a host plant of the Colorado beetle, *Leptinotarsa decemlineata, J. Chem. Ecol.,* 5, 13, 1979.

Vogel, S., Low-speed wind tunnels for biological investigations, in *Experiments in Physiology and Biochemistry,* G. A. Kerkut, ed., Academic Press, London, 1969, 410.

Volz, H. A., Monoterpenes governing host selection in the bark beetles *Hylurgops palliatus* and *Tomicus piniperda, Entomol. Exp. Appl.,* 47, 31, 1988.

Voyksner, R., Pack, T., Smith, C., Swaisgood, H. and Chen, D., Techniques for enhancing structural information from high-performance liquid chromatography/mass spectrometry, *ACS Symp. Ser.,* 420, 14, 1990.

Wadhams, L. J., The coupled gas chromatography—single cell recording technique, in *Techniques in Pheromone Research,* H. E. Hummel and T. A. Miller, eds., Springer-Verlag, New York, 1984, 464.

Wadhams, L. J., Angst, M. E. and Blight, M. M., Responses of the olfactory receptors of *Scolytus scolytus* (F.) (Coleoptera: Scolytidae) to the stereoisomers of 4-methyl-3-heptanol, *J. Chem. Ecol.,* 8, 477, 1982.

Waladde, S. M., Chemoreception of adult stem borers: tarsal and ovipositor sensilla on *Chilo partellus* and *Eldana saccharina, Insect Sci. Appl.,* 4, 159, 1983.

Waladde, S. M., Hassanali, A. and Ochieng, S. A., Taste sensilla responses to limnids, natural insect antifeedants, *Insect Sci. Appl.,* 10, 301, 1989.

Waladde, S. M., Ochieng, S. A. and Kahoro, H. M., Responses of *Chilo partellus* larvae to host plant materials: behaviour and electrophysiological bioassay, *Insect Sci. Appl.,* 11, 79, 1990.

Wearing, C. H. and Hutchings, R. F. N., 2-Farnesene, a naturally occurring oviposition stimulant for the codling moth, *Laspeyresia pomonella, J. Insect Physiol.,* 19, 1251, 1973.

Wearing, C. H., Connor, P. J. and Ambler, K. D., Olfactory stimulation of oviposition and flight activity of the codling moth, *Laspeyresia pomonella* using apples in an automated olfactometer, *N. Z. J. Sci.,* 16, 697, 1973.

Weissling, T. J., Meinke, L. J., Trimnell, D. and Golden, K. L., Behavioral responses of *Diabrotica* adults to plant-derived semiochemicals encapsulated in a starch borate matrix, *Entomol. Exp. Appl.,* 53, 219, 1989.

Wheeler, G. S. and Slansky, Jr., F., Effect of constituline and herbivore-induced extractables from susceptible and resistant soybean foliage on nonpest and pest noctuid caterpillars, *J. Econ. Entomol.,* 84, 1068, 1991.

White, J. D. and Richmond, J. A., Two olfactometers for observing orientation of the southern pine beetle to host odors, *J. Georgia Entomol. Soc.,* 14, 99, 1979.

White, P. P. and Chapman, R. F., Tarsal chemoreception in the polypagous grasshopper, *Schistocerca americana:* behavioural assay, sensilla distributions and electrophysiology, *Physiol. Entomol.,* 15, 105, 1990.

Wilson, D. D., Severson, R. F., Son, K-C. and Kays, S. J., Oviposition stimulants in sweet potato periderm for the seed potato weevil, *Cylas formicarius elegantulus* (Coleoptera: Curculionidae), *Environ. Entomol.,* 17, 691, 1988.

Withycombe, D. A., Lindsay, R. C. and Stuiber, D. A., Isolation and identification of volatile components from wild rice grain (*Zizania aquatica*), *J. Agric. Food Chem.,* 26, 816, 1978.

Wolfson, J. L. and Murdock, L. L., Method for applying chemicals to leaf surfaces for bioassay with herbivorous insects, *J. Econ. Entomol.,* 80, 1334, 1987.

Woodhead, S., Surface chemistry of *Sorghum bicolor* and its importance in feeding by *Locusta migratoris, Physiol. Entomol.,* 8, 345, 1983.

Woodhead, S. and Bernays, E. A., The chemical basis of resistance of *Sorghum bicolor* to attack of *Locusta migratoris, Entomol. Exp. Appl.,* 24, 123, 1978.

Wotiz, H. H. and Clark, S. J., *Gas Chromatography in the Analysis of Steroid Hormones,* Plenum Press, New York, 1966, 288.

Woodhead, S. and Padgham, D. E., The effect of plant surface characteristics on resistance of rice to the brown planthopper, *Nilaparvata lugens, Entomol. Exp. Appl.,* 47, 15, 1988.

Yamamoto, R. T. and Fraenkel, G., Assay of the principle gustatory stimulant for the tobacco hornworm, *Protoparce sexta* (Hohan.) from Solanaceous plants, *Ann. Entomol. Soc. Am.,* 53, 499, 1960.

Yoshihara, T., Sogawa, K., Pathak, M. D., Juliano, B. O. and Sakamura, S., Soluble silicic acid as a sucking inhibitory substance in rice against the brown planthopper (Delphacidae, Homoptera), *Entomol. Exp. Appl.,* 26, 314, 1979.

Yoshihara, T., Sogawa, K., Pathak, M. D., Juliano, B. O. and Sakamura, S., Oxalic acid as a sucking inhibitor of the brown planthopper in rice (Delphacidae, Homoptera), *Entomol. Exp. Appl.,* 27, 149, 1980.

Zalkow, L. H., Gordon, M. M. and Lanir, N., Antifeedants from rayless goldenrod and oil of pennyroyal: toxic effects for the fall armyworm, *J. Econ. Entomol.,* 72, 812, 1979.

CHAPTER 5

Evaluation of the Morphological Bases of Insect Resistance

There is ample evidence to suggest that plant morphological characters such as tissue toughness, pubescence, spines, or bristles act as barriers to normal feeding and oviposition by insects. Similarly, factors such as the lignification of cell walls, the presence of silica, tissue sclerotization, and arrangement and number of vascular bundles affect the feeding and establishment of insects on plants.

Since the insect resistance in many genetically improved crop plants involves morphological factors, it is important to elucidate the roles of such factors in insect resistance and susceptibility. In many cases the role of plant morphology in insect resistance can be easily determined. In some cases, however, it is difficult to demonstrate if a certain morphological character contributes towards insect resistance. Simple bioassays, involving insect responses on plant varieties of differing morphological characters, may not provide sufficient evidence to prove the role of plant morphology in insect resistance because such resistance may also be imparted by allelochemicals.

This chapter has been prepared to serve as a comprehensive overview of the entomological techniques developed during the past 40 years to better understand the morphological bases of insect resistance in plants.

5.1. TRICHOMES

Trichomes, also known as hairs or pubescence, are one of the more important morphological parameters of plant resistance to insects. In numerous plant species, a negative correlation has been established between trichome density on the plant surface and insect feeding and oviposition (Gallun et al. 1973, Levin 1973, Webster 1975, Hoxie et al. 1975, Roberts and Foster 1983,

Khan et al. 1986, Sosa, Jr. 1988, Tingey 1991). Long and dense trichomes reportedly hinder normal feeding and oviposition by various insect pests. According to Norris and Kogan (1980), the roles of plant pubescence have been studied in 58 insect species in at least 19 crops belonging to 11 plant families. Trichomes reportedly contribute to plant resistance against 32 insects and susceptibility to 14 insects. No trichome effect was found for five insect species and no consistent relationship was found with seven other species.

Nonglandular, simple erect trichomes on plants deter insect feeding and oviposition and may also adversely affect insect egg hatchability and larval development (Johnson and Hollowell 1935, Sikka et al. 1966, Ringlund and Everson 1968, Schillinger, Jr. and Gallum 1968, Broersma et al. 1972, Gallun et al. 1973, Hoxie et al. 1975, Roberts et al. 1979, Lyman and Cardona 1982, Roberts and Foster 1983, Khan et al. 1986, Schultz and Coffelt 1987). Hooked trichomes on the foliage of the green bean, *Phaseolus vulgaris* L., impale nymphs of the potato leafhopper, *Empoasca fabae* (Harris), and the aphid, *Aphis craccivora* Koch during movement on the bean plant leaf (Johnson 1955, Pillemer and Tingey 1976, 1978).

Glandular trichomes contained on leaves, stems, and reproductive structures of several plant species adversely affect the feeding, growth, survival, and oviposition of several insect pests (Shade et al. 1979, Johnson et al. 1980, Duffey and Isman 1981, Johnson et al. 1980, MacLean and Byers 1983, Shade and Kitch 1983, Lapointe and Tingey 1984, Tingey and Laubengayer 1986, Farrar, Jr. and Kennedy 1987, Neal et al. 1991, Tingey 1991). Upon contact, glandular trichomes release a viscous substance which accumulates on tarsi and mouthparts, entrapping small insects, and decreasing the feeding and oviposition of larger insects. Trichome exudates are composed of various aldehydes, alkanes, esters, ketones, sesquiterpenes, etc. (Triebe et al. 1981, Lin et al. 1987). The biochemical aspects of glandular trichome-mediated insect resistance was reviewed by Steffens and Walters (1991).

The relative contribution of trichome- and nontrichome-based plant resistance to insects may not be well understood unless trichomes are removed to detect insect resistance. Without removing trichomes, the effects of plant allelochemicals could also be mistakenly ascribed to trichome-based resistance.

5.1.1. Removal of Trichomes

In order to remove nonglandular trichomes from pubescent leaves, electric razors can be used following the methods described by Khan et al. (1986), Baur et al. (1991) and Navon et al. (1991) (Figure 5.1).

Glandular trichome exudates can be removed from plants by wiping or by freezing. If exudates are removed by wiping, plant foliage is first gently wiped with a dry cotton ball, next with a cotton ball soaked in a 0.1% aqueous solution of Triton X-155 (a non-phytotoxic emulsifying agent), then rinsed in 0.1% Triton X-155, and then rinsed in water (Farrar and Kennedy

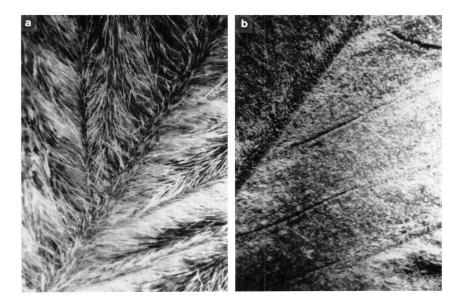

FIGURE 5.1. Trichomes on the abaxial surface of the uppermost juvenile leaf of PI227687 soybean plant (a); and the same leaf after trichomes were removed using an electric razor (b). (From Khan, Z. R. et al., *Entomol. Exp. Appl.*, 42, 109, 1986. With permission of Kluwer Academic Publishers.)

1987). Tingey and Laubengayer (1986) removed glandular trichomes by vigorously wiping the abaxial leaflet surface using a cotton gauge pad moistened with 95% ethanol. Yerger et al. (1992) developed a procedure to recover glandular trichome exudate by freezing and treatment with dry ice. Plant tissues containing trichomes are quick frozen in liquid nitrogen, and then shaken with finely powdered dry ice. Trichomes are sheared from the supporting plant tissues, sieved from the trichome-free tissues, and then extracted in the solvent to be used in bioassays.

5.1.2. Bioassays

5.1.2.1. Feeding

The roles of trichomes in resistance to feeding by chewing insects can be evaluated in choice (Khan et al. 1986, Bauer et al. 1991) or no-choice (Navon et al. 1991) tests. Khan et al. (1986) studied the roles of soybean foliar trichomes in soybean resistance to the cabbage looper, *Trichoplusia ni* (Hübner), using choice feeding bioassays conducted with disks from trichome-shaven vs. unshaven leaves (Figure 5.2). Leaf disks are presented for 8 hr to 10 third-instar *T. ni* larvae, that have been starved for 2 hr with access

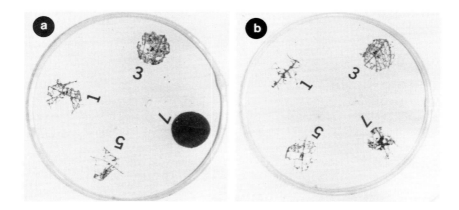

FIGURE 5.2. Results from a choice test bioassay of *Trichoplusia ni* larval feeding on leaf disks from trifoliate leaf (TL) 1 (1), TL 3 (3), TL 5 and TL 7 (7) of a susceptible soybean variety Davis. In one set of experiment, (a) dense trichomes were not shaven from the uppermost juvenile TL7; whereas in another set (b), TL7 trichomes were shaven using an electric razor. (Photograph provided by Z. R. Khan.)

to water. Disks are positioned with their abaxial sides up and arranged equidistantly on a circle inside a petri dish arena. Khan et al. (1986) reported that in the youngest and uppermost leaf of the susceptible soybean variety Davis, *T. ni* resistance was attributed to trichomes, whereas in the uppermost leaf of the resistant soybean variety PI227687, resistance involved both trichomes and chemical factors.

From dual choice tests, Baur et al. (1991) demonstrated that trichomes of a grey alder, *Alnue incana* (L.), significantly deterred the feeding behavior of larvae and adults of a chrysomelid beetle, *Agelastica alni* L. From a no-choice test, Navon et al. (1991) also reported that removal of trichomes from the leaves of the pubescent cotton cultivar "Texas-172" did not increase feeding by *Heliothis armigera* (Hübner).

The role of trichomes in plant resistance to sucking insects can be evaluated using an electronic feeding monitor (described in Chapter 2.2.2.1.5). Tingey and Laubengayer (1986) recorded the feeding behavior of the potato leafhopper, *E. fabae,* on intact and trichome-free leaves of resistant potato hybrids. The waveforms produced by fourth instar nymphs are recorded on a strip chart recorder operating at a constant speed of 6 cm/min. Feeding parameters such as the number of probes, the total probing time, and the time to initiate the first probe are determined through the analysis of strip chart recording. Tingey and Laubengayer (1986) demonstrated that *E. fabae* acceptance and feeding success on the resistant potato was significantly improved following the removal of glandular trichomes. The adult feeding of the leafminer, *Liriomyza trifoli* (Burgess), on a wild tomato species,

Lycopersicon pennellii (Carr), was significantly increased following rinsing of glandular trichomes on foliage with ethanol (Hawthorne et al. 1992). Resistant attributes of *L. pennellii* were transferred to a susceptible tomato species by application of exudates of *L. pennellii* foliage to the susceptible plant.

5.1.2.2. Growth

In this type of test, a preweighed test insect is caged singly on an intact or a trichome-shaven leaf. After 24 to 48 hr of feeding, the weight of the insect is recorded and weight gain due to feeding is calculated following methods of Navon et al. (1991), who reported that cotton leaf trichomes had no effect on weight gain by *H. armigera* fed on the resistant cotton variety Texas 172.

Neal et al. (1989) measured the influence of glandular trichomes of a resistant wild potato, *Solanum berthaultii* Hawkes, on the growth of the Colorado potato beetle, *Leptinotarsa decemlineata* (Say), by incorporating *Solanum* spp. fresh leaves or lyophilized leaf powder into an artificial diet, which eliminates the physical feeding barrier caused by the trichomes. Beetle growth was similar on artificial diets containing either pubescent *S. berthaultii* or susceptible smooth *S. tuberosum* L. leaves, indicating the role of glandular trichomes of *S. berthautii* in resistance to *L. decemlineata*.

5.1.2.3. Mortality

The relative importance of trichome and non-trichome-based resistance factors to larval insect mortality can be studied by rearing the test insect on foliage with either intact or removed trichomes, and recording insect mortality after 48 or 72 hr.

Farrar and Kennedy (1987) reported that the removal of glandular trichomes from the resistant wild tomato, *Lycopersicon hirsutum f. glabratum* C. M. Mull, and the susceptible tomato, *L. esculentum*, significantly increased the suitability of each for *H. zea* survival. On the resistant *L. hirsutum* f. *glabratum*, trichome removal significantly reduced larval mortality.

5.2. SILICA

Soluble silica absorbed by plant roots is translocated and deposited in the tissues of a number of plants. Several studies have demonstrated a relationship between silica deposits in certain plant tissues and the resistance of these plants to insect attack (Ponnaiya 1951, Djamin and Pathak 1967, Hanifa et al. 1974, Lanning et al. 1980). Djamin and Pathak (1967) and Hanifa et al. (1974) demonstrated that the presence of greater amounts of silica in resistant plants interfered with the feeding and boring of insects and caused severe wear to the mandibles. However, because studies relating to silica and insect resistance have been conducted on different susceptible and resistant plant varieties, it is possible that presumed silica-based resistance could be due to plant allelochemicals. Evaluating plants for the role of silica in insect

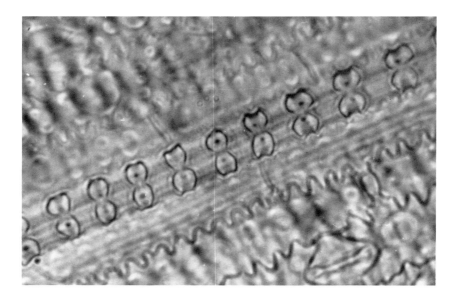

FIGURE 5.3. Silica cells deposited on leaf surface of TKM6 plants grown in a culture solution with 100 ppm silica. (Photo courtesy of B. Mandras, Visayas State College of Agriculture, Philippines.)

resistance necessitates that all test varieties be grown with and without silica for bioassays, following the methods described by Sasamoto (1953, 1955) and Mandras (1991).

Mandras (1991) evaluated the role of silica in susceptible and resistant rice varieties as a resistance factor to the yellow stem borer, *Scirpophaga incertulas* (Walker), by growing susceptible and resistant rice varieties in culture solutions with 0, 5, 10, 25, 50, or 100 ppm silica. Seeds of test varieties are washed and soaked in distilled water for 24 hr, then incubated for 48 hr at a constant temperature of 30° C. The germinated seeds are transferred to styropore trays supported underneath by a fiber glass net. The trays are placed in plastic pans containing a silica-free culture solution (Yoshida et al. 1976) for 2 weeks, and then transferred to similar trays with multiple holes and placed in plastic pans containing culture solutions with or without silica. Culture solution is maintained at a pH of 5.0 and replaced after every 5 days.

Silica cells in the resulting plants can be examined by boiling the plant tissue for 1 min in 20 ml phenol containing 5 drops of 0.1% aqueous safranin solution (Ramachandran and Khan 1991). The plant tissue is then mounted on a glass slide in glycerine and examined under a light microscope (Figure 5.3). The following are some of the bioassay techniques which can be used to understand the role of silica in insect resistance.

FIGURE 5.4. Yellow stem borer damage to susceptible Rexoro plants grown in culture solutions with (a) no silica, and with (b) 10 ppm, (c) 25 ppm, (d) 50 ppm, and (e) 100 ppm silica. With increase in silica level a decrease in insect damage was noted. (Photo courtesy of B. Mandras, Visayas State College of Agriculture, Philippines.)

5.2.1. Plant Damage

Plants grown in silica culture solutions are infested with neonate larvae of the test insect and observed for 3 to 4 weeks until appearance of damage symptoms. Mandras (1991) reported that regardless of the rice variety, plants grown in silica-free culture solution incurred significantly more damaged tillers from *S. incertulas* (Figure 5.4), compared to plants grown in cultivars with silica. Damage to susceptible Rexoro plants grown in 100 ppm silica solution was comparable to damage to resistant TKM6 plants grown in culture solution without silica. The reduced damage of resistant TKM6 plants may also be the result of the antibiotic effects of allelochemicals.

5.2.2. Larval Penetration Time

Larval penetration time of stem borers on susceptible and resistant plants, grown with or without silica, can also be measured following the method described by Mandras (1991) for *S. incertulas.* A 5 cm × 5 cm Parafilm® sachet (similar to Figure 2.18) is attached to the rice stem, at the point where stem borers generally penetrate. One newly emerged larva is introduced into each sachet. The time spent by each larva from the first attempt until it has completely penetrated the stem is recorded. Mandras (1991) reported that penetration of

S. incertulas larvae into the stems of rice varieties was greatly affected by the addition of silica in the culture solution on all varieties tested. The penetration time was shortest in plants devoid of silica, and longest in those grown on 100 ppm silica solution.

5.2.3. Feeding and Food Utilization

Feeding and utilization of food by both chewing and sucking insects on plants grown on different levels of silica can be determined by the methods described in Chapters 2.2.2 and 2.2.3.

In an attempt to learn more about the specific mechanisms involved in silica-based plant resistance in the southern armyworm, *Spodoptera eridania* (Cramer), Peterson et al. (1988) amended artificial diets with 0, 0.5, 2.5, 5, 10, or 20% (dry weight) of silicic acid. Indices of nutrition and growth (calculated as described in Chapter 7.2.3) indicated that silica at levels of 10% dry weight and above reduced *S. eridania* digestibility and increased consumption rates.

5.3. TISSUES TOUGHNESS AND THICKNESS

The toughness and thickness of various plant tissues has been reported to be correlated with resistance to insect pests. Toughness of stems, foliage, and pods of various crops adversely affects penetration and feeding by insects (Issac 1939, Khanna et al. 1947, Tanton 1962, Agarwal 1969, Chiang and Jackai 1988, Kang et al. 1990, Springer et al. 1990). Stems thickened by increased layers of epidermal cells deter or limit the entrance of stem damaging insects of several crops (Patanakamjorn and Pathak 1967, Wallace et al. 1974, Martin et al. 1975, Fiori and Dolan 1981). A thick cortex layer in the stems of wild tomato, *Lycopersicon hirsutum* var. *typicum,* is correlated with resistance to the aphid, *Macrosiphum euphorbiae* Thomas, because the proboscis is not sufficiently long to reach the phloem (Quiros et al. 1977).

Although adverse effects of tissue toughness and thickness on feeding activities of phytophagous insects have been discussed by many researchers, the relative importance of these morphological resistance factors has not been compared with allelochemical resistance factors affecting insect feeding.

5.3.1. Weight-Type Penetrometer

Various types of penetrometers have been used for quantitative measurements of the toughness of plant tissues. Williams (1954) and Tanton (1962) developed a simple penetrometer based on the weight of sand required to push a sharp pin through a leaf blade. This principle was further modified with a metal rod to punch out a leaf disk by a shearing and tearing action (Feeny 1970). Each of these methods measured toughness as force (weight of sand) required to penetrate a leaf.

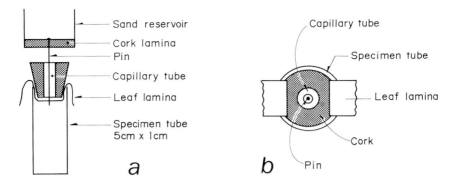

FIGURE 5.5. The "Weight type" penetrometer (a); cross section of the cork and the tube (b). (From Tanton, M. T., *Entomol. Exp. Appl., 5,* 74, 1962. With permission of Kluwer Academic Publishers.)

The penetrometer described by Williams (1954) and Tanton (1962) consists of a pin attached to a sand reservoir (Figure 5.5). A strip of leaf is placed across the mouth of a specimen tube and a snug fitting cork, having a capillary tube fitted in its center, is pushed into the tube. The pin is placed in the capillary tube so that the point rests on the leaf surface. Dry silver sand is poured into the sand reservoir until the pin just penetrates the leaf. The reservoir and the sand are weighed. On the basis of weight needed for the pin to penetrate the various leaves, a toughness scale of T1 to T12 was developed, where T1 is the least tough and T12 is the toughest (equivalent to weights of 7.0 g and 10.6 g, respectively).

5.3.2. Spring-Type Penetrometer

In the spring type penetrometer developed by Cherrett (1968), increasing pressure is applied to a needle by means of an extension spring. The degree of extension required to produce penetration by the needle point is read on a millimeter scale (Figure 5.6). This penetrometer consists of a lever (G) with a control pivot, one end of which is connected to a spring (E) and the other end to a needle (I). To zero the apparatus, the capstan (M) is turned until the indicator knob (B) is opposite zero on the scale. The capstan is then slackened off to raise the needle and the leaf tissue to be measured is positioned on the table. The capstan is then rotated at a slow speed until the needle has just emerged through the lower side of the leaf. The reading of the indicator knob (B) is then noted. By balancing against weights, extension readings for the spring are expressed as grams of pressure.

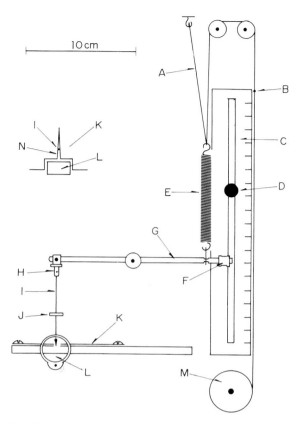

FIGURE 5.6. The "Spring type" penetrometer; A, fine nylon cord; B, indicator knob; C, graduated sliding scale; D, locking unit; E, interchangeable spring; F, counterweight; G, stainless steel lever with central pivot; H, needle holder; I, needle; J, relieved guide; K, stainless steel table; L, magnifying glass; M, friction-damped take-up capstan; N, V-shaped notch in table. (From Cherrett, J. M., *J. Econ. Entomol.,* 61, 1736, 1968. With permission of the Entomological Society of America.)

5.3.3. Strain Gauge Transducer Penetrometer

A penetrometer developed by Beckwith and Helmers (1976), is based on a strain gauge transducer, miniature lathe, and recorded printout. The penetrometer is a small lathe in which the tailstock (1) supports a strain gauge transducer (2), and the carriage (3) supports the leaf (4). The penetrometer needle (5) is fitted to the transducer. Plant material is clamped between two pieces of plastic (6) that are part of the carriage. Carriage movement is controlled by a lead screw (7) which is turned by a motor (8). The transducer signal is measured by a strain gauge bridge and noted graphically by a recorder (Figure 5.7).

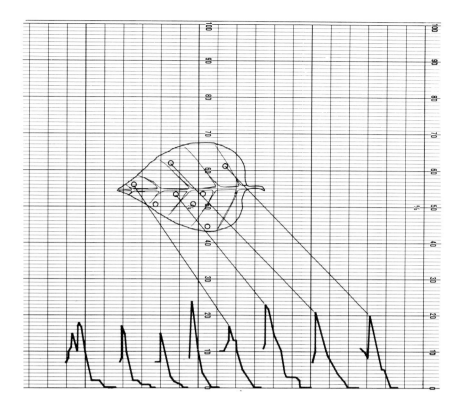

FIGURE 5.7. Recorder deflection of "Strain gauge transducer" penetrometer showing variance at different positions in a white birch leaf. (Redrawn from Beckwith, R. C. & Helmers, A. E., *Environ. Entomol.*, 5, 291, 1976. With permission of the Entomological Society of America.)

5.3.4. King Penetrometer

King (1988) designed a penetrometer to measure leaf toughness of *Pelargonium* sp. and *Schefflera* sp. The major components of the penetrometer are two pressure valves and a pressure gauge from a sphygmomanometer, a plastic syringe, and a metal rod (Figure 5.8). A cylindrical hole is drilled through two smooth, hardwood blocks (15 × 10 cm). A leaf is placed between two blocks which are held firmly together by two wing nuts and bolts. The upper block has a hole about 5 mm in diameter lined by a plastic sleeve through which is passed a metal rod (2 mm diameter) to puncture the leaf. The rod is attached to an inverted 10-cc plastic hypodermic syringe. A hypodermic needle is attached to the upper end of the syringe along with a T-shaped plastic tubing connector. Automatic pressure release valves are attached to each outlet of the T-tube (Figure 5.8). The upper valve is fixed to an

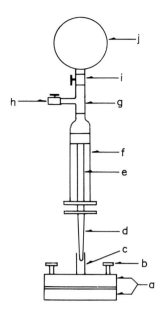

FIGURE 5.8. The King penetrometer; a, wood block assembly; b, wing nut and bolt; c, plastic sleeve; d, brass punching rod; e, plunger of plastic syringe; f, plastic syringe; g, T-shaped plastic connector; h and i, plastic release valves; j, aneroid pressure gauge. (From King, B. L., *Va. J. Sci., 39,* 405, 1988. With permission.)

aneroid pressure gauge. This gauge allows air into the gauge but not out and effectively converts it into a stop gauge. The side valve allows air into the system but not out.

To use the King penetrometer, a test leaf is placed between the wooden blocks. As force is applied to the leaf with the punching rod, the hypodermic plunger is compressed and pressure increase in the system is registered on the pressure gauge. When the metal rod pushes completely through the leaf, the pressure increase ceases and the pressure reading at that time is the measure of leaf toughness. Although measurements recorded by this penetrometer and the penetrometer described by Feeny (1970) were similar for fairly tough leaves, subsequent trials on very tender leaves have indicated that the King penetrometer may not be as sensitive without modification (B. L. King, personal communication).

5.3.5. Portable Penetrometer

Sands and Brancatini (1991) described a portable penetrometer and tongs for measuring leaf toughness while the leaves remain attached to the plant. The penetrometer consists of a dial gram gauge with removable probes

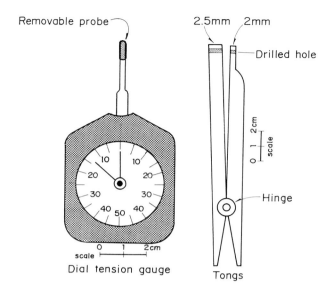

FIGURE 5.9. The "Portable" penetrometer and the tongs. (From Sands, D. P. A. and Brancatini, V. A., *Proc. Entomol. Soc. Wash.*, 93, 786, 1991. With permission of Entomological Society of Washington.)

(Figure 5.9). The probes are made from stainless steel pins or rods with the apex ground flat. The gauge is calibrated for 0 to 50 g, but gauges with higher capacities can be used on plants with greater leaf toughness.

The tongs are constructed from two rectangular polyacrylamide plates (160 × 33 × 5 mm) with tapered apices and holes are drilled (2 mm diameter upper; 2.5 mm diameter lower plate) to accommodate the penetrometer probe. The penetrometer is calibrated by clamping it in a boss head clamp which is pivoted on a retort stand so that the probe rests under pressure on an open pan of an electronic balance. Weights registered on the tension gauge are corrected for error by comparing readings with those on the electronic balance. Sands and Brancatini (1991) reported that when used on leaves of the vine, *Stephania japorica* (Thunberg), penetrometer readings had coefficients of variance less than 5%.

5.4. SURFACE WAXES

The plant surface is protected against desiccation, insect feeding, and disease by a layer of surface waxes over the epicuticle. Chemically, plant waxes are esters formed by the linkage of a long chain fatty acid and an aliphatic alcohol. For a more detailed discussion of the structural aspects of plant surface waxes, see Jeffree (1986).

Foliar wax coatings play an important role in the resistance of some crop cultivars to insect attack when sense organs on the insect tarsi and mouthparts receive negative chemical and tactile stimuli from the leaf surface. The raspberry species, *Rubus phoenicolasius,* has heavy wax secretions that serve as a resistance barrier against the raspberry beetle, *Byturus tomentosus* Barber, and the rubus aphid, *Amphorophora rubi* (Kaltenbach) (Lupton 1967). Wax blooms on the leaves of some cruciferous crops deter feeding of the cabbage flea beetle, *Phyllotreta albionica* (LeConte), and diamondback moth, *Plutella xylostella* (L.) (Anstey and Moore 1954, Eigenbrode and Shelton 1990, Eigenbrode et al. 1991, Stoner 1990). Wax blooms on the leaves of brussels sprouts consist of vertical rods and dendritic plates that interfere with adhesion of the tarsal setae of the mustard beetle, *Phaedon cochleariae,* to the leaf surface (Stork 1980). In contrast, the waxy leaves of kale stimulate feeding of the cabbage aphid, *Brevicoryne brassicae* (L.), and the cabbage whitefly, *Aleurodes brassicae* (Walker), more than glossy-leaved cultivars (Thompson 1963).

Foliar surface waxes of other plants contain allelochemicals that also affect the behavior of insects. Sulfur compounds from the surface waxes of onions stimulate oviposition by the leek moth, *Acrolepiopsis assectella* (Zeller) (Thibout et al. 1982). Linear furanocoumarins from carrot foliage waxes have a similar effect on oviposition by the carrot rust fly, *Psilia rosae* (F.) (Städler and Buser 1982). Hydrocarbon- and carbonyl-containing fractions of the wax of rice cultivars resistant to the brown planthopper, *Nilaparvata lugens* (Stål), adversely alter hopper feeding behavior by causing hoppers to move away from their preferred feeding site (Woodhead and Padgham 1988).

In sorghum, a parallel exists to that in cruciferous plants. Wax on sorghum leaves deters feeding by the migratory locust, *Locusta migratoides* (R & F) (Atkin and Hamilton 1982). Wax from young plants is more deterrent than that from older leaves, apparently because of the wax "blooms" of younger, faster growing foliage.

5.4.1. Responses to Surface Waxes

The roles of surface wax-based insect resistance factors can be evaluated by studying insect behavior on foliage with surface waxes either intact or removed (Woodhead and Padgham 1988, Eigenbrode and Shelton 1990). According to Woodhead and Padgham (1988), surface waxes of rice plants can be extracted by immersing intact rice tillers in ^{n}hexane (approximately 20 ml/tiller) for 30 sec. Eigenbrode and Shelton (1990) removed surface waxes of *Brassica oheracea* L. by 10 sec dips in three consecutive baths of dichloromethane.

In bioassays described by Woodhead and Padgham (1988), plant waxes are redissolved in appropriate volumes of hexane to yield a desired concentration. Extracts from each cultivar are then painted onto tillers of other rice cultivars to cover approximately 10 cm of tillers just above the water level.

Control plants are painted with solvent only. After allowing approximately 2 min for evaporation of the solvent from plant foliage, insects are placed on plants and observed for timing, number, distance, and direction of movements in the first 30 min. All movements are recorded on sketches of the plant. Plants are randomly positioned with respect to cultivar. Movements off of the treated area are also noted. Woodhead and Padgham (1988) demonstrated that *N. lugens* movement on susceptible IR22 rice plants could be increased by painting the plants with wax from resistant IR46 plants.

According to Eigenbrode and Shelton (1990), movement rates of neonate larvae of *P. xylostella* can be recorded on waxed and dewaxed leaves of resistant glossy leaved 8329 and susceptible normal bloom "Round-Up" cauliflowers. Removal of epicuticular waxes with dichloromethane increases *P. xylostella* larval movement rates on "Round-Up" and decreases them on 8329 so that movement rates on dewaxed leaves of both genotypes are similar.

5.5. PLANT CUTICLE

The cuticle of plants is a heterogenous membrane in which wax, pectin, and cellulose in varying proportions are contained in a cutin framework (Holloway and Baker 1968). Plant cuticle plays a significant role in the probing and oviposition behavior of insect pests, and young plant cuticle is normally preferred over thick mature cuticle (Zettler et al. 1969, Walker and Aitken 1985, Walker 1987, 1988).

5.5.1. Isolation of Plant Cuticle

The effect of leaf cuticle on probing and oviposition behavior by insects can be studied by removing the cuticles from young and mature plants and using the isolated cuticle as a membrane over artificial media (Walker 1988). Plant cuticle can be isolated by reacting plant leaves with either a zinc chloride-hydrochloric acid solution or with pectinase enzymes. Holloway and Baker (1968) developed a method to obtain leaf cuticle. Leaf disks, (2 cm^2) which have already been dewaxed by washing in chloroform, are immersed at room temperature in a solution containing 1 g zinc chloride in 1.7 ml of concentrated hydrochloric acid, using 5 ml of reagent per disk. The reagent reacts immediately with disks and the solution often becomes highly pigmented. Membranes begin to separate after 1 hr and the reaction is considered to be complete if membranes are released when a test disk is placed into distilled water. Depending on the plant tested, this normally occurs within 2 to 12 hr.

Walker (1988) developed an enzymatic method to obtain plant cuticle. Leaves are laid abaxial surface down and their adaxial cuticles are abraded off with sand paper. A 6-cm^2 disk is cut from the abraded portion of the leaf with a cork borer. The disk is floated, abraded surface down, on a pectinase solution (in 40% glycerol diluted 1:3 with acetate buffer at pH 4:0) for 24 hr at room temperature. The surface abaxial cuticle faces up and does not contact the enzyme

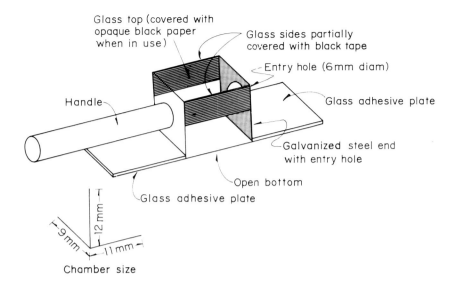

FIGURE 5.10. An observation chamber for observing probing behavior of *Parabemesia myricae* on experimental substrates. (From Walker, G. P., *Ann. Entomol. Soc. Am.*, 81, 365, 1988. With permission of the Entomological Society of America.)

solution. After 24 hr, the disks are carefully lifted off in such a way that the solution does not come in contact with the abaxial surface and are floated on distilled water. The cuticle then separates from the rest of the leaf disk.

5.5.2. Bioassays with Plant Cuticle

The effect of leaf cuticle on the probing behavior of sucking insects can be studied using the method described by Walker (1988) for bayberry whitefly, *Parabemisia myricae* (Kuwana). A 9-cm disk of white Whatman no. 1 filter paper is laid in a petri dish containing a hot aqueous solution of 1% agar, 15% sucrose, and 0.1 M phosphate buffer (pH 7.0). After the agar has solidified and has encapsulated the paper in a thin agar film, the paper is cut into rectangles (4.0 cm × 1.3 cm) and the rectangles are submerged in buffered 15% sucrose. Cuticle from different plants floating on buffered 15% sucrose are attached to the rectangles and the rectangles are attached to the bottom of an observation chamber described by Walker (1987, 1988) (Figure 5.10). Using a stereomicroscope, data on the proportion of insects that probe on first contact with leaf cuticle and the probe duration of each probe are recorded.

Walker (1988) reported that the duration of probes by *P. myricae* on sucrose solutions covered by young cuticle was significantly greater than on those covered by mature thick cuticle. It was concluded that since the media beneath the cuticle was the same for both young and mature cuticles, the inability of *P. myricae* to feed on mature lemon leaves was due to the properties of leaf cuticle.

5.6. COLOR

Specific color-related insect resistance in plants does exist, but genetic manipulation of plant color usually has an affect on some fundamental physiological plant process (Norris and Kogan 1980). Red cotton plants are less attractive to the boll weevil, *Anthonomus grandis* Boheman, than green plants (Stephens 1957). The imported cabbage worm, *Pieris rapae* (L.), is less attracted to the foliage of red brussels sprout varieties than to green-leaved varieties (Dunn and Kempton 1976). Some cucurbit cultivars with silver-colored leaves reflect more blue and ultraviolet wavelengths of light than normal cultivars and are resistant to aphids (Shifriss 1981). A purple pigment in the stem epidermis of wild soybean, *Glycine soja* Sieb and Zuec, has been used as an indicator of soybean stem resistance to the bean fly, *Ophiomyia centrosematis* (de Meijere) (Chiang and Norris 1984). Similarly, the color intensity of leaf supernatants of birch trees has been used to determine the degree of resistance to oviposition by the birch leafminer, *Fenusa pusila* (Lepeletier) (Fiori and Craig 1987). Foliar color influences oviposition by the cabbage root fly, *Delia radicum* (L.) (Roessingh and Städler 1990), onion fly *Delia antiqua* (Meigen) (Harris and Miller 1983), carrot fly, *Psila rosae* (F.) (Städler 1977), and Hessian fly, *Mayetiola destructor* (Say) (Harris and Rose 1990). However, the occurrence of color differences between susceptible and resistant plants does not always conclusively demonstrate that plant color is involved in conditioning insect resistance. It is, therefore, important to evaluate insect responses to various colors in the absence of other plant variables. The following are some of the bioassay techniques which can be used to understand insect color preferences.

5.6.1. Color Preference of Sucking Insects

The color preferences of sucking insects can be determined using a rearing chamber designed by Cartier and Auclair (1964) to test the color preference of aphids fed artificial diets. The chamber (Figure 5.11) consists of the top of a petri dish (10 cm diameter) with two ventilation holes on the flat surface and with the vertical wall painted black. The dish is closed by a piece of circular cardboard supporting six feeding units. Each feeding unit is made of a piece of glass tubing (23 mm long × 25 mm diameter) that supports the diet at one end enclosed between two Parafilm® membranes. The units are set into holes in the cardboard so that the pockets are flush with the ceiling of the chamber. The other end of the unit is covered with a colored rubber membrane acting as exterior light filter. The spectral transmittance characteristics of the colored rubber membrane can be determined with a spectrophotometer. Thirty to 150 aphids are released in the test chamber and their settling responses are recorded on feeding units transmitting different colors of light.

Cartier and Auclair (1964) demonstrated that pea aphid, *Acyrthosiphon pisum* (Harris), preferred to feed in orange or yellow (615 μm and 595 μm, respectively) lights, or both. Aphids confined to chemical diets backlighted blue or white survived only a few days, whereas aphids lived longer and grew rapidly on the same diets in yellow or orange light.

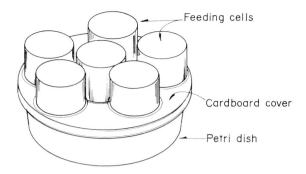

FIGURE 5.11. A color preference chamber used for aphids and other sucking insects. (Redrawn from the photograph in Cartier, J. J. and Auclair, J. L., *Can. Entomol.*, 96, 1240, 1964. With permission from the Entomological Society of Canada.)

5.6.2. Color Preference of Chewing Insects

The color preferences of chewing lepidopterous larvae can be determined following the methods of Meisner and Ascher (1973) and Saxena and Onyango (1987). Meisner and Ascher (1973) determined the influence of edible colors on the feeding behavior of cotton leafworm, *Spodoptera littoralis* (Boisduval), larvae in a choice or no-choice test. A range of commercial food colors are offered to test insects on styropor (foamed polystyrene) lamellae with 3% dry sucrose residues as phagostimulants. In choice tests, styropor lamellae treated with different colors are arranged in a large petri dish (15 cm diameter) (Figure 5.12) and test insects are released in the center. The weight of styropor consumed during the test period indicates the degree of phagostimulation. In no-choice tests, styropor lamellae of only one color is offered in each dish for insect feeding. In choice tests, *S. littoralis* larvae preferred to feed on green and yellow lamellae. Lamellae colored red and blue were inactive or even repellent.

Saxena and Onyango (1987) used papers of different colors to evaluate color preference of the sorghum stem borer, *Chilo partellus* (Swinhoe). A filter paper disk (15 cm diameter) is placed on an inverted glass petri dish (15 cm diameter) which is placed on a glass plate under a uniform illumination (1050 to 1150 lux). A cone (5 cm high, 5 cm base diameter) of the test color paper is placed upright on the filter paper disk, occupying its central area (5 cm diameter). About 5 cm from the basal edge of the cone, 20 first-instar larvae are released along a circular track around the cone. The number of larvae that move to the cone or out of the filter paper disk are recorded for 15 min and then they are removed. Different colored papers are tested one after

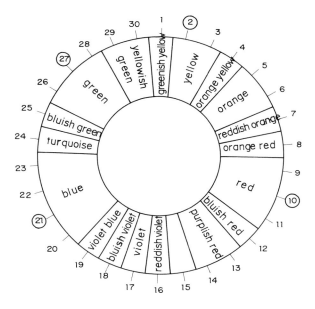

FIGURE 5.12. The color preference circle used for lepidopterous larvae; the numbers representing the four primary colors are encircled. (From Meisner, J. and Ascher, K. R. S., *Nature,* 242, 332, 1973. With permission.)

another, and the greater the percentage of larvae reaching a color cone, the greater its attractions. An attractancy index (Saxena and Onyango 1987) was calculated as follows:

$$\text{Attractancy Index} = 100 \times \frac{\% \text{ larvae reaching the cone} - \% \text{ larve reaching blank area}}{100 - \% \text{ larvae reaching blank area}}$$

Saxena and Onyango (1987) reported that bice green and light green colors elicited maximum attraction to *C. partellus* larvae, followed by foliage green and cadmium yellow colors.

5.6.3. Color Preference of Ovipositing Females

The effects of foliar color on the oviposition behavior of insects can be determined by presenting females with colored paper leaf models or filter paper strips (Harris and Rose 1990, Roessingh and Städler 1990). Paper models or paper strips of different colors are prepared either from commercially available colored cardboard or by dipping into diluted vegetable dyes and allowing the papers to dry. The models are coated with a thin layer of paraffin wax

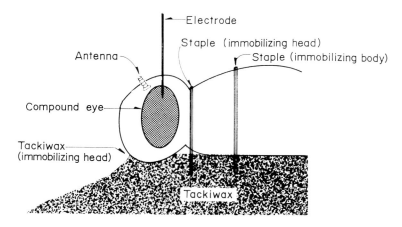

FIGURE 5.13. Side view of compound eye of a mounted insect showing typical site for insertion of electrode for making electroretinograms. (From Agee, H. R., 1977, courtesy of the U.S. Department of Agriculture.)

by short immersion in warm water (50° C) with a floating layer of melted wax on top. Sometimes, in order to provide sufficient oviposition stimulation, each paper model or paper strip is sprayed with extract of a susceptible host plant. Color choice bioassays are conducted under uniform illumination, and at the end of the experiment, the number of eggs on each paper model or paper strip are recorded.

Harris and Rose (1990) reported that Hessian fly, *Mayetiola destructor* (Say), females laid more eggs on green, yellow, and orange filter paper strips than on blue and red strips. Roessingh and Städler (1990) reported similar results for the cabbage root fly, *D. radicum* (L.).

5.6.4. Electroretinograms (ERG)

Electroretinograms (ERGs) measure an insect's visual response spectrum and can provide information concerning an insect's perception of both monochromatic light and color. The ERG is an oscilloscopic display of the voltage changes that occur in the insect eye when it is stimulated with flashes of light. Techniques for determining ERGs of insects have been described in detail by Agee (1977) and Colwell and Page (1989).

The test insect is temporarily anesthetized with CO_2 and is mounted on a stand with wax in such a way that one eye is positioned in the light beam from the insect vision analyzer and the other eye is entirely in the shadow of the illuminated eye (Figure 5.13). The ERG recording procedure involves inserting a recording electrode 20 to 50 μm just beneath the lens of the illuminated eye, with the tip of the electrode close to the rhabdom. A micromanipulator is generally used to insert the electrode into the eye. A reference

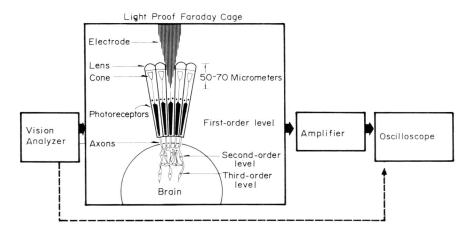

FIGURE 5.14. A flow diagram of an electroretinogram system showing electrode position in compound eye of a typical insect. (From Agee, H. R., 1977, courtesy of the U.S. Department of Agriculture.)

electrode is inserted into the nonilluminated eye to detect only the resting potential of the eye. The electrode is connected to an amplifier and an oscilloscope (Figure 5.14).

The light chopper of a vision analyzer generates short pulses of light that are beamed onto the eye. When the insect eye is stimulated with a particular wavelength of light, electrical changes occur in the eye that can be detected by the electrodes. These electrical changes are amplified by the amplifier and are displayed on the oscilloscope for measurement. For a more detailed description of instrumentation and techniques for ERG, the reader should refer to Agee (1977).

5.7. SHAPES AND SIZES

Shape and size perception is likely to elicit some generalized behavioral patterns by insects, but no resistance mechanisms among crop plants have been reported using these parameters as selection criteria (Norris and Kogan 1980). Certain plant shapes and sizes may, however, be linked with other resistance factors. For example, the frego-bract and okra leaf traits of cotton exhibit resistance to the boll weevil, *Anthonomus grandis grandis* Boheman (Niles 1980). The efficacy of frego-bract cotton in weevil resistance is due mainly to modifications of weevil behavior (Mitchell et al. 1973), whereas weevil mortality in squares under the canopy of okra leaf cotton is increased because of dessication (Reddy 1974). Although other allelochemical factors may also be involved in resistance associated with shape and size, plant shape

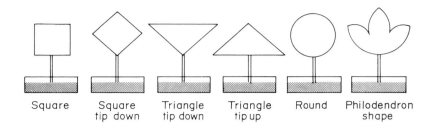

Square   Square   Triangle   Triangle   Round   Philodendron
          tip down   tip down   tip up   shape

FIGURE 5.15. Various leaf models made of paper used for testing the influence of foliar shapes on oviposition behavior of *Delia radicum*. (From Roessingh, P. and Städler, E., *Entomol. Exp. Appl.*, 57, 93, 1990. With permission of Kluwer Academic Publishers.)

and size could be used as a selection criterion in resistance evaluation programs, following the methods of Prokopy and Bush (1973), Roessingh and Städler (1990), and Pittara and Katsoyannos (1992).

Prokopy and Bush (1973) evaluated the oviposition response of apple maggot flies, *Rhagoletis pomonella* (Walsh), on artificial fruits of different sizes. Four different sizes of hemispheres and spheres (10, 20, 40 and 70 mm diameter), made of solid agar (2% in water) dyed red with food coloring and wrapped in a single layer of Parafilm® were tested in no-choice experiments. The number of eggs deposited inside each artificial fruit is closely correlated to the size of artificial fruit preferred.

Roessingh and Städler (1990) and Koštál (1993) investigated the effects of different shapes and sizes of paper models on the oviposition behavior of the cabbage fly, *D. radicum*. To test the effects of different shapes, ovipositing females are given a choice of flat leaves of equal surface area (90 cm^2): oval, long vertical, long horizontal, square, square with tip down, triangular with tip up, triangular with tip down, round and philodendron-shaped (Figure 5.15). To test the effect of size, two sizes of the standard flat surrogate leaf (6.5 × 6.5 cm and 13 × 13 cm) are offered to gravid females for oviposition. Different model shapes have little effect on *D. radicum* oviposition. However, about twice as many *D. radicum* eggs were laid in the sand around the larger surrogate leaf model.

Pittara and Katsoyannos (1992) studied the effects of shape and size on the selection of oviposition sites by the fly, *Chaetorellia australis* Hering, using artificial substrates made of yellow-colored paraffin wax mimicking the natural fly oviposition sites. Wax spheres, cubes, cylinders, and cones are presented in a choice test to eight gravid females. For determining the effect of size, 5, 10, 15, and 20 mm diameter spheres are compared in a choice test. Oviposition preference for different shapes and sizes is determined by recording the number of females present on the objects at 5 min intervals for 1 hr. Pittara and Katsoyannos (1992) reported that spheres and cones are visited more frequently than cylinders and cubes of the same surface area. Also, females more frequently visited small (5 and 10 mm diameter) spheres than large (15 and 20 mm diameter) spheres. For a detailed review on insect response to different shapes and sizes, the reader is referred to Katsoyannos (1989).

5.8. CONCLUSIONS

The behavior of insects in selecting a host plant for food and shelter is affected by a wide array of physical stimuli. Although most researchers are primarily concerned with plant chemicals that influence insect behavior, it is also important to recognize the critical involvement of physical factors in the insect host selection process. Several morphological characteristics of potential resistant host plants present barriers to insect feeding and oviposition. However, many assumptions have been made in even the most thoroughly studied plant-insect system. It is still necessary to identify additional morphological characteristics using proper bioassay techniques to help explain the complex behavioral events in insect host plant selection.

REFERENCES

Agarwal, R. A., Morphological characteristics of sugarcane and insect resistance, *Entomol. Exp. Appl.,* 12, 767, 1969.

Agee, H., Instrumentation and Techniques for Measuring the Quality of Insect Vision with the Electroretinogram, Agricultural Research Service, U.S. Department of Agriculture, ARS-S-162, 1977, 13.

Anstey, T. H. and Moore, J. F., Inheritance of glossy foliage and cream petals in green sprouting broccoli, *J. Hered.,* 45, 39, 1954.

Atkin, D. S. J. and Hamilton, R. J., The effects of plant waxes on insects, *J. Natural Products,* 45, 694, 1982.

Baur, R., Binder, S. and Benz, G., Nonglandular leaf trichomes as short-term inducible defense of grey alder, *Alnus indicana* (L.), against the chrysomelid beetle, *Agelastica alni* L., *Oecologie,* 87, 219, 1991.

Beckwith, R. C. and Helmers, A. E., A penetrometer to quantify leaf toughness in studies of defoliators, *Environ. Entomol.,* 5, 291, 1976.

Broersma, D. B., Bernard, R. L. and Luckmann, W. H., Some effects of soybean pubescence on populations of the potato leafhopper, *J. Econ. Entomol.,* 65, 78, 1972.

Cartier, J. J. and Auclair, J. L., Pea aphid behaviour: colour preference on a chemical diet, *Can. Entomol.,* 96, 1240, 1964.

Cherrett, J. M., A simple penetrometer for measuring leaf toughness in insect feeding studies, *J. Econ. Entomol.,* 61, 1736, 1968.

Chiang, H.-S. and Jackai, L. E. N., Tough pod wall: a factor involved in cowpea resistance to pod sucking bugs, *Insect Sci. Applic.,* 9, 389, 1988.

Chiang, H.-S. and Norris, D. M., "Purple stem", a new indicator of soybean stem resistance to bean flies (Diptera: Agromyzidae), *J. Econ. Entomol.,* 77, 121, 1984.

Colwell, C. S. and Page, T. L., The electroretinogram of the cockroach *Leucophaea maderae, Comp. Biochem. Physiol.,* 92A, 117, 1989.

Djamin, A. and Pathak, M. D., Role of silica in resistance to Asiatic rice borer, *Chilo suppressalis* (Walker), in rice varieties, *J. Econ. Entomol.,* 60, 347, 1967.

Duffey, S. S. and Isman, M. B., Inhibition of insect larval growth by phenolics in glandular trichomes of tomato leaves, *Experientia,* 37, 574, 1981.

Dunn, J. A. and Kempton, D. P. H., Varietal differences in the susceptibility of Brussel sprouts to lepidopterous pests, *Ann. Appl. Biol.,* 82, 11, 1976.

Eigenbrode, S. D. and Shelton, A. M., Behavior of neonate diamondback moth larvae (Lepidoptera: Plutellidae) on glossy-leaved resistant *Brassica oleracea* L., *Environ. Entomol.,* 19, 1566, 1990.

Eigenbrode, S. D., Stoner, K. A., Shelton, A. M. and Kain, W. C., Characteristics of glossy leaf waxes associated with resistance to diamondback moth (Lepidoptera: Plutellidae) in *Brassica oleracea, J. Econ. Entomol.,* 84, 1609, 1991.

Farrar, R. R., Jr. and Kennedy, G. G., Growth, food consumption and mortality of *Heliothis zea* larvae on foliage of the wild tomato *Lycopersicon hirsutum* f. *glabratum* and the cultivated tomato, *L. esculentum, Entomol. Exp. Appl.,* 44, 213, 1987.

Feeny, P., Seasonal changes in oak leaf tannins and nutrients as a cause of spring feeding by winter moth caterpillar, *Ecology,* 51, 565, 1970.

Fiori, B. J. and Dolan, D. D., Field tests for *Medicago* resistance against the potato leafhopper (Homoptera: Cicadellidae), *Can. Entomol.,* 113, 1049, 1981.

Fiori, B. J. and Craig, D. W., Relationship between color intensity of leaf supernatants from resistant and susceptible birchtree and rate of oviposition by the birch leafminer (Hymenoptera: Tenthredinidae), *J. Econ. Entomol.,* 80, 1331, 1987.

Gallun, R. L., Roberts, J. J., Finney, R. E. and Patterson, F. L., Leaf pubescence of field grown wheat: a deterrent to oviposition by the cereal leaf beetle, *J. Environ. Quality,* 2, 333, 1973.

Hanifa, A. M., Subramaniam, T. R. and Ponnaiya, B. W. X., Role of silica in resistance to the leafroller, *Cnaphalocrocis medinalis* Guenée, in rice, *Indian J. Exp. Biol.,* 12, 463, 1974.

Harris, M. O. and Miller, J. R., Color stimuli and oviposition behavior of the onion fly, *Delia antiqua* (Meigen), (Diptera: Anthomyiidae), *Ann. Entomol. Soc. Am.,* 76, 766, 1983.

Harris, M. O. and Rose, S., Chemical, color, and tactile cues influencing oviposition behavior of the Hessian fly (Diptera: Cecidomyiidae), *Environ. Entomol.,* 19, 303, 1990.

Hawthorne, D. J., Shapiro, J. A., Tingey, W. M. and Mutschler, M. A., Trichome-borne and artificially applied acylsugars of wild tomato deter feeding and oviposition of the leafminer *Liriomyza trifoli, Entomol. Exp. Appl.,* 65, 65, 1992.

Holloway, P. J. and Baker, E. A., Isolation of plant cuticles with zinc chloride-hydrochloric acid solution, *Plant Physiol.,* 43, 1878, 1968.

Hoxie, R. P., Wellso, S. G. and Webster, J. A., Cereal leaf beetle response to wheat trichome length and density, *Environ. Entomol.,* 4, 365, 1975.

Issac, P. V., How mid-rib hardness affords resistance to the sugarcane top-borer *Scirpophaga nivella* F., in India, *Current Sci.,* 8, 211, 1939.

Jeffree, C. E., The cuticle, epicuticular waxes and trichomes of plants, with reference to their structure, function and evolution, in *Insects and the Plant Surface,* B. E. Juniper and T. R. E. Southwood, eds., Edward Arnold, London, 1986, 360.

Johnson, B., The injurious effects of the hooked epidermal hairs of French beans (*Phaseolus vulgaris* L.) on *Aphid craccivora* Koch., *Bull. Entomol. Res.,* 44, 779, 1955.

Johnson, H. W. and Hollowell, E. A., Pubescent and glabrous characters of soybeans as related to resistance to injury by the potato leafhopper, *J. Agric. Res.,* 51, 371, 1935.

Johnson, K. J. R., Sorensen, E. L. and Horber, E. K., Resistance in glandular-haired annual *Medicago* species to feeding by adult alfalfa weevils (*Hypera postica*), *Environ. Entomol.,* 9, 133, 1980.

Kang, M. S., Sosa, O., Jr. and Miller, J. D., Genetic variation and advance for rind hardness, flowering and sugar yield trials in sugarcane, *Field Crops Res.,* 23, 69, 1990.

Katsoyannos, B. I., Responses to shape, size and color, in *Fruit Flies, Their Biology Natural Enemies and Control,* Vol. 3A, A. S. Robinson and G. Hooper, eds., Elsevier, New York, 1989, 372.

Khan, Z. R., Ward, J. T. and Norris, D. M., Role of trichomes in soybean resistance to cabbage looper, *Trichoplusia ni, Entomol. Exp. Appl.,* 42, 109, 1986.

Khanna, K. L., Sharma, S. L. and Ramanathan, K. R., Studies in the association of plant characters and pest incidence: III. Hardness of leaf midrib and top borer infestation in sugarcane, *Indian J. Entomol.,* 9, 115, 1947.

King, B. L., Design and evaluation of a simple penetrometer for measuring leaf toughness in studies of insect herbivory, *Virginia J. Sci.,* 39, 405, 1988.

Koštàl, V., Physical and chemical factors influencing landing and oviposition by the cabbage root fly on host-plant models, *Entomol. Exp. Appl.,* 66, 109, 1993.

Lanning, F. C., Hopkins, T. L. and Loera, J. C., Silica and ash content and depositional patterns in tissues of mature *Zea mays* L. plants, *Ann. Bot.,* 45, 549, 1980.

Lapointe, S. L. and Tingey, W. M., Feeding response of the green peach aphid (Homoptera: Aphididae) to potato glandular trichomes, *J. Econ. Entomol.,* 77, 386, 1984.

Levin, D. A., The role of trichomes in plant defense, *Q. Rev. Biol.,* 48, 3, 1973.

Lin, S. Y. H., Trumble, J. T. and Kumamoto, J., Activity of volatile compounds in glandular trichomes of *Lycopersicon* species against two insect herbivores, *J. Chem. Ecol.,* 13, 837, 1987.

Lupton, F. G. H., The use of resistant varieties in crop protection, *World Rev. Pest Control,* 6, 47, 1967.

Lyman, J. M. and Cardona, C., Resistance in lima beans to a leafhopper, *Empoasca kraemeri, J. Econ. Entomol.,* 75, 281, 1982.

MacLean, P. S. and Byers, R. A., Ovipositional preferences of the alfalfa blotch leafminer (Diptera: Agromyzidae) among some simple and glandular-haired *Medicago* species, *Environ. Entomol.,* 12, 1083, 1983.

Mandras, B. T., Resistance of Deepwater Rice Varieties and Elongating Wild Rices to Yellow Stem Borer, *Scirpophaga incertulas* (Walker) (Lepidoptera: Pyralidae), Ph.D. Thesis, University of the Philippines at Los Baños, Philippines, 1991, 107.

Martin, G. A., Richard, C. A. and Hensley, S. D., Host resistance to *Diatraea saccharalis* (F.): relationship of sugarcane node hardness to larval damage, *Environ. Entomol.,* 4, 687, 1975.

Meisner, J. and Ascher, K. R. S., Attraction of *Spodoptera littoralis* larvae to colours, *Nature,* 242, 332, 1973.

Mitchell, H. C., Cross, W. H., McGovern, W. L. and Dawson, E. M., Behavior of the boll weevil on frego bract cotton, *J. Econ. Entomol.,* 66, 677, 1973.

Navon, A., Melamed-Madjar, V., Zur, M. and Ben-Moshe, E., Effects of cotton cultivars on feeding of *Heliothis armigera* and *Spodoptera littoralis* larvae and on oviposition of *Bemisia tabaci, Agric. Ecosyst. Environ.,* 34, 73, 1991.

Neal, J. J., Plaisted, R. L. and Tingey, W. M., Feeding behavior and survival of Colorado potato beetle, *Leptinotarsa decemlineata* (Say), larvae on *Solanum berthaultii* Hawkes and an F_6 *S. tuberosum* × *S. berthaultii* hybrid, *Am. Potato J.,* 68, 649, 1991.

Neal, J. J., Steffens, J. C. and Tingey, W. M., Glandular trichomes of *Solanum berthaultii* and resistance to the Colorado potato beetle, *Entomol. Exp. Appl.,* 51, 133, 1989.

Niles, G. A., Breeding cotton for resistance to insect pests, in *Breeding Plants Resistant to Insects,* F. G. Maxwell and P. R. Jennings, eds., Chapter 15, John Wiley & Sons, New York, 1980, 683.

Norris, D. M. and Kogan, M., Biochemical and morphological bases of resistance, in *Breeding Plants Resistance to Insects,* F. G. Maxwell and P. R. Jennings, eds., Chapter 3, John Wiley & Sons, New York, 1980, 683.

Patanakamjorn, S. and Pathak, M. D., Varietal resistance of rice to Asiatic rice borer, *Chilo suppressalis* (Lepidoptera: Crambidae), and its association with various plant characters, *Ann. Entomol. Soc. Am.,* 60, 287, 1967.

Peterson, S. S., Scriber, J. M. and Coors, J. G., Silica, cellulose and their interactive effects on the feeding performance of the southern armyworm, *Spodoptera eridania* (Cramer) (Lepidoptera: Noctuidae), *J. Kans. Entomol. Soc.,* 61, 169, 1988.

Pillemer, E. A. and Tingey, W. M., Hooked trichomes: a physical plant barrier to a major agricultural pest, *Science,* 193, 482, 1976.

Pillemer, E. A. and Tingey, W. M., Hooked trichomes and resistance of *Phaseolus vulgaris* to *Empoasca fabae* (Harris), *Entomol. Exp. Appl.,* 24, 83, 1978.

Pittara, I. S. and Katsoyannos, B. I., Effect of shape, size and color on selection of oviposition sites by *Chaetorellia australis, Entomol. Exp. Appl.,* 63, 105, 1992.

Ponnaiya, B. W. X., Studies in the genus *Sorghum:* the cause of resistance in sorghum to the insect pest *Antherigona indica* M., *Madras Univ. J.,* Section B, 21, 203, 1951.

Prokopy, R. J. and Bush, G. L., Ovipositional response to different sizes of artificial fruit by flies of *Rhagoletis pomonella* species group, *J. Econ. Entomol.,* 66, 627, 1973.

Quiros, C. F., Stevens, M. A., Rick, C. M. and Kok-Yokomi, M. L., Resistance in tomato to the pink form of the potato aphid (*Macrosiphum euphorbiae* Thomas): the role of anatomy, epidermal hairs, and foliage composition, *J. Am. Soc. Hort. Sci.,* 102, 166, 1977.

Ramachandran, R. and Khan, Z. R., Mechanisms of resistance in wild rice *Oryza brachyantha* to rice leaffolder *Cnaphalocrocis medinalis* (Guenée) (Lepidoptera: Pyralidae), *J. Chem. Ecol.,* 17, 41, 1991.

Reddy, P. S. C., Effects of three leaf shape genotypes of *Gossypium hirsutum* L. and row types on plant microclimate, boll weevil survival, boll rot and important agronomic characters, Ph.D. Dissertation, Louisiana State University, Baton Rouge, 1974.

Ringland, K. and Everson, E. H., Leaf pubescence in common wheat, *Triticum aestivum* L., and resistance to the cereal leaf beetle, *Oulema melanopus* (L.), *Crop Sci.,* 8, 707, 1968.

Roberts, J. J. and Foster, J. E., Effect of leaf pubescence in wheat on the bird cherry oat aphid (Homoptera: Aphidae), *J. Econ. Entomol.,* 76, 1320, 1983.

Roberts, J. J., Gallun, R. L., Patterson, F. L. and Foster, J. E., Effects of wheat leaf pubescence on the Hessian fly, *J. Econ. Entomol.,* 72, 211, 1979.

Roessingh, P. and Städler, E., Foliar form, color and surface characteristics influence oviposition behaviour in the cabbage root fly, *Delia radicum, Entomol. Exp. Appl.,* 57, 93, 1990.

Sands, D. P. A. and Brancatini, V. A., A portable penetrometer for measuring leaf toughness in insect herbivory studies, *Proc. Entomol. Soc. Wash.,* 93, 786, 1991.

Sasamoto, K., Studies on the relation between insect pests and silica content in rice plant (II). On the injury of the second generation larvae of rice stem borer, *Oyo-Kontyu,* 9, 108, 1953.

Sasamoto, K., Studies of the relation between insect pests and silica content in rice plant (III). On the relation between some physical properties of silicified rice plant and injuries by rice stem borer, rice plant skipper and rice stem maggot, *Oyo-Kontyu,* 11, 66, 1955.

Saxena, K. N. and Onyango, J. D., Attraction of the stem borer *Chilo partellus* (Swinhoe) larvae (Lepidoptera: Pyralidae) by certain colored surfaces for trapping, *Appl. Entomol. Zool.,* 22, 493, 1987.

Schillinger, Jr., J. A. and Gallun, R. L., Leaf pubescence of wheat as a deterrent to the cereal leaf beetle, *Oulema melanopus, Ann. Entomol. Soc. Am.,* 61, 900, 1968.

Schultz, P. B. and Coffelt, M. A., Oviposition and nymphal survival of the hawthorn lace bug (Hemiptera: Tingidae) on selected species of *Cotoneaster* (Rosaceae), *Environ. Entomol.,* 16, 365, 1987.

Shade, R. E. and Kitch, L. W., Pea aphid (Homoptera: Aphididae) biology on glandular-haired *Medicago* species, *Environ. Entomol.,* 12, 237, 1983.

Shade, R. E., Doskocil, M. J. and Maxon, N. P., Potato leafhopper resistance in glandular-haired alfalfa species, *Crop Sci.,* 19, 287, 1979.

Shifriss, O., Do *Cucurbita* plants with silvery leaves escape virus infection? *Cucurbit Gen. Coop. Rep.,* 4, 42, 1981.

Sikka, S. M., Sahni, V. M. and Butani, D. K., Studies on jassid resistance in relation to hairiness of cotton leaves, *Euphytica,* 15, 383, 1966.

Sosa, O., Jr., Pubescence in sugarcane as a plant resistance character affecting oviposition and mobility by the sugarcane borer (Lepidoptera: Pyralidae), *J. Econ. Entomol.,* 81, 663, 1988.

Springer, T. L., Kindler, S. D. and Sorensen, E. L., Comparison of pod-wall characteristics with seed damage and resistance to the alfalfa seed chalcid (Hymenoptera: Eurytomidae) in *Medicago* species, *Environ. Entomol.,* 19, 1614, 1990.

Städler, E., Host selection and chemoreception in the carrot rust fly *(Psila rosae)* F., Dipt. Psilidae: extraction and isolation of oviposition stimulants and their perception by the female, in *Comportement des Insectes et Milieu Trophique,* C.N.R.S. Editions, 1977, 493.

Städler, E. and Buser, H. R., Oviposition stimulation for the carrot fly in the surface wax of carrot leaves, in *Proc. 5th Int. Symp. Insect-Plant Relationships,* J. H. Visser and A. K. Minks, eds., Wageningen Centre for Agricultural Publishing and Documentation, Wageningen, The Netherlands, 1982, 464.

Steffens, J. C. and Walters, D. S., Biochemical aspects of glandular trichome-mediated insect resistance in the *Solanaceae,* in *Naturally Occurring Pest Bioregulators,* P. A. Hedin, ed., ACS Symp. Ser. 449, Chapter 10, American Chemical Society, Washington, DC, 1991, 456.

Stephens, S. G., Sources of resistance of cotton strains to the boll weevil and their possible utilization, *J. Econ. Entomol.,* 50, 415, 1957.

Stoner, K. A., Glossy leaf wax and plant resistance to insects in *Brassica oleracea* under natural infestation, *Environ. Entomol.,* 19, 730, 1990.

Stork, N. E., Role of waxblooms in preventing attachment to brassicas by the mustard beetle, *Phaedon cochleariae, Entomol. Exp. Appl.,* 28, 99, 1980.

Tanton, M. T., The effect of leaf "toughness" on the feeding of larvae of the mustard beetle *Phaedon cochleariae* Fab., *Entomol. Exp. Appl.,* 5, 74, 1962.

Thibout, E., Auger, J. and Lecomte, C., Host plant chemicals responsible for attraction and oviposition in *Acrolepiopis assectella,* in *Proc. 5th Int. Symp. Insect-Plant Relationships,* J. H. Visser and A. K. Minks, eds., Wageningen Centre for Agricultural Publishing and Documentation, Wageningen, The Netherlands, 1982, 464.

Thompson, K. F., Resistance to the cabbage aphid *(Brevicoryna brassicae)* in *Brassica* plants, *Nature,* 198, 209, 1963.

Tingey, W. M., Potato glandular trichomes, in *Naturally Occurring Pest Bioregulators,* P. A. Hedin, ed., ACS Symp. Ser. 449, Chapter 9, American Chemical Society, Washington, DC, 1991, 456.

Tingey, W. M. and Laubengayer, J. E., Glandular trichomes of a resistant hybrid potato alter feeding behavior of the potato leafhopper (Homoptera: Cicadellidae), *Entomol.,* 79, 1230, 1986.

Triebe, D. C., Meloan, C. E. and Sorensen, E. L., The chemical identification of the glandular hair exudate for *Medicago scutellata,* in Proceedings of 27th Alfalfa Improvement Conference, ARM-NC-19, U.S. Department of Agriculture, 1981, 52.

Walker, G. P., Probing and oviposition behavior of the bayberry whitefly (Homoptera: Aleyrodidae) on young and mature lemon leaves, *Ann. Entomol. Soc. Am.,* 80, 524, 1987.

Walker, G. P., The role of leaf cuticle in leaf age preference by bayberry whitefly (Homoptera: Aleyrodidae) on lemon, *Ann. Entomol. Soc. Am.,* 81, 365, 1988.

Walker, G. P. and Aitken, D. C. G., Oviposition and survival of bayberry whitefly, *Parabemisia myricae* (Homoptera: Aleyrodidae) on age, *Environ. Entomol.,* 14, 254, 1985.

Wallace, L. E., McNeal, H. and Berg, M. A., Resistance to both *Oulema melanopus* and *Cephus cinctus* in pubescent-leaved and solid stemmed wheat selections, *J. Econ. Entomol.,* 67, 105, 1974.

Webster, J. A., Association of plant hairs and insect resistance. An annotated bibliography, USDA-ARS Misc. Pub. No. 129, 1975, 18.

Williams, L. H., The feeding habits and food preferences of *Acrididae* and the factors which determine them, *Trans. R. Entomol. Soc. London,* 105, 423, 1954.

Woodhead, S. and Padgham, D. E., The effect of plant surface characteristics on resistance of rice to the brown planthopper, *Nilaparvata lugens, Entomol. Exp. Appl.,* 47, 15, 1988.

Yerger, E. H., Grazzini, R. A., Hesk, D., Cox-Foster, D. L., Craig, R. and Mumma, R. O., A rapid method for isolating grandular trichomes, *Plant Physiol.,* 99, 1, 1992.

Yoshida, S., Forno, D. A., Cock, J. H. and Gomez, K. A., Laboratory Manual for Physiological Studies of Rice, *International Rice Research Institute,* Los Baños, Laguna, Philippines, 1976, 83.

Zettler, F. W., Smyly, M. O. and Evans, I. R., The repellency of mature citrus leaves to probing aphids, *Ann. Entomol. Soc. Am.,* 62, 399, 1969.

CHAPTER 6

Techniques for Identifying Insect Biotypes

The protective properties of insect resistant cultivars may be overcome by the development of resistance-breaking insect populations or biotypes that possess an inherent genetic capability to overcome plant resistance (Smith 1989). Typically, insect biotypes occur in nature as products of a survival mechanism for the persistence of an insect species and develop as a result of selection from the parent population in response to exposure to resistant cultivars. Therefore, failure to recognize the existence of insect biotypes may lead to severe infestation of crop plants even on formerly resistant cultivars. The study of insect biotypes may become a significant part of insect resistance programs and can provide tools for the analysis of insect plant relationships that serve as the bases for breeding resistant plants (Saxena and Barrion 1985, 1987). The identification of new biotypes is of utmost importance in accurately identifying genetic sources of resistance in crop plants.

Identifying insect biotypes may be a long and difficult process. The types of methods used to determine differences vary considerably. In many insects, biotypes may be detected by the response of a group of differential host varieties to an insect population. In others, electrophoretic, morphometric, or cytological methods may be used to detect differences between biotypes. The most useful method of differentiation, however, is the one that gives the most precise and efficient delineation of biotypes in an insect population. The following methods have been used to identify insect biotypes.

6.1. DIFFERENTIAL CULTIVAR REACTION

Differential sets of plant cultivars with different resistance genes are commonly used to detect biotypes of insect pests. In this method, biotype identification is based on cultivar reaction to insect attack. Cultivars susceptible to a particular biotype are killed upon infestation with nymphs or adults while those resistant are not damaged. Both sympatric and allopatric biotypes can be monitored on the basis of differential cultivar reaction in greenhouse or field tests (Gallun et al. 1961, Oka 1978, Seshu and Kauffman 1980, IRRI 1982, Heinrichs and Rapusas 1985, Saxena and Barrion 1985, Lidell and Schuster 1990). To monitor allopatric biotypes of an insect pest, a uniform set of host cultivars are tested in different countries or geographic regions. If particular cultivars are resistant in one area but susceptible in another, insects at the two locations may be different biotypes. However, this method directly measures insect biotype presence by cultivar damage rather than by insect characteristics. Furthermore, biotype identification is based on the response of a population of insects and no individual insect type can be identified. In the greenhouse, the evaluation of uniform sets of cultivars in a "seed box test," has been developed by Oka (1978) and Seshu and Kauffman (1980) as a common method for detecting brown planthopper, *Nilaparvata lugens* (Stål), biotypes in rice.

A uniform set of rice cultivars (TN1, no resistance gene; Mudgo, *Bph*1 resistance gene; ASD7, *Bph*2 resistance gene; Rathu Heenati, *Bph*3 resistance gene; Babawee, *Bph*4 resistance gene etc.) are seeded in rows 5 cm apart in $60 \times 45 \times 10$ cm seed boxes. Seven days after seeding, the test cultivars are thinned to contain about 30 to 40 seedlings per row, and the seed boxes are then placed in a galvanized iron tray inside a fine mesh screen cage. The tray contains about 5 cm of standing water, providing high humidity for insect survival and eliminating the need for watering the plants. Each seedling is then infested with 5 to 10 second- and third-instar *N. lugens* nymphs, an infestation which kills the susceptible cultivars in about 7 days. The scoring of individual seedlings begins as soon as the susceptible check plants die, using a standard evaluation system (IRRI 1988) (Figure 6.1). In Indonesia, Oka (1978) used the seedbox test to show that if susceptible TN1 and Pelita I-1 cultivars are rated as susceptible and cultivars with the *Bph*1 gene are resistant, then *N. lugens* biotype 1 prevailed in that area. Susceptible reactions of rice cultivars with no resistance gene and with the *Bph*1 gene, but resistant ratings on rice cultivars with the *Bph*2 gene indicated the presence of biotype 2. If TN1 and Pelita I-1 are susceptible and cultivars with the *Bph*1 gene are moderately susceptible while cultivars with the *Bph*2 are moderately resistant, the *N. lugens* populations are a mixture of biotypes 1 and 2. Lidell and Schuster (1990) also used the seedbox evaluation method in North Central Texas to demonstrate yearly variations in Hessian fly, *Mayetiola destructor* (Say), biotypes.

Field evaluation of differential cultivars to detect insect biotypes may also be undertaken following the methods described by Seshu and Kauffman (1980) and IRRI (1982) for *N. lugens*. Several border rows of susceptible TN1 rice are planted one month earlier than the differential cultivars and high levels of nitrogen are applied to induce rapid insect population increases.

FIGURE 6.1. Differential reactions of rice varieties with different resistance genes to *Nilaparvata lugens* biotype 1 (a), biotype 2 (b), and biotype 3 (c). (Photographs provided by M. D. Pathak.)

Sometimes it may be necessary to spray the susceptible border rows with a sublethal dose of insecticide that induces resurgence of the *N. lugens* population. When the border TN1 plants show the initiation of hopper burn, the rows are cut on alternate days so that the *N. lugens* population moves to the test cultivars. Test entries are scored when the susceptible check is killed, using a standard evaluation system (IRRI 1988). Seshu and Kauffman (1980) reported the results from International Rice Brown Planthopper Nursery evaluations in several Asian countries over a period of 5 years, which demonstrated a major distinction between the hopper populations present in Bangladesh, India and Sri Lanka and those in the rest of Asia.

6.2. PLANT-MEDIATED DIFFERENTIAL INSECT RESPONSES

Insect biotypes can be separated from each other by their differential behavioral responses to plants of known genotypes. The following responses of an insect can be used to determine the biotype to which it belongs.

6.2.1. Feeding

The duration of feeding and quantity of food intake in pest insects on resistant and susceptible cultivars can serve to identify biotypes (Saxena and Barrion 1985, Khan and Saxena 1990). Biotypes incapable of feeding adequately on resistant genotypes fail to establish populations as large as those on susceptible genotypes. A number of techniques have been devised and used for the quantitative determination of the duration of feeding and the amount of food intake by insect biotypes on differential host cultivars.

6.2.1.1. Gravimetric Technique

Pathak and Painter (1958), Saxena and Pathak (1977), Pathak et al. (1982), and Khan and Saxena (1990) described gravimetric techniques for the quantitative determination of differential amounts of food intake by different insect biotypes on susceptible and resistant plants.

Pathak and Painter (1958) passed sorghum leaves through transparent plastic boxes (7.3 × 5.1 × 0.5 cm), and placed a starved, preweighed corn leaf aphid, *Rhopalosiphum maidis* (Fitch), in each box. After 12 or 24 hr of feeding, aphids were weighed again and their percent loss or gain is determined. Pathak and Painter (1958) reported that all of the four biotypes of *R. maidis* gained significantly more weight on susceptible plants than on resistant plants. Biotype KS-2 differed from other biotypes in its ability to take significantly more material from resistant sorghum plants.

Pathak et al. (1982) and Khan and Saxena (1990) used a Parafilm® sachet, developed by Saxena and Pathak (1977), for collection and quantitative determination of the honeydew secreted by *N. lugens* biotypes on susceptible and resistant rice cultivars. See the detailed description of this method in Chapter

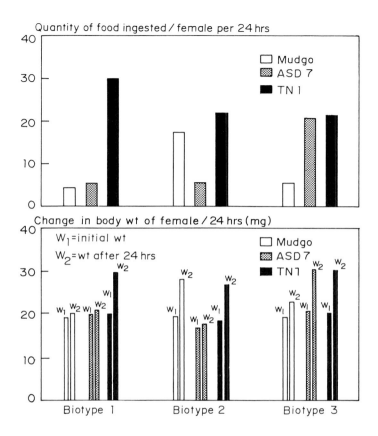

FIGURE 6.2. Quantity of food ingested and change in body weight of three biotypes of *Nilaparvata lugens* on three rice varieties with different levels of resistance. (From, Saxena, R. C. and Pathak, M. D., 1977, courtesy of the Pest Control Council of the Philippines.)

2.2.2.1.1. The quantity of honeydew secreted per insect in 24 hr was used as a parameter to compare *N. lugens* feeding on susceptible and resistant rice cultivars. This technique also permits the quantitative determination of the utilization of ingested food (Khan and Saxena 1990). Insect weight gains are smaller on resistant cultivars than on susceptible cultivars because of reduced food assimilation.

Saxena and Pathak (1977), Pathak et al. (1982) and Khan and Saxena (1990) reported that *N. lugens* biotypes could be separated from each other by their differential ingestion, weight increase, and assimilation of food on susceptible and resistant rice cultivars (Figure 6.2).

6.2.1.2. Filter Paper Technique

The relative amounts of feeding by different insect biotypes can also be assessed by measuring the area of honeydew excreted on a filter paper disk. The honeydew of *N. lugens* biotypes 1, 2, and 3 excreted on filter paper disks is stained with 0.001% ninhydrin solution in acetone and quantified by measuring the area (mm^2) of the honeydew spot (Paguia et al. 1980). A detailed description of this technique appears in Chapter 2.2.2.1.2. The filter paper method, based on differential feeding activity, is relatively simple, rapid, and sufficiently precise for biotype identification.

6.2.1.3. Electronic Recording of Insect Probing

An electronic feeding monitor, developed by McLean and Kinsey (1964, 1968) can be used to identify insect biotypes by their differential probing response to and duration of feeding on susceptible and resistant plants. A detailed description of the electronic monitoring device is presented in Chapter 2.2.2.1.5. The technique has been used to study differential probing responses of four biotypes of the spotted alfalfa aphid, *Therioaphis maculata* (Buckton), on alfalfa clones (Nielson and Don 1974), two biotypes of greenbug, *Schizaphis graminum* (Rondani) on wheat genotypes (Niassy et al. 1987, Ryan et al. 1987), and three biotypes of *N. lugens* on rice cultivars (Khan and Saxena 1988) (Figure 6.3). Various researchers have demonstrated that electronically recorded waveforms corresponding to probing, salivation, and ingestion differ significantly between aphids housed on susceptible and resistant cultivars (Figure 6.3). Virulent biotypes probe more readily and ingest longer on their respective susceptible cultivars than on cultivars which resist them.

6.2.1.4. Diagnostic Feeding Lesions

Feeding damage caused by several plant sap-sucking bugs and aphids is characterized by necrotic lesions (Miles 1972, Dreyer and Campbell 1984) which occur as a phytotoxic response to pectic enzymes from the saliva of the pest insect (Ma et al. 1990).

Using diagnostic feeding lesions, Puterka and Peters (1988, 1989) differentiated four biotypes of greenbug, *Schizaphis graminum* (Rondani), feeding on susceptible and resistant wheat cultivars. Red-brown necrotic lesions are caused by virulent biotypes of *S. graminum* on susceptible wheat plants (Puterka and Peters 1988); however, such lesions do not appear on resistant plants.

The study of lesion formation is facilitated by a clip-on cage, described by Puterka and Peters (1988) (Figure 6.4a), constructed from clear plastic drinking straws (6 mm diameter), hair curling clips, and a white felt pad. The clip arms are shortened to 12 mm, and 6 mm of the 2 center posts of the upper clip arm are bent at a 90° angle. A 1-cm length of plastic straw is glued to the center posts. One end of the straw piece is positioned flush with the bottom clip arm before gluing to the upper clip arm. This allows a good seal with a piece of felt (0.8 × 0.8 cm) that is glued to the bottom clip arm. Foam corks

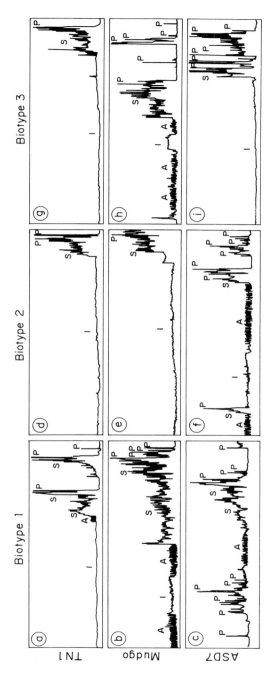

FIGURE 6.3. Waveforms recorded during probing of the three biotypes of *Nilaparvata lugens* on different resistant and susceptible rice varieties using an electronic monitoring device; charts are to be read from right to left: P, probes; S, salivation; I, ingestion; A, "A waveform". (From Khan, Z. R. and Saxena, R. C., *J. Econ. Entomol.*, 81, 1338, 1988. With permission of the Entomological Society of America.)

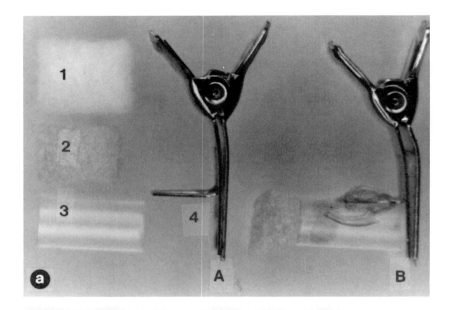

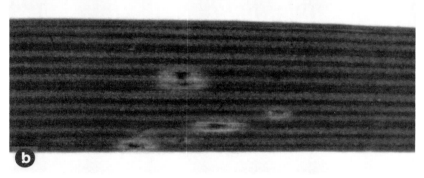

FIGURE 6.4. (a) A clip-on cage used to study lesion formation on plants, (A) cage parts: 1, foam plug; 2, felt pad; 3, straw piece; 4, shortened clip with bent center posts; (B) a constructed cage; (b) brown lesions with yellow halos on a susceptible sorghum plant due to *Schizaphis graminum* feeding. (From Puterka, G. J. and Peters, D. C., *J. Econ. Entomol.,* 81, 396, 1988. With permission of the Entomological Society of America. Photo courtesy of G. J. Puterka.)

are cut with a no. 3 cork borer (7 mm diameter) from a sheet of white, dense, resilient foam (1 cm thick). The completed clip-on cage is positioned on a leaf, infested from the top of the cage, and plugged with the foam cork. The clip cages minimize the time required for *S. graminum* to locate and settle on the leaf, and feeding lesions are easily located within the confined area of the cage. Puterka and Peters (1988) reported that *S. graminum* feeding lesions appear on the leaves of susceptible plants 72 hr after a 4-hr feeding exposure (Figure 6.4b). Lesions do not form on cultivars resistant to each biotype.

6.2.2. Insect Biology

Biotypes incapable of feeding and assimilating food adequately on a resistant cultivar will have poorer growth and fecundity, and therefore fail to establish in the large numbers which develop on a susceptible cultivar (Saxena and Barrion 1985). With resistant cultivars, insect survival (Gallun et al. 1961, Sato and Sogawa 1981), growth (Peters et al. 1988, Kerns et al. 1989), adult longevity (Khan and Saxena 1990), and fecundity (Pathak and Painter 1958, Cartier 1959, Gallun et al. 1961, Sato and Sogawa 1981, Niassy et al. 1987, Ryan et al. 1987, Kerns et al. 1989, Khan and Saxena 1990) are inhibited. Population increases on resistant and susceptible cultivars are important criterion for differentiating insect biotypes (Medrano and Heinrichs 1984, Heinrichs and Rapusas 1985, Kerns et al. 1989) since population increases represent the combined effects of feeding rates, the nutritional value of food, ovipositional rate, growth, adult survival, and fecundity.

Using these biological parameters, the biotypes of several insects have been differentiated, including *M. destructor* (Say) (Gallun et al. 1961), *R. maidis* (Pathak and Painter 1958), *Acyrthosiphon pisum* (Harris) (Cartier 1959), *S. graminum* (Niassy et al. 1987, Ryan et al. 1987, Peters et al. 1988, Kerns et al. 1989), *Nephotettix cincticeps* (Uhler) (Sato and Sogawa 1981), *Nephotettix virescens* (Distant) (Heinrichs and Rapusas 1985), and *N. lugens* (Medrano and Heinrichs 1984, Khan and Saxena 1990).

6.3. MORPHOMETRIC VARIATIONS

In-depth evaluations of morphological and morphometric differences among insect biotypes can provide an additional tool for the identification of both sympatric and allopatric biotypes. A biotype or race may not appear to be morphologically distinct from others using univariate statistical methods. However, the two populations may differ if a combined analysis of characters is considered in a multivariate space (Fargo et al. 1986). This technique has been used for identification of different populations and biotypes of European corn borer, *Ostrinia nubilalis* (Hübner) (Kim et al. 1967), *N. lugens* (Saxena and Rueda 1982, Saxena et al. 1983), *S. graminum* (Fargo et al. 1986, Inayatullah et al. 1987), yellow peach moth, *Conogethes punctiferalis* (Guenée) (Honda and Mitsuhashi 1989), and *Aulacorthum solani* (Kaltenbach) (Damsteegt and Voegtlin 1990).

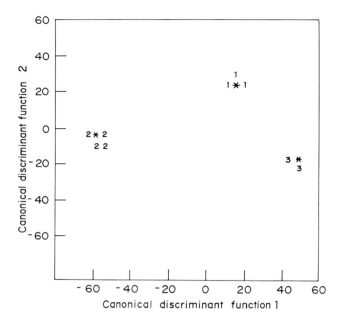

FIGURE 6.5. Discriminant scores of three biotypes of *Nilaparvata lugens* based on adult rostral, legs and antennal characters of brachypterous females. (From Saxena, R. C. and Rueda, L. M., *Insect Sci. Appl.*, 3, 193, 1982. With permission of ICIPE Science Press.)

A large number of quantitative morphological characters, especially those involved in host plant discrimination such as the rostrum, legs, and antennae may be morphometrically analyzed. These characters are analyzed separately in males and females and in different morphs. Measurements of body parts are made using a microscope equipped with a linear, graduated ocular micrometer, and discriminant analysis is run on the combined morphological characters to classify biotype populations. The relationship between two groups is determined by the Mahalanobis distance (Rao 1952) between their centroids in a scatter diagram. The smaller this distance, the more closely related the groups are and vice versa. Classified by leg and antennal characters, Saxena and Rueda (1982) reported a 100% probability of correct morphological identification of the three biotypes of *N. lugens* (Figure 6.5). The scatter diagram based on computed discriminant scores of the three biotypes of *S. graminum* were classified into their appropriate groups (Inayatullah et al. 1987).

6.4. CYTOLOGICAL VARIATIONS

Chromosome cytology has been used to determine differences between insect biotypes (Mayo and Starks 1972, Saxena and Barrion 1984, 1985, Mayo et al. 1988). Variations in the number, morphology, morphometry, and behavior of autosomes and sex chromosomes can be used to differentiate insect biotypes (Saxena and Barrion 1985).

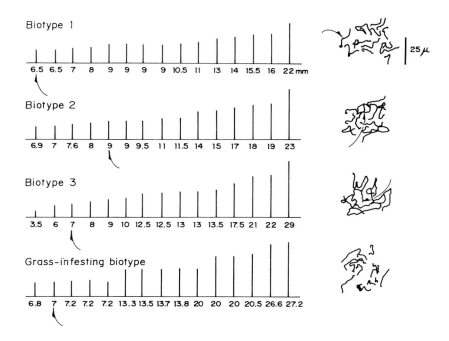

FIGURE 6.6. Karyotypes and idiograms of the bivalent pachytene chromosomes of *Nilaparvata lugens* biotypes; sex chromosomes are indicated by arrows. (From Saxena, R. C. and Barrion, A. A., *Insect Sci. Appl.*, 6, 271, 1985. With permission of ICIPE Science Press.)

Saxena and Barrion (1982) described a simple and rapid technique for preparing meiotic chromosomes for studying *N. lugens* biotypes. Fifth-instar nymphs and newly emerged males of *N. lugens* are fixed in glass vials containing carnoy's fluid (1 part 99.7% glacial acetic acid and 3 parts 95% ethyl alcohol) for at least 2 min. A fixed insect is then dissected in a drop of Ringer's solution on a clean glass slide. The head and thorax are discarded and the abdomen is dorsally incised to extract the tiny, translucent testes. The testes are submerged in a drop of 2% aceto-orcein or carmine solution for 2 min. Each testes is macerated with a fine-tipped needle on a clean slide with a drop of 2% aceto-orcein. The cells are kept from drying out by adding a drop of 45% acetic acid. A clean cover slip is placed over the macerated testes. Most of the chromosome measures are recorded close to their most condensed (metaphase) stage, but substantial variations among biotypes can also be observed during other stages. Chromosome measurements are made directly from the slide under 100× magnification. Data for each biotype can be analyzed using the discriminant analyses procedure following the method of Mayo et al. (1988).

Saxena and Barrion (1985) reported that at pachynema, the karyotypes and idiograms of bivalent autosomes and sex chromosomes manifested distinct variations in their relative mean lengths among four biotypes of *N. lugens* (Figure 6.6). The clumping and clustering behavior of highly condensed autosomes during first metaphase also differed among the four *N. lugens* biotypes

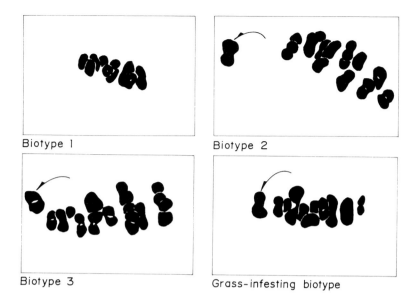

Biotype 1

Biotype 2

Biotype 3

Grass-infesting biotype

FIGURE 6.7. First metaphase autosomes and sex chromosomes of *Nilaparvata lugens* biotypes; sex chromsomes are indicated by arrows. (From Saxena, R. C. and Barrion, A. A., *Insect Sci. Appl.* 6, 271, 1985. With permission of ICIPE Science Press.)

(Figure 6.7). Mayo et al. (1988) found significant differences in the total lengths of chromosome sets among five biotypes of *S. graminum*. The relationship among biotypes based on cytological studies was similar to those reported by plant-damage studies.

6.5. ENZYME POLYMORPHISM

Gel electrophoresis of isozymes and their mobility variants (allozymes or electromorphs), whose position on a gel are revealed using appropriate stains and which are phenotypic expressions of gene coding, has been widely used to detect insect biotypes and populations (Sogawa 1978, Simon et al. 1982, Saxena and Mujer 1984, Wellso et al. 1988, Williams and Shambaugh 1988, Abid et al. 1989, Black et al. 1990). Isozymes, or multiple molecular forms of enzymes, are enzymes that share a common substrate but differ in electrophoretic mobility (Market and Moller 1959). They are revealed when tissue extracts are subjected to electrophoresis in various types of gels and are subsequently submerged in solutions containing enzyme-specific stains

(Wendel and Weeden 1989). When electrophoretic gels are submerged in enzyme-specific staining solutions, one or more regions of enzyme activity may be revealed. Those enzymes which are coded by more than a single gene display complex phenotypes with several bands (polymorphic), in which more than one isozyme is present. On the other hand, "nonsegregating isozymes" remain common to all members of the population (monomorphic). The most commonly used measures of intrapopulation variation are the percent of polymorphic loci, the number of alleles per locus, and the mean proportion of loci heterozygous per individual.

Horizontal starch gels or vertical polyacrylamide gels are generally used for surveying enzyme polymorphism among insect biotypes. The procedure involves crude enzyme preparation, electrophoresis, and histochemical staining. Immediately prior to electrophoresis, single insect individuals are homogenized in 15–20 µl cold, homogenizing solution, containing a few crystals of bromephenol blue as a tracking dye during electrophoresis. Insect individuals are homogenized, using glass pasteur pipettes with flame-sealed tips in rows of small wells (3 mm diameter × 8 mm deep) drilled into a flat perspex box.

For starch gel electrophoresis, Whatman no. 3 filter paper bits (4 mm × 6 mm) are used to absorb the crude extract and are inserted directly into the gel. In polyacrylamide gel electrophoresis (PAGE), 10 µl samples are injected into the pockets of a large pore gel with a microsyringe. For detailed descriptions on the two electrophoretic procedures and gel staining techniques, refer to Brewer and Sing (1970), Loxdale et al. (1983), Soltis and Soltis (1989), and Hames and Rickwood (1990).

Using a PAGE technique, several enzymes (Table 6.1) have been investigated to differentiate insect biotypes. Initial studies on the electrophoretic variations in esterases among *N. lugens* biotypes were reported by Sogawa (1978). *N. lugens* biotype 2 is differentiated from the other biotypes by the involvement of the electrophoretic variant type D in its population. Saxena and Mujer (1984) reported polymorphism in 5 out of 11 enzymes of *N. lugens* bioytypes, including catalase, esterase, malate dehydrogenase, isocitric dehydrogenase, and glucose phosphate isomerase. Isozyme analysis of *A. pisum,* however, did not disclose variation among biotypes and only the superoxide dismutase enzyme varied for one biotype (Simon et al. 1982). Similarly, Wellso et al. (1988) reported that genetic differentiation among *M. destructor* biotypes is low and that electrophoresis does not identify biotypes. However, Black et al. (1990) detected polymorphism in 13 of 47 enzymes examined from 18 *M. destructor* populations. Williams and Shambaugh (1988) reported clear banding differences in leucine aminopeptidase, phosphoglucose mutase, and malate dehydrogenase (= malic enzyme) of two biotypes of grape phylloxera, *Daktulosphaira vitifoliae* (Fitch), reared on seven different grape

**Table 6.1. Enzyme Loci Examined for Identification
of Biotypes of Selected Insect Pests**

Enzyme	GB	GP	HF	PA	BPH
Acid phosphatase	+[a]		+[c]	+[e]	+[f]
Aconitate hydrogenase (= Aconitase)			+[c,d]		
Adenylate kinase			+[c,d]		
Alcohol hydrogenase	+[a]		++[c,d]		+[f]
Aldehyde oxidase			+[c]	+[e]	
Aldolase			+[c]		
Alkaline phosphatase	+[a]		+[c]	+[e]	
Amylase			+[c]		
Carbonic anhydrase			++[c]		
Catalase			+[c]		++[f]
Diaphorase			+[c]		
Esterase	+[a]		+[c]	+[e]	++[f,g]
Fructose-1,6-biphosphate			+[c]		
Fumarate hydrogenase (= Fumarase)			++[c]		
Glucose dehydrogenase			+[c]		
Glucose oxidase			+[c]		
Glucose-6-phosphate dehydrogenase (= Phosphoglucoisomerase)			+[c]		+[f]
Glutamate dehydrogenase			+[c]		
Glutamate oxaloacetate transaminase (= Asparate aminotransferase)	+[a]		++[c]		
Glutamate oxaloacetate mutase					+[f]
Glutamic acid dehydrogenase			+[c]		
Glyceraldehyde-3-phosphate dehydrogenase			+[c]		
Glycerol-3-phosphate dehydrogenase			+[c]		
β-glycosidase	+[a]				
Hexokinase			++[c,d]	+[e]	
Hydroxyacid dehydrogenase			+[c]		
Isocitrate dehydrogenase	+[a]		++[c,d]		++[f]
Lactic acid dehydrogenase			+[c]		
Lactate dehydrogenase					+[f]
Leucine aminopeptidase		++[b]			+[f]
Malate dehydrogenase (= Malic enzyme = Malic acid dehydrogenase)	+[a]	++[b]	++[c,d]	+[e]	++[f]
Mannose-6-phosphate dehydrogenase			++[c]		
Octanol dehydrogenase			+[c]		
Peroxidase					+[f]
Phenol oxidase			+[c]		
Phosphatase					
Phosphoglucoisomerase			++[c,d]		++[f]
Phosphoglucomutase	+[a]	++[b]	++[c,d]	+[e]	

Table 6.1. Continued

Enzyme	GB	GP	HF	PA	BPH
Phosphogluconate dehydrogenase	++[a]		+[d]		+[f]
6-Phosphogluconic acid dehydrogenase			++[c]		
Phosphohexose isomerase	+[a]				
Shikimate dehydrogenase					+[f]
Sorbitol dehydrogenase			+[c]		
Succinate dehydrogenase			+[c]		
Sucrase			+[c]		
Superoxide dimutase				++[e]	
Tetrazolium oxidase					+[f]
Trehalase			+[c]		
Triose phosphate isomerase			+[c,d]		
Xanthine dehydrogenase			+[c]		

GB, Greenbug, *Schizaphis graminum* (Rondani); GP, Grape phylloxera, *Daktulosphaira vitifoliae* (Fitch); HF, Hessian fly, *Mayetiola destructor* (Say); PA, Pea aphid, *Acyrthosiphon pisum* (Harris); BPH, Brown planthopper, *Nilaparvata lugens* (Stål).

[+], Enzyme loci examined.
[++], Polymorphism detected.
[a]Abid et al. 1989.
[b]Williams and Shambaugh 1988.
[c]Black et al. 1990.
[d]Wellso et al. 1988.
[e]Simon et al. 1982.
[f]Saxena and Mujer 1984.
[g]Sogawa 1978.

species (Figure 6.8). Isozyme characterization of *S. graminum* revealed differences in the phosphogluconate dehydrogenase enzyme among some biotypes (Abid et al. 1989).

6.6. MITOCHRONDRIAL DNA VARIATIONS

Nucleotide differences among mitochondrial genomes from insect populations can provide estimates of genetic divergence and a possible rapid identification of insect biotypes (Powers et al. 1989). Because of the rapid accumulation of nucleotide polymorphism in the mitochondrial genomone, and since in most insect species mitochondrial DNA (mt DNA) is inherited intact from the female parent without being altered by recombination or meiotic segregation (Dawid and Blackler 1972, Lansman et al. 1981, Brown 1983, Zehnder et al. 1992), the analysis of mt DNA may provide information regarding the genetic structure of populations not obtained through other methods (Zehnder et al. 1992). Mitochondrial DNA exhibits considerable variation among individuals both within and between populations.

The methodology involves comparison of mt DNA fragments generated by digestion with restriction endonucleases, followed by electrophoretic fractionation on agarose or polyacrylamide gels. Polymorphisms in mt DNA from individual insects are examined to determine genetic differences in biotypes.

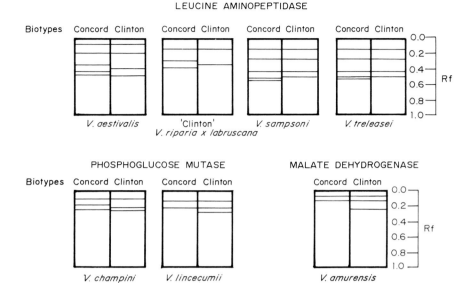

FIGURE 6.8. Isozymes of three enzymes in 'Concord' and 'Clinton' biotypes of *Daktulosphaira vitifoliae* when reared on seven different *Vitis* species. (From Williams, R. N. and Shambaugh, G. F., *Ann. Entomol. Soc. Am.,* 81, 1, 1988. With permission of the Entomological Society of America.)

Pairwise comparisons of mt DNA fragments from biotypes are conducted following the formula of Powers et al. (1989): $F = (2Nxy/(Nx + Ny))$, where Nx is the number of fragments in genotype x; Ny is the number of fragments in genotype y; and Nxy is the number of common fragments. F values are converted to estimates of percent nucleotide sequence divergence (p) following the equation of Nei and Li (1979). Powers et al. (1989) examined the relationships among *S. graminum* biotypes B, C, E, and F by comparing mt DNA restriction patterns. The mitochondrial data indicated a close relationship between biotype C and biotype E, but the affinities of biotype F were not clear. For detailed explanations of the methodology involved in mt DNA isolation and mt DNA restriction analysis, see Powers et al. (1989), Martinez et al. (1992), and Zehnder et al. (1992).

6.7. CUTICULAR HYDROCARBONS

The surface lipids of insects have a number of physiological and behavioral roles in relation to regulation of water loss and as a barrier limiting the entry of chemicals and microorganisms (Espelie and Bernays 1989). Previously, gas chromatography of cuticular hydrocarbons was used to identify specimens of morphologically indistinguishable members of species

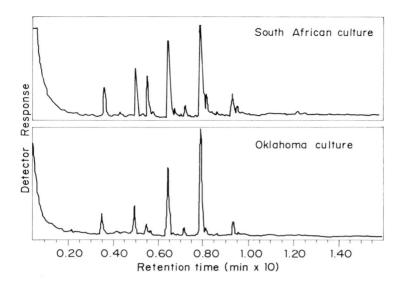

FIGURE 6.9. Cuticular hydrocarbons of South African and Oklahoma biotypes of *Diuraphis noxia,* analyzed on a RSL-150 capillary column. (Bergman, D. K. et al., *Southwest. Entomol.,* 15, 91, 1990. With permission.)

complexes, such as *Anopheles* sp. (Carlson and Service 1980), *Simiulium* sp. (Phillips et al. 1985) and *Blatella* sp. (Carlson and Brenner 1988). In addition, cuticular hydrocarbons have also been used to investigate phylogenic relationships in *Drosophila* sp. (Bartelt et al. 1986). Recently it has been demonstrated that cuticular hydrocarbons can also serve as biotypic markers (Bergman et al. 1990, Dillwith et al. 1991).

Extraction and analysis of the cuticular hydrocarbons of insect biotypes is conducted using the method described by Bergman et al. (1990) for the Russian wheat aphid, *Diuraphis noxia* (Mordvilko). Cuticular lipids are extracted from a pooled sample of 20 to 50 individuals of a biotype with 5 to 7 ml of hexane for 30 sec. Hydrocarbons are isolated using column chromatography (Dillwith et al. 1991) and hydrocarbon structures are determined by gas chromatography-mass spectroscopy. Data on hydrocarbon composition are subjected to multivariate and discriminate analysis followed by canonical and cluster analysis.

Bergman et al. (1990) demonstrated that the cuticular hydrocarbons of *D. noxia* populations from South Africa and Oklahoma differed significantly (Figure 6.9). Oklahoma populations have less 3-methylheptacosane and more hentriacontane in cuticular extracts than *D. noxia* from South Africa. Similarly, Dillwith et al. (1991) reported that cuticular hydrocarbon analyses of *S. graminum* can be used for identifying biotypes. A 3-D canonical plot based on n-alkane composition showed distinct groupings for biotypes F, G, and H, and the Idaho isolates of *S. graminum* (Figure 6.10).

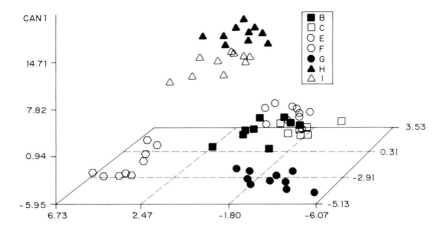

FIGURE 6.10. A 3-D Canonical plot based on *n*-alkane composition of *Schizaphis graminum* biotypes; biotypes B, C, E, F, G, H, and Idaho (I) are identified by different symbols. (From Dillwith, J. W. et al., courtesy of the U.S. Department of Agriculture and Oklahoma State University.)

6.8. CONCLUSIONS

Despite the problem of continually developing new (virulent) biotypes, plant resistance to insects will continue to play the dominant role in crop plant IPM in the future. The continuous identification of new genes and donors for insect resistance are of the utmost importance in varietal resistance programs. Pure line populations of insect biotypes will be needed for identifying new genetic sources of crop plants (Saxena and Barrion 1985). Any mixtures in biotype populations may provide conflicting results in studies of the genetic analysis of insect-resistant plants and insect virulence. Differential varietal reactions will probably remain the most commonly used method to detect resistance-breaking insect biotypes. However, in-depth investigations into pest insect behavior, sensory, and regulatory physiology, ecology, and genetics will all be very useful in developing techniques to complement or supplant this type of biotype identification.

REFERENCES

Abid, H. S., Kindler, S. D., Jensen, S. G., Thomas-Compton, M. A. and Spomer, S. M., Isozyme characterization of sorghum aphid species and greenbug biotypes (Homoptera: Aphididae), *Ann. Entomol. Soc. Am.,* 82, 303, 1989.

Bartelt, R. J., Armold, M. T., Schaner, A. M. and Jackson, L. L., Comparative analysis of cuticular hydrocarbons in the *Drosophila virilis* species group, *Comp. Biochem. Physiol.,* 83B, 731, 1986.

Bergman, D. K., Dillwith, J. W., Campbell, R. K. and Eikenbary, R. D., Cuticular hydrocarbons of the Russian wheat aphid, *Southwest. Entomol.,* 15, 91, 1990.

Black IV, W. C., Hatchett, J. H. and Krchma, L. J., Allozyme variation among populations of the Hessian fly *(Mayetiola destructor)* in the United States, *J. Heredity,* 81, 331, 1990.

Brewer, G. J. and Sing, C. F., *An Introduction to Isozyme Techniques,* Academic Press, New York, 1970, 186.

Brown, W. M., Evolution of animal mitochondrial DNA, in *Evolution of Genes and Proteins,* M. Nei and R. K. Koehn, eds., Sinauer, Sunderland, MA, 1983, 331.

Carlson, D. A. and Brenner, R. J., Hydrocarbon-based discrimination of three North American *Blattella* cockroach species (Orthoptera: Blattellidae) using gas chromotography, *Ann. Entomol. Soc. Am.,* 81, 711, 1988.

Carlson, D. A. and Service, M. W., Identification of mosquitoes of *Anopheles gambiae* species complex A and B by analysis of cuticular components, *Science,* 207, 1089, 1980.

Cartier, J. J., Recognition of three biotypes of the pea aphid from southern Quebec, *J. Econ. Entomol.,* 52, 293, 1959.

Damsteegt, V. D. and Voegtlin, D. J., Morphological and biological variation among populations of *Aulacorthum solani* (Homoptera: Aphididae), the vector of soybean dwarf virus, *Ann. Entomol. Soc. Am.,* 83, 949, 1990.

Dawid, I. B. and Blackler, A. W., Material and cytoplasmic inheritance of mitochondrial DNA in *Xenopus Dev. Biol.,* 29, 152, 1972.

Dillwith, J. W., Bergman, D. K., Fargo, W. S., Puterka, G. J. and Peters, D. C., Cuticular hydrocarbons of greenbug biotypes, in *Proceedings Aphid-Plant Interactions: Population to Molecules,* D. C. Peters and J. A. Websters, eds., USDA-Agricultural Research Service–Oklahoma State University, 1991, 355.

Dreyer, D. L. and Campbell, B. C., Association of the methylation of intercellular pectin with plant resistance to aphids and with induction of aphid biotypes, *Experienta,* 40, 224, 1984.

Espelie, K. E. and Bernays, E. A., Diet-related differences in the cuticular lipids of *Manduca sexta* larvae, *J. Chem. Ecol.,* 15, 2003, 1989.

Fargo, W. S., Inayatullah, C., Webster, J. A. and Holbert, D., Morphometric variation within apterous females of *Schizaphis graminum* biotypes, *Res. Popul. Ecol.,* 28, 163, 1986.

Gallun, R. L., Deay, H. O. and Cartwright, W. B., Four races of Hessian fly selected and developed from an Indiana population, *Purdue Univ. Res. Bull.,* 732, 1961, 8.

Hames, B. D. and Rickwood, D., *Gel Electrophoresis of Proteins: A Practical Approach,* 2nd Ed., Oxford Press, New York, 1990, 383.

Heinrichs, E. A. and Rapusas, H. R., Cross-virulence of *Nephotettix virescens* (Homoptera: Cicadellidae) biotypes among some rice cultivars with the same major-resistance gene, *Environ. Entomol.,* 14, 696, 1985.

Honda, H. and Mitsuhashi, W., Morphological and morphometrical differences between the fruit- and pinaceae-feeding types of yellow peach moth, *Conogethes punctiferalis* (Guenée) (Lepidoptera: Pyralidae), *Appl. Entomol. Zool.,* 24, 1, 1989.

Inayatullah, C., Webster, J. A. and Fargo, W. S., Morphometric variations in the alates of greenbug (Homoptera: Aphididae) biotypes, *Ann. Entomol. Soc. Am.,* 80, 306, 1987.

IRRI (International Rice Research Institute), Levels of Resistance of Rice Varieties to Biotypes of the Brown Planthopper, *Nilaparvata lugens,* in South and Southeast Asia, Report of the 1979 International Collaborative Project on Brown Planthopper Resistance, *IRRI Res. Paper Ser.,* 72, 1982, 14.

IRRI (International Rice Research Institute), *Standard Evaluation System for Rice,* IRRI, Los Baños, Laguna, Philippines, 1988, 54.

Khan, Z. R. and Saxena, R. C., Probing behavior of three biotypes of *Nilaparvata lugens* (Homoptera: Delphacidae) on different resistant and susceptible rice varieties, *J. Econ. Entomol.,* 81, 1338, 1988.

Khan, Z. R. and Saxena, R. C., Purification of biotype 1 population of brown planthopper, *Nilaparvata lugens* (Homoptera: Delphacidae), *Insect Sci. Appl.,* 11, 55, 1990.

Kerns, D. L., Puterka, G. J. and Peters, D. C., Intrinsic rate of increase for greenbug (Homoptera: Aphididae) biotypes E, F, G, and H on small grain and sorghum varieties, *Environ. Entomol.,* 18, 1074, 1989.

Kim, K. C., Chiang, H. C. and Brown, Jr., B. W., Morphometric differences among four biotypes of *Ostrinia nubilalis* (Lepidoptera: Pyralidae), *Ann. Entomol. Soc. Am.,* 57, 155, 1967.

Lansman, R. A., Shade, R. O., Shapira, J. F. and Avise, J. C., The use of restriction endonucleases to measure mitochondrial DNA sequence relatedness in natural populations. III. Techniques and potential applications, *J. Mol. Evol.,* 17, 214, 1981.

Lidell, M. C. and Schuster, M. F., Effectiveness of wheat genes for Hessian fly (Diptera: Cecidomyiidae) resistance in Texas, *J. Econ. Entomol.,* 83, 1135, 1990.

Loxdale, H. D., Castanera, P. and Brookes, C. P., Electrophoretic study of enzymes from cereal aphid populations. I. Electrophoretic techniques and staining systems for characterizing isozymes from six species of cereal aphid (Hemiptera: Aphididae), *Bull. Entomol. Res.,* 73, 645, 1983.

Ma, R., Reese, J. C., Black IV, W. C. and Bramel-Cox, P., Detection of pectinesterase and polygalacturonase from salivary secretions of living greenbugs, *Schizaphis graminum* (Homoptera: Aphididae), *J. Insect Physiol.,* 36, 507, 1990.

Market, C. L. and Moller, F., Multiple forms of enzymes: tissue, ontogenetic and species specific patterns, *Natl. Acad. Sci. U.S.A.,* 45, 753, 1959.

Martinez, D., Moya, A., Latorre, A. and Fereres, A., Mitochondrial DNA variation in *Rhopalosiphum padi* (Homoptera: Aphididae) populations from four Spanish locations, *Ann. Entomol. Soc. Am.,* 85, 241, 1992.

Mayo, Z. B. Jr., and Starks, K. J., Sexuality of the greenbug, *Schizaphis graminum,* in Oklahoma, *Ann. Entomol. Soc. Am.,* 65, 671, 1972.

Mayo, Z. B., Starks, K. J., Banks, D. J. and Veal, R. A., Variation in chromosome length among five biotypes of greenbug (Homoptera: Aphididae), *Ann. Entomol. Soc. Am.,* 81, 128, 1988.

McLean, D. L. and Kinsey, M. G., A technique for electronically recording aphid feeding and salivation, *Nature (London),* 202, 1358, 1964.

McLean, D. L. and Kinsey, M. G., Probing behavior of pea aphid, *Acyrthosiphon pisum.* II. Comparison of salivation and ingestion in host and nonhost plant leaves, *Ann. Entomol. Soc. Am.,* 61, 730, 1968.

Medrano, F. and Heinrichs, E. A., A method for purifying brown planthopper (BPH) *Nilaparvata lugens* biotypes, *Int. Rice Res. Newsl.,* 9(4), 16, 1984.

Miles, P. W., The saliva of Hemiptera, *Adv. Insect Physiol.,* 9, 183, 1972.

Nei, M. and Li, W. H., Mathematical model for studying genetic variations in terms of restriction endonucleases, *Proc. Natl. Acad. Sci. U.S.A.,* 76, 5269, 1979.

Niassy, A., Ryan, J. D. and Peters, D. C., Variations in feeding behavior, fecundity, and damage of biotype B and E of *Schizaphis graminum* (Homoptera: Aphididae) on three wheat genotypes, *Environ. Entomol.,* 16, 1163, 1987.

Nielson, M. W. and Don, H., Probing behavior of biotypes of the spotted alfalfa aphid on resistant and susceptible alfalfa cloves, *Entomol. Exp. Appl.,* 17, 477, 1974.

Oka, I. N., Quick method for identifying brown planthopper biotypes in the field, *Int. Rice Res. Newsl.,* 3(6), 11, 1978.

Paguia, P., Pathak, M. D. and Heinrichs, E. A., Honeydew excretion measurement techniques for determining differential feeding activity of biotypes of *Nilaparvata lugens* on rice varieties, *J. Econ. Entomol.,* 73, 35, 1980.

Pathak, M. D. and Painter, R. H., Differential amounts of material taken up by four biotypes of corn leaf aphids from resistant and susceptible sorghums, *Ann. Entomol. Soc. Am.,* 51, 250, 1958.

Pathak, P. K., Saxena, R. C. and Heinrichs, E. A., Parafilm sachet for measuring honeydew excretion by *Nilaparvata lugens* on rice, *J. Econ. Entomol.,* 75, 194, 1982.

Peters, D. C., Kerns, D., Puterka, G. J. and McNew, R., Feeding behavior, development and damage by biotypes B, C, and E of *Schizaphis graminum* (Homoptera: Aphididae) on 'Wintermalt' and 'Post' barley, *Environ. Entomol.,* 17, 503, 1988.

Phillips, A., Walsh, J. F., Garms, R., Molyneux, D. H., Milligan, P. and Ibrahim, G., Identification of adults of the *Simulium damnosum* complex using hydrocarbon analysis, *Trop. Med. Parasit.,* 36, 97, 1985.

Powers, T. O., Jense, S. G., Kindler, S. D., Stryker, C. J. and Sandall, L. J., Mitochondrial DNA divergence among greenbug (Homoptera: Aphididae) biotypes, *Ann. Entomol. Soc. Am.,* 82, 298, 1989.

Puterka, G. J. and Peters, D. C., Rapid technique for determining greenbug (Homoptera: Aphididae) biotypes B, C, E, and F, *J. Econ. Entomol.,* 81, 396, 1988.

Puterka, G. J. and Peters, D. C., Inheritance of greenbug, *Schizaphis graminum* (Rondani), virulence to *Gb2* and *Gb3* resistance genes in wheat, *Genome,* 32, 109, 1989.

Rao, C. R., *Advanced Statistical Methods in Biometric Research,* John Wiley & Sons, New York, 1952.

Ryan, J. D., Dorschner, K. W., Girma, M., Johnson, R. C. and Eikenbary, R. D., Feeding behavior, fecundity, and honeydew production of two biotypes of greenbug (Homoptera: Aphididae) on resistant and susceptible wheat, *Environ. Entomol.,* 16, 757, 1987.

Sato, A. and Sogawa, K., Biotype variations in the green rice leafhopper, *Nephotettix cincticeps* Uhler (Homoptera: Deltocephalidae), in relation to rice varieties, *Appl. Entomol. Zool.,* 16, 55, 1981.

Saxena, R. C. and Barrion, A. A., A technique for preparation of brown planthopper chromosomes, *Int. Rice. Res. Newsl.,* 7(2), 8, 1982.

Saxena, R. C. and Barrion, A. A., Comparative cytology of brown planthopper populations infesting *Leersia hexandra* Swartz and rice in the Philippines, *Int. Rice Res. Newsl.,* 9(1), 23, 1984.

Saxena, R. C. and Barrion, A. A., Biotypes of the brown planthopper *Nilaparvata lugens* (Stål) and strategies in deployment of host plant resistance, *Insect Sci. Appl.,* 6, 271, 1985.

Saxena, R. C. and Barrion, A. A., Biotypes of insect pests of agricultural crops, *Insect Sci. Appl.,* 8, 453, 1987.

Saxena, R. C. and Mujer, C. V., Detection of enzyme polymorphism among populations of brown planthopper (BPH) biotypes, *Int. Rice Res. Newsl.,* 9(3), 18, 1984.

Saxena, R. C. and Pathak, M. D., Factors affecting resistance of rice varieties to the brown planthopper, *Nilaparvata lugens* (Stål), Paper presented at the 9th Annual Conference of the Pest Control Council of the Philippines, Bacolod City, 18–20 May, 1977.

Saxena, R. C. and Rueda, L. M., Morphological variations among three biotypes of brown planthopper *Nilaparvata lugens* in the Philippines, *Insect Sci. Appl.,* 3, 193, 1982.

Saxena, R. C., Velasco, M. V. and Barrion, A. A., Morphological variations between brown planthopper biotypes on *Leersia hexandra* and rice in the Philippines, *Int. Rice Res. Newsl.,* 8(3), 3, 1983.

Seshu, D. V. and Kauffman, H. E., Differential response of rice varieties to the brown planthopper in international screening tests, *IRRI Res. Paper Ser.,* 52, 1980, 13.

Simon, J. P., Parent, M. A. and Auclair, J. L., Isozyme analysis of biotypes and field populations of the pea aphid, *Acyrthosiphon pisum, Entomol. Exp. Appl.,* 32, 186, 1982.

Smith, C. M., *Plant Resistance to Insects: A Fundamental Approach,* John Wiley & Sons, New York, 1989, 286.

Sogawa, K., Electrophoretic variations in esterase among biotypes of the brown planthopper, *Int. Rice Res. Newsl.,* 3, 8, 1978.

Soltis, D. E. and Soltis, P. S., *Isozymes in Plant Biology,* Advances in Plant Sciences Series, Vol. 4, Dioscorides Press, Portland, 1989, 268.

Wellso, S. G., Howard, D. J., Adams, J. L. and Arnold, J., Electrophoretic monomorphism in six biotypes of the Hessian fly (Diptera: Cecidomyiidae), *Ann. Entomol. Soc. Am.,* 81, 50, 1988.

Wendel, J. F. and Weeden, N. F., Visualization and interpretation of plant isozymes, in *Isozymes in Plant Biology,* Advances in Plant Sciences Series, Vol. 4, D. E. Soltis and P. S. Soltis, eds., Dioscorides Press, Portland, 1989, 268.

Williams, R. N. and Shambaugh, G. F., Grape phylloxera (Homoptera: Phylloxeridae) biotypes confirmed by electrophoresis and host susceptibility, *Ann. Entomol. Soc. Am.,* 81, 1, 1988.

Zehnder, G. W., Sandall, L., Tisler, A. M. and Powers, T. O., Mitochondrial DNA diversity among 17 geographic populations of *Leptinotarsa decemlineata* (Coleoptera: Chrysomelidae), *Ann. Entomol. Soc. Am.,* 85, 234, 1992.

CHAPTER 7

Use of Tissue Culture and Artificial Diets for Evaluating Insect Resistance

7.1. TISSUE CULTURE

The use of undifferentiated plant tissue (callus) from plant tissue culture as a method to evaluate the resistance of plants against insect attack provides new possibilities for understanding the nature and causes of resistance to several chewing insects. In general, tissue culture offers an ideal way of evaluating resistance under a controlled environment free of almost any physical, environmental, or morphological factor. Thus, for cases where there is evidence of an allelochemical factor involved in plant resistance, tissue culture will provide additional information if the expression of resistance can be correlated to the results of field experiments.

The evaluation of insect resistance using plant callus is generally based on larval feeding preferences and larval growth on callus initiated from susceptible and resistant plant varieties. Callus tissues of several insect-resistant plants exhibit resistance to insect feeding and growth. Maize callus tissue exhibits resistance to the fall armyworm, *Spodoptera frugiperda* (J. E. Smith), the southwestern corn borer, *Diatraea grandiosella* Dyar, European corn borer, *Ostrinia nubilalis* Hübner, and the corn earworm, *Helicoverpa zea* (Boddie), similar to the resistance shown by whole plant foliage (Williams and Davis 1985, Williams et al. 1983, 1985, 1987a,b). Callus tissues from fall armyworm-resistant bermudagrass cultivars also exhibit resistance equivalent to whole plant foliage (Croughan and Quisenberry 1989a). Caballero et al. (1988) used callus tissue from rice plants to evaluate their resistance to the yellow stem borer, *Scirpophaga incertulas* (Walker), the striped stem borer, *Chilo suppressalis* (Walker), and the rice leaffolder, *Cnaphalocrocis medinalis* (Guenée).

7.1.1. Production of Callus

Mature seed kernels are dipped in 70% ethanol in a beaker for 1.5 min. Then they are placed in a sterile beaker with 2 g laboratory detergent in 100 ml 5.25% sodium hypochlorite solution and swirled for 20 min and then rinsed 5 to 7 times in sterile, distilled water. Seeds are maintained 24 to 48 hr at approximately 25 ° C on moistened filter paper in petri plates. Embryos are excised with sterile dissecting needles swirled for 5 min in 5.25% sodium hypochlorite and rinsed 5 to 7 times in sterile, distilled water. Embryos are placed in petri dishes on a cell culture medium.

Four different media, MS-3 (Murashige and Skoog 1962), N6 (Chu et al. 1981), B5 (Gamborg et al. 1968), and LS (Linsmaier and Skoog 1965) are generally used for callus induction. The composition of different cell culture media is given in Table 7.1. Macro- and micro-nutrients, required for each one liter of each medium, are measured and stock solutions are prepared separately in one-liter size beakers. Eight grams of agar and 30 g of sucrose are added to each stock solution with a known volume of distilled water. Solutions are heated in an oven to melt the agar completely. The resulting mixtures are measured separately using a graduated glass flask, and an additional volume of distilled water is added to obtain one liter of each medium. The media is further heated and thoroughly mixed with a magnetic stirrer. The hydrogen ion concentration (pH) of each medium is determined using a pH meter and the required volume of sodium hydroxide or hydrochloric acid is added to attain a pH of 5.8. The media is steam sterilized in an autoclave at a temperature of 120°C for 20 min. After autoclaving and cooling, vitamins and 2,4-D (2,4 dichlorophenoxyacetic acid), both filter sterilized, are added. The developing callus are transferred onto fresh medium at 3-week intervals using a sterile spatula.

7.1.2. Disinfection of Insect Eggs

Gravid female moths are generally allowed to lay their eggs on sterilized filter papers or cheese cloth if possible. The eggs are disinfected by dipping in 70% ethanol for 15 sec and subsequently in 0.1% mercuric chloride solution for 15 min. Sometimes eggs are disinfected with 0.25% sodium hypochlorite prior to emergence. Disinfected eggs are washed with sterile water and kept in a petri dish lined with moist sterilized filter paper and incubated at 27° C until they hatch. Newly hatched larvae are then placed in petri dishes with plant callus.

7.1.3. Bioassays

7.1.3.1. Orientation and Settling Responses

To determine whether neonate larvae feed preferentially on callus initiated from susceptible or resistant plants, larval orientation and settling responses are measured. Medium is distributed in a plastic container ($130 \times 130 \times 15$ mm) (Williams et al. 1985) or in a 15-cm-diameter petri dish (Croughan and

Table 7.1. Composition of Different Cell Culture Media Used for Callus Induction in Selected Rice Varieties[a]

	Constituents of different media[b]			
	N_6	B_5	MS_3	LS
Inorganic (mg/ml/l)				
KNO_3	2830	2500	1900	1900
$(NH_4)_2SO_4$	463	134	—	—
NH_4NO_3	—	—	1650	1650
$MgSO_4 \cdot 7H_2O$	185	250	370	370
$CaCl_2 \cdot 2H_2O$	166	150	440	440
KH_2PO_4	400	—	170	170
$NaH_2PO_4 \cdot H_2O$	—	150	—	—
$FeSO_4 \cdot 7H_2O$	27.8	—	27.8	27.8
Sequestrene 330 Fe	—	28	—	—
$Na_2EDTA \cdot 2H_2O$	37.3	—	37.3	37.3
$MnSO_4 \cdot 4H_2O$	4.4	—	22.3	22.3
$MnSO_4 \cdot H_2O$	—	10	—	—
H_3BO_3	1.6	3	6.2	6.2
$ZnSO_4 \cdot 7H_2O$	1.5	2	8.6	8.6
$CuSO_4 \cdot 5H_2O$	—	0.025	0.025	0.025
$NaMoO_4 \cdot 2H_2O$	—	0.25	0.25	0.25
KI	0.8	0.75	0.83	0.83
$CaCl_2 \cdot 6H_2O$	—	0.025	0.025	0.025
Organic				
Myo-inositol	—	100	100	100
Nicotinic acid	0.5	1	0.5	0.5
Glycine	2	—	2	2
Pyridoxine . HCl	0.5	1	0.5	0.5
Thiamine . HCl	1.0	10	0.1	0.4
Kinetin	—	—	—	0.03
Others				
3, Indoleacetic acid	—	—	—	2
2,4-D	2	2	2	2
Agar	8	8	8	10
Sucrose	30	30	30	30
pH	5.8	5.5	5.5–5.8	5.8

[a]Laboratory experiment, IRRI, 1990.
[b]N_6 – Chu et al. (1981); B_5 – Gamborg et al. (1968); MS – Murashige and Skoog (1962); LS – Linsmair and Skoog (1965).

Quisenberry 1989a, Sharma 1991, Thapa 1991). Five hundred mg of callus of each test variety are placed in the corners of each container or in a circular manner equidistantly from the center of a petri dish. Fifty to one hundred blackhead stage eggs or freshly hatched larvae are transferred carefully to the center of each container or petri dish. All containers are kept in complete darkness to minimize the effects of light on larval orientation. Petri dishes are sealed with a Parafilm® membrane to prevent escape of larvae and to avoid contamination. The number of larvae present on each callus is recorded at 1, 6, 12, 24, and 48 hr after introduction of larvae and the mean number of larvae present on each callus is then compared.

When given a choice of callus of four maize hybrids, more than twice as many *S. frugiperda* larvae preferred callus from the susceptible hybrid than callus from resistant hybrids (Williams et al. 1985). Croughan and Quisenberry (1989a) also reported differences among susceptible and resistant bermudagrass varieties to *S. frugiperda.* Similarly, in a choice test, significantly more *D. grandiosella, D. saccharalis* and *O. nubilalis* larvae preferred the callus of maize hybrids which are susceptible to leaf feeding (Williams et al. 1987b). Significantly fewer larvae of *S. incertulas* and *C. medinalis* settled on callus of resistant rice varieties as compared to susceptible ones (Sharma 1991, Thapa 1991).

7.1.3.2. Consumption and Utilization of Food

In a typical experiment of this type, third-instar larvae (in this example Lepidoptera), are reared on callus of a susceptible plant and starved (but water satiated) for 4 hr. Larvae are weighed individually and allowed to feed only on callus of the test variety (Sharma 1991). After 24 to 48 hr, larvae are deprived of food but water satiated for 4 hr so that all food matter is excreted from the gut. Individual larval weights are recorded and the percentage of increase in body weight is calculated for callus from each test plant. Sharma (1991) reported that the increase in the weights of *C. medinalis* larvae fed on callus of susceptible rice varieties for 48 hr was significantly greater than the weight gains of larvae fed on resistant varieties.

7.1.3.3. Growth and Development

To determine the growth and development of larvae feeding on callus initiated from susceptible and resistant plants, petri dishes with approximately 600 mg to 1 g of callus are infested with 3 to 5 neonate larvae. The larvae are maintained until they pupate in a growth chamber at 27° C and 12:12 (L:D) (Williams and Davis 1985, Williams et al. 1985, Williams et al. 1987b, Caballero et al. 1988, Sharma 1991, Thapa 1991). Larval growth and development is measured by recording percent larval survival, larval weight after 7 and 15 days, length of larval development period, percent larvae becoming pupae, and pupal weight. An insect growth index (the ratio of percentage of larvae developing into pupae and the mean growth period) has been developed by Saxena et al. (1974). The higher the growth index, the more suitable the callus is for insect growth. Williams et al. (1983) reported that *D. grandiosella* larvae reared for 7 days on callus of a highly resistant maize genotype were smaller than larvae reared on callus from susceptible maize genotypes. Similar differences were also reported in weights of *S. frugiperda* larvae reared for 7 days on resistant and susceptible maize hybrids (Williams et al. 1985). *D. grandiosella, D. saccharalis,* and *O. nubilalis* larvae reared for 7 days on callus initiated from two *D. grandiosella* resistant maize hybrids and two susceptible hybrids weighed significantly more on susceptible hybrids than on resistant ones (Williams et al. 1987b).

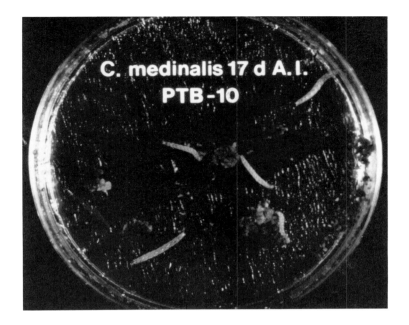

FIGURE 7.1. *Cnaphalocrocis medinalis* larvae reared for 17 days on callus originated from Ptb 10, a susceptible rice cultivar. (Photo courtesy of P. Caballero.)

Caballero et al. (1988) reported that the development of *S. incertulus, C. suppressalis,* and *C. medinalis* on callus from Rexoro (susceptible) rice was normal and similar to their development on Rexoro foliage. At 20 days after infestation, *S. incertulus* larvae developed to fifth instar on Rexoro callus, to only the third instar on callus of the resistant Chianan 2 cultivar, and failed to survive on callus of a resistant wild rice, *Oryza ridleyi* (acc. no. 100821). At 15 days after infestation, *C. suppressalis* developed to fifth instar on Rexoro, to fourth instar on callus of the resistant cultivar Yabami M. 47, and to only second instar on *O. ridleyi* callus. At 17 days after infestation, *C. medinalis* larvae developed to fifth instar on susceptible Rexoro and Ptb 10 calli (Figure 7.1), to fourth instar on moderately resistant IR 5865-26-1 callus, and to only second instar on resistant *O. ridleyi* callus.

Isenhour and Wiseman (1988) incorporated callus of susceptible and resistant maize varieties into an artificial diet to measure resistance to *S. frugiperda,* but callus-diet mixtures failed to confer the degree of resistance of foliage-diet mixtures.

Callus has many potential uses in developing insect-resistant crop plants. Plants derived from callus tissue may possess sufficient genetic variability to contain insect resistance. Such variation in corn, sugarcane, rice, and oat callus has been used to select for disease resistance, agrochemical tolerance, and improved agronomic traits. Attempts to identify insect resistance in somaclonal variants (plants generated from the variation among cultured cells or tissue) have yet to identify such differences. White and Irvine (1987) produced 2000 somaclonal variants of sugarcane callus tissue from a cultivar susceptible to the sugarcane borer, *Diatraea saccharalis* (F.). Regenerated plants grown and infested in field plots exhibited only random variation in the amount of borer damage incurred and increased levels of borer resistance were detected.

Tissue culture methods were used to develop regenerated plants from three bermudagrass cultivars (Croughan and Quisenberry 1989b). The regenerated plants were compared with the original genotypes for resistance to *S. frugiperda*. Of the seven regenerated plants tested, two lines (Brazos-R3 and OSU LCB W26-R2) appeared to have increased resistance. The other five lines were slightly more resistant or exhibited no increased resistance to the fall armyworm than the parent from which they were developed.

Sorghum genotypes regenerated from tissue culture were evaluated under field and laboratory conditions for resistance to leaf-feeding by *S. frugiperda* (Isenhour et al. 1991). Two regenerated lines were identified with higher resistance than their nonregenerated resistant parent or the susceptible parent.

Isenhour and Wiseman (1991) successfully regenerated *S. frugiperda* resistant maize genotypes via somatic embryogenesis. Progeny from these regenerates were evaluated under field and laboratory conditions and significant differences in resistance were observed between regenerated and nonregenerated lines.

Cell culture may also be used to select insect-resistant plants from cell populations. Several homopteran insects inject cell-digesting enzymes into plants during feeding (Miles 1972). Insects such as the greenbug, *Schizaphis graminum* (Rondani), the lygus bug, *Lygus disponsi* Linnavuori, the squash bug, *Anasa tristis* (De Geer), and the spotted alfalfa aphid, *Therioaphis maculata* (Buckton), also inject enzymetic toxins into host tissues (Wadley 1929, Beard 1940, Paschke and Sylvester 1957, Hori et al. 1987, Reese et al. 1990, Ma et al. 1991). If the toxins from these insects can be identified, they may be used to treat plant cell cultures and select insect-resistant plant types. Shukle et al. (1985) have partially characterized the enzymes of Hessian fly, *Mayetiola destructor* (Say), larvae in order to use these enzymes in cell culture selection in wheat for fly resistance.

7.2. ARTIFICIAL DIETS

To detect the presence of larval insect growth inhibitors in plants, artificial diets have been widely used. Artificial diets have played a particularly important role in the bioassay of individual factors involved as resistance mechanisms (Reese et al. 1989). Fresh or dried plant materials, plant extracts, or a particular plant allelochemical is thoroughly blended with a known amount of artificial diet. First-instar larvae are fed the amended diets and comparisons of insect growth on such diets incorporated with plant tissues or extracts of different susceptible or resistant varieties can be used effectively to assay for antibiosis resistance (Wiseman and Isenhour 1990). If the tissue extract is from a susceptible plant, the larvae that are fed the mixture will have a normal development compared with larvae that are fed on a control diet. However, when the tissue is from a resistant plant, larvae fed the diet mixture develop abnormally. This abnormal development can be expressed as high mortality, reduced larval growth, reduced larval and pupal size, extended mean generation time, or reduced fecundity. Such bioassays are more discriminating when antibiosis results from the presence of a toxin or a strong antifeedant (Buckley et al. 1991). However, even subtle effects of such compounds can be detected if the concentration of nutrients in the artificial diet is reduced to a level equivalent to that found in the host plant.

7.2.1. Isolation of Antifeedants Using Diets

Eucoubas et al. (1992) developed a quick bioassay-guided isolation method to detect antifeedants from a Japanese plant, *Skimmia japonica* (Rutaceae) against *Spodoptera littoralis* (Boisduval). The method uses the concept of bioautography, where insect feeding on thin layer chromatography plates (TLC) indicates chemical activity. This method can be applicable to any chewing insect for which an artificial diet has been developed. TLC silica gel plates with fluorescent indicator are cut into 5 × 10-cm pieces. The sample containing candidate compound is applied as a band, and the plate is developed in an appropriate solvent mixture in order to give the best possible separation. The solvent is then completely evaporated under reduced pressure. The plate is then coated with hot insect artificial diet and left to solidify for a few minutes. The diet used quenches the fluorescence, making the coated plate appear dark under UV light (254 nm). A second plate developed under the same conditions, but not coated with any diet, serves as a reference plate for measuring Rf values. Each plate is laid in a petri dish, lined with moist filter paper, with test insect larvae. The larvae are allowed to feed for 16 to 20 hr in the dark. Under these conditions, the larvae consume most of the diet layer on the control plate (without antifeedants), leaving only the fluorescent silica gel. The plates are dried with a hair dryer and then examined under UV light to visualize the uneaten areas. The location of these areas is compared with the Rf values of compounds detected on a reference plate.

7.2.2. Incorporation of Plant Materials into Diets

Oven-dried or lyophilized plant materials or plant extracts may be incorporated into an artificial diet to determine the effects of these constituents on plant resistance to insects. Intact plant tissues of susceptible and resistant plants are harvested at the developmental stage when resistance is expressed, dried at 40 to 42° C, ground in a Wiley mill using a 1-mm screen and stored at −10° C (Smith and Fisher 1983, Khan et al. 1989, Diawara et al. 1991, Wiseman et al. 1991, Wiseman and Isenhour 1991). Similarly, intact plant tissues may be harvested and placed in plastic freezer bags and frozen at −20° C. These tissues are later lyophilized and ground to a fine powder in a laboratory mill with a 1-mm mesh screen (Straub and Fairchild 1970, Wilson and Wissink 1986, Williams et al. 1990, Buckley et al. 1991, Ramachandran and Khan 1991). Oven-dried or lyophilized plant tissues are added to the artificial diet by blending together appropriate ratios of the plant material and the artificial diet. No plant tissue is added to one set of the diet which serves as a control. Ten to 15 g of either treatment or control diet is placed in 30-ml plastic cups and allowed to cool and solidify at room temperature. One neonate larva is placed on top of the diet and the cup is capped. The cups are maintained in environmental chambers at 27 ± 2° C, 75 ± 5% RH and 12:12 (L:D) photoperiods until larvae pupate. Ten to 15 days after infestation, depending on the species of insect, larval survival and larval weights are recorded.

To test crude plant extracts or plant allelochemicals, they are generally admixed with the dry portion of the artificial diet, and the carrier solvent is removed under vacuum in a desiccator (Ohigashi and Koshimizu 1976, Reese and Beck 1976, Dowd et al. 1983, Jenkins et al. 1983, Khan et al. 1986, Salloum and Isman 1989). Controls consist of artificial diets similarly treated with the carrier solvent alone.

Chan et al. (1978) and Reese et al. (1989) have developed techniques to bioassay phytochemicals in an artificial diet. The plant extracts or allelochemicals are coated onto a carrier such as alphacel, and this mixture is dissolved in a small beaker with a minimum volume of solvent such as diethyl ether, n-hexane, ethylacetate, acetone, chloroform methanol, or ethanol. Once the carrier is added to the artificial diet, care should be taken to add allelochemicals or extracts to the carrier in amounts that will equal their biological concentration in living plant tissues. Alphacel to 5% of the final diet weight is generally added as a carrier. The solvent is dried off with a stream of nitrogen and traces of solvent are removed in a vacuum desiccator. If the substance is water soluble it can be added directly in the diet and the water removed by freeze-drying. Phytochemical-coated alphacel is mixed similarly with the dry portion of an artificial diet. Controls consist of artificial diets similarly incorporated with untreated alphacel.

Zhou et al. (1984) developed a technique for bioassaying water soluble plant extracts. Artificial diets are poured into petri dishes and diet plugs weighing about 3 g are cut from the diet. A corkborer (0.5 cm in diameter) is then used to drill a hole in the center of each plug. The drilled plugs are frozen at −10° C for 24 hr and lyophilized. The shriveled, lyophilized plugs of diet are very porous, having lost about 80% of their weight. The plugs are dipped in plant water extractables and are allowed to absorb extracts for 12 hr at 4°C. After each of the diet plugs has thoroughly absorbed the extract, the surplus extract on the outside of each plug is removed with filter paper and they are infested with larvae of test insects.

7.2.3. Bioassays

By placing eggs or newly emerged larvae on control and treated diets, differences in feeding, weight gain, survival, and growth are detected. The following criteria are used to evaluate allelochemicals in plants.

7.2.3.1. Nutritional Indices

Nutritional indices of insect feeding have been developed by Waldbauer (1964, 1968), Beck and Reese (1976), and Reese et al. (1989), based on the development of neonate larvae feeding on control and treated diets. The procedure involves weighing larvae and placing them individually on weighed cylinders of treated or control diet and allowing larvae to feed *ad libitum* for approximately 10 days. Larvae, feces, and uneaten diet are carefully separated after 10 days, weighed, and dried to a constant weight. Fresh control and treated diets are also weighed and dried to a constant weight for calculating the initial percentage dry matter. These measurements are then used to calculate:

$$\text{Approximate Digestibility (AD)} = \frac{\text{Amount ingested (mg)} - \text{feces (mg)}}{\text{Amount ingested (mg)}} \times 100$$

$$\text{Efficiency of Conversion of Digested Food (ECD)} = \frac{\text{Weight gain (mg)}}{\text{Amount ingested (mg)} - \text{feces (mg)}} \times 100$$

$$\text{Efficiency of Conversion of Ingested Food (ECI)} = \frac{\text{Weight gain (mg)}}{\text{Amount ingested (mg)}} \times 100$$

Reese and Beck (1976) calculated nutritional indices of black cutworm, *Agrotis ipsilon* (Hufnagel), feeding on artificial diet incorporated with catechol, chlorogenic acid and dopamine. Catechol did not inhibit survival or food ingestion, but did inhibit larval growth by reducing ECD and ECI. Chlorogenic acid and dopamine had no apparent effect on survival, growth, pupation, and pupal weight of *A. ipsilon*. Salloum and Isman (1989) reported severe growth inhibition (extremely low ECD and ECI) of the variegated cutworm, *Peridroma saucia* (Hübner), fed on artificial diet amended with the ethanolic extract of an asteraceous weed, *Artemisia tridentata* Nutt.

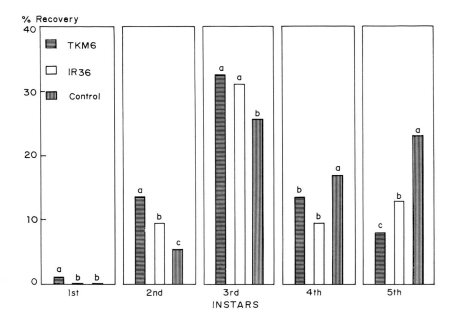

FIGURE 7.2. Different larval instars of *Cnaphalocrocis medinalis* recovered after 15 days of infestation on an artificial diet and diet mixed with leaf powders of susceptible IR36 and resistant TKM6; means followed by the same letter were not significantly different. (From Khan, Z. R. et al., *Entomol. Exp. Appl.*, 52, 7, 1989. With permission of Kluwer Academic Publishers.)

7.2.3.2. Growth

Larval development on control and treated diets is routinely measured by recording larval survival (Ohigashi and Koshimizu 1976, Zhou et al. 1984, Khan et al. 1989), weights from 5 to 15 days after infestation (Reese and Beck 1976, Dreyer et al. 1979, Jenkins et al. 1983, Reynolds et al. 1984, Zhou et al. 1984, Broadway et al. 1986, Salloum and Isman 1989, Williams et al. 1990, Wiseman and Isenhour 1991, Ramachandran and Khan 1991, Buckley et al. 1991, Diawara et al. 1991), developmental time to pupation (Ramachandran and Khan 1991, Wiseman et al. 1991), weight of pupae (Reese and Beck 1976, Ramachandran and Khan 1991, Diawara et al. 1991, Wiseman et al. 1991, Diawara et al. 1991), larval stadia (Khan et al. 1989, Wiseman et al. 1991), and larval length (Wilson and Wissink 1986).

Steam distillate extracts from resistant PI 227687 soybean plants, when incorporated in an artificial diet (25 g of artificial diet containing 25, 50, or 100 mg distillate), were more toxic to the cabbage looper, *Trichoplusia ni* (Hübner), than extractables from susceptible Davis plants (Khan et al. 1986). Larval mortality on diets treated with Davis extracts was low but significantly more than that on solvent-treated (control) diet.

Rice leaffolder, *C. medinalis* (Guenée), larval survival was significantly lower on diets containing leaf powders of the leaffolder-resistant wild rices, *Oryza perennis* or *Oryza punctata,* than on control diets or the diet containing leaf powder of the susceptible rice variety IR36 (Khan et al. 1989). Similarly, incorporation of leaf powder of the leaffolder-resistant variety TKM6 into artificial diet significantly delays *C. medinalis* larval development as compared to development on control diet or diet containing susceptible IR36 leaf powder (Figure 7.2). Larval development on the artificial diet with IR36 leaf powder was also delayed compared to that on the control diet. This finding supports the concept that most susceptible plants growing under natural conditions are not necessarily the optimal host for the involved insect (de Ponti 1982). Similarly, Reese and Field (1986) reported that black cutworm, *A. ipsilon,* larvae grew more slowly on susceptible corn seedlings than a control artificial diet.

Wiseman and Isenhour (1990) demonstrated that low concentrations of resistant (antibiotic) maize silk incorporated into an artificial diet reduced larval growth of the corn earworm, *H. zea,* and extended the life cycle by 3 days. High concentrations of this silk also reduced *H. zea* growth and extended the life cycle by as much as 20 days. Wiseman et al. (1991) determined that larval and pupal weights of *H. zea* larvae fed on silks of a resistant (Zapalote Chico) maize were lower than weights of larvae fed on susceptible (Stowell's Evergreen) maize silks. Larval growth and larval and pupal weights are inversely related to the amount of resistant silk added to the diet, and as the concentration of resistant silk increases in the diet, *H. zea* stadial length also significantly increases.

Although artificial diets could be very helpful in bioassays intended to assess the economic importance of plant allelochemicals, there are some problems with this procedure. Not all insects can be reared on artificial diets, making this procedure unavailable to some researchers. In many instances, diets are suboptimal and are themselves a source of stress for developing insect larvae. On the other hand, some artificial diets act as "super diets" and often support faster luxuriant larval growth (Beck 1974, Reese and Field 1986). Such luxuriant growth makes larvae less susceptible to allelochemicals, and can mask the effects of allelochemicals from resistant plant that occur at low concentrations and have subtle effects on usual growth and development (Rose et al. 1988).

7.3. CONCLUSIONS

Recent advances in insect bioassay techniques using tissue culture and artificial diets to evaluate insect resistance have greatly improved the understanding of allelochemical bases of insect resistance.

The use of tissue cultures to investigate the allelochemical bases of mechanisms of insect resistance promises to be an exciting area of research. However, a major constraint is that all cultivars respond differently to the composition of the callus culturing media, and calli cannot be obtained from some plant species. Realistically, even if all cultivars can be grown in a general medium, screening tissue culture will not totally replace screening under field conditions, since field experiments involve a complex set of insect-plant-environment interactions. However, if similar allelochemicals can be identified in both callus and plant tissue and their occurrence correlated to field conditions, tissue culture may be used in biotechnology programs to produce large quantities of insect toxins at a low cost. There is also a need for an improved understanding of foliar nutrient composition, in order for artificial insect diets to contain a nutrient content similar to that of the plant. Most artificial diets used to rear insects contain excessive amounts of nutrients that can inhibit the expression of the effects of allelohemicals on the pest insect. This improvement will be necessary in order to more fully understand the contributions of allelochemicals in plant resistance to insects.

REFERENCES

Beard, R. L., The biology of *Anasa tristis* DeGeer with reference to the Tachinid parasite, *Trichopoda pennipes* Fabr., *Bull. Conn. Exp. Stn.,* 440, 593, 1940.

Beck, S. D., Theoretical aspects of host plant specificity in insects, in *Proceedings of the Summer Institute on Biological Control of Plants, Insects and Diseases,* F. G. Maxwell and F. A. Harris, eds., University Press of Mississippi, Jackson, 1974, 647.

Beck, S. D. and Reese, J. C., Insect-plant interactions: nutrition and metabolism, *Rec. Adv. Phytochem.,* 10, 41, 1976.

Broadway, R. M., Duffey, S. S., Pearce, G. and Ryan, C. A., Plant proteinase inhibitors: A defense against herbivorous insects? *Entomol. Exp. Appl.,* 41, 33, 1986.

Buckley, P. M., Davis, F. M. and Williams, W. P., Identifying resistance in corn to corn earworm (Lepidoptera: Noctuidae) using a laboratory bioassay, *J. Agric. Entomol.,* 8, 67, 1991.

Caballero, P., Shin, D. H., Khan, Z. R., Saxena, R. C., Juliano, B. O. and Zapata, F. J., Use of tissue culture to evaluate rice resistance to lepidopterous pests, *Int. Rice Res. Newsl.,* 13(5), 14, 1988.

Chan, B. G., Waiss, Jr., A. C., Stanley, W. L. and Goodban, A. E., A rapid diet preparation method for antibiotic phytochemical bioassay, *J. Econ. Entomol.,* 71, 366, 1978.

Chu, C. C., Wang, C. S. and Sun, C. C., The N6 medium and its application to another culture of cereal crops, in *Symposium on Plant Tissue Culture (1978: Peking, China),* Pitman Advanced Publishing Program, Boston, 1981, 531.

Croughan, S. S. and Quisenberry, S. S., Evaluation of cell culture as a screening technique for determining fall armyworm (Lepidoptera: Noctuidae) resistance in bermudagrass, *J. Econ. Entomol.,* 82, 232, 1989a.

Croughan, S. S. and Quisenberry, S. S., Enhancement of fall armyworm (Lepidoptera: Noctuidae) resistance in bermudagrass through cell culture, *J. Econ. Entomol.,* 82, 236, 1989b.

de Ponti, O. M. B., Plant resistance to insects: a challenge to plant breeders and entomologists, in *Proceedings 5th International Symposium on Insect-Plant Relationships,* J. H. Visser and A. K. Minks, eds., Centre for Agricultural Publishing and Documentation, Wageningen, 1982, 464.

Diawara, M. M., Wiseman, B. R. and Isenhour, D. J., Bioassays for screening plant accessions for resistance to fall armyworm (Lepidoptera: Noctuidae) using artificial diets, *J. Entomol. Sci.,* 26, 367, 1991.

Dowd, P. F., Smith, C. M. and Sparks, T. C., Influence of soybean leaf extracts on ester cleavage in cabbage and soybean loopers (Lepidoptera: Noctuidae), *J. Econ. Entomol.,* 76, 700, 1983.

Dreyer, D. L., Binder, R. G., Chan, B. G., Waiss, Jr., A. C., Hartwig, E. E. and Beland, G. L., Pinitol, a larval growth inhibitor for *Heliothis zea* in soybeans, *Experientia,* 35, 1182, 1979.

Escubas, P., Fukushi, Y., Lajide, L. and Mizutani, J., A new method for fast isolation of insect antifeedant compounds from complex mixtures, *J. Chem. Ecol.,* 18, 1819, 1992.

Gamborg, O. L., Miller, R. A. and Ojima, K., Nutrient requirements of soybean root cells, *Exp. Cell. Res.,* 50, 151, 1968.

Hori, K., Torikura, H. and Kumagi, M., Histological and biochemical changes in the tissue of pumpkin fruit injured by *Lygus disponsi* Linnavuori (Hemiptera: Miridae), *Appl. Entomol. Zool.,* 22, 259, 1987.

Isenhour, D. J. and Wiseman, B. R., Incorporation of callus tissue into artificial diet as a means of screening corn genotypes for resistance to the fall armyworm and the corn earworm (Lepidoptera: Noctuidae), *J. Kans. Entomol. Soc.,* 61, 303, 1988.

Isenhour, D. J. and Wiseman, B. R., Fall armyworm resistance in progeny of maize plants regenerated via tissue culture, *Fla. Entomol.,* 74, 221, 1991.

Isenhour, D. J., Duncan, R. R., Miller, D. R., Waskom, R. M., Hanning, G. E., Wiseman, B. R. and Nabors, M. W., Resistance to leaf-feeding by the fall armyworm (Lepidoptera: Noctuidae) in tissue culture derived sorghums, *J. Econ. Entomol.,* 84, 680, 1991.

Jenkins, J. N., Hedin, P. A., Parrott, W. L., McCarty, Jr., J. C. and White, W. H., Cotton allelochemics and growth of tobacco budworm larvae, *Crop Sci.,* 23, 1195, 1983.

Khan, Z. R., Norris, D. M., Chiang, H. S., Weiss, N. E. and Oosterwyk, A. S., Light-induced susceptibility in soybean to cabbage looper, *Trichoplusia ni* (Lepidoptera: Noctuidae), *Environ. Entomol.,* 15, 803, 1986.

Khan, Z. R., Rueda, B. R. and Caballero, P., Behavioral and physiological responses of rice leaffolder, *Cnaphalocrocis medinalis* to selected wild rices, *Entomol. Exp. Appl.,* 52, 7, 1989.

Linsmaier, E. M. and Skoog, F., Organic growth factor requirements of tobacco tissue cultures, *Physiologia plantarum,* 18, 100, 1965.

Ma, R., Reese, J. C., Black IV, W. C. and Bramel-Cox, P., Detection of plant cell wall-degrading enzymes in greenbug saliva, in *Proceedings, Aphid-Plant Interactions: Populations to Molecules,* D. C. Peters, J. A. Webster and C. S. Chlouber, eds., Oklahoma State University, Stillwater, 1991, 335.

Miles, P. W., The saliva of Hemiptera, *Adv. Insect Physiol.,* 9, 183, 1972.

Murashinge, T. and Skoog, F., A revised medium for rapid growth and bioassays with tobacco tissue cultures, *Physiologia plantarum*, 15, 473, 1962.

Ohigashi, H. and Koshimizu, K., Chavicol, as a larva-growth inhibitor, from *Viburnum japonicum* Spreng., *Agric. Biol. Chem.*, 40, 2283, 1976.

Paschke, J. D. and Sylvester, E. S., Laboratory studies on the toxic effects of *Therioaphis maculata* (Buckton), *J. Econ. Entomol.*, 50, 742, 1957.

Ramachandran, R. and Khan, Z. R., Mechanisms of resistance in wild rice *Oryza brachyantha* to rice leaffolder *Cnaphalocrocis medinalis* (Guenée) (Lepidoptera: Pyralidae), *J. Chem. Ecol.*, 17, 41, 1991.

Reese, J. C. and Beck, S. D., Effects of allelochemics on the black cutworm, *Agrotis ipsilon;* effects of p-benzoquinone, hydroquinone, and duroquinone on larval growth, development, and utilization of food, *Ann. Entomol. Soc. Am.*, 69, 58, 1976.

Reese, J. C. and Field, M. D., Defense against insect attack in susceptible plants: black cutworm (Lepidoptera: Noctuidae) growth on corn seedlings and artificial diet, *Ann. Entomol. Soc. Am.*, 79, 372, 1986.

Reese, J. C., Waiss, Jr., A. C. and Legacion, D. M., Biochemical methodologies and approaches for determining the basis of maize resistance to insects, in *Towards Insect Resistance Maize for the Third World: Proceedings of the International Symposium on Methodologies for Developing Host Plant Resistance to Maize Insects,* CIMMYT, Mexico, D.F., 327, 1989.

Reese, J. C., Bramel-Cox, P., Dixon, A. G. O., Schmidt, D. J., Ma, R., Noyes, S., Margolies, D. C. and Black, W. C., IV, Novel approaches to the development of *Sorghum* germplasm resistant to greenbugs, *Symp. Biol. Hung.*, 39, 523, 1990.

Reynolds, G. W., Smith, C. M. and Kester, K. M., Reductions in consumption, utilization, and growth rate of soybean looper (Lepidoptera: Noctuidae) larvae fed foliage of soybean genotype PI227687, *J. Econ. Entomol.*, 77, 1371, 1984.

Rose, R. L., Sparks, T. C. and Smith, C. M., Insecticide toxicity to the soybean looper and the velvetbean caterpillar (Lepidoptera: Noctuidae) as influenced by feeding on resistant soybean (PI227687) leaves and coumestrol, *J. Econ. Entomol.*, 81, 1288, 1988.

Salloum, G. S. and Isman, M. B., Crude extracts of asteraceous weeds, growth inhibitors for variegated cutworm, *J. Chem. Ecol.*, 15, 1379, 1989.

Saxena, K. N., Gandhi, J. R. and Saxena, R. C., Patterns of relationship between certain leafhoppers and plants. I. Responses to plants, *Entomol. Exp. Appl.*, 17, 303, 1974.

Sharma, N. R., Use of Tissue Culture for Studying Resistance to the Leaffolder, *Cnaphalocrocis medinalis* (Guenée) in Rice, M.S. Thesis, University of the Philippines at Los Baños, Philippines, 1991, 111.

Shukle, R. H., Murdock, L. L. and Gallun, R. L., Identification and partial characterization of a major gut enzyme from larvae of the Hessian fly *Mayetiola destructor* (Say) (Diptera: Cecidomyiidae), *Insecta Biochem.*, 15, 93, 1985.

Smith, C. M. and Fisher, N. H., Chemical factors of an insect resistant soybean genotype affecting growth and survival of the soybean looper, *Entomol. Exp. Appl.*, 33, 343, 1983.

Straub, R. W. and Fairchild, M. L., Laboratory studies of resistance in corn to the corn earworm, *J. Econ. Entomol.,* 63, 1901, 1970.

Thapa, R. B., Use of Tissue Culture for Studying Rice Resistance to Yellow Stem Borer, *Scirpophaga incertulas* (Walker), Ph.D. Thesis, University of the Philippines at Los Baños, Philippines, 1991, 101.

Wadley, F. M., Observations on the injury caused by *Toxoptera graminum* Rond (Homoptera: Aphididae), *Proc. Entomol. Soc. Wash.,* 31, 130, 1929.

Waldbauer, G. P., The consumption, digestion and utilization of solanaceous and non-solanaceous plants by larvae of the tobacco hornworm, *Protoparce sexta* (Johan.) (Lepidoptera: Sphingidae), *Entomol. Exp. Appl.,* 7, 253, 1964.

Waldbauer, G. P., The consumption and utilization of food by insects, *Adv. Insect Physiol.,* 5, 229, 1968.

White, W. H. and Irvine, J. E., Evaluation of variation in resistance to sugarcane borer (Lepidoptera: Pyralidae) in a population of sugarcane derived from tissue culture, *J. Econ. Entomol.,* 80, 182, 1987.

Williams, W. P. and Davis, F. M., Southern corn borer larval growth on corn callus and its relationship with leaf feeding resistance, *Crop Sci.,* 25, 317, 1985.

Williams, W. P., Buckley, P. M. and Taylor, V. N., Southwestern corn borer growth on callus initiated from corn genotypes with different levels of resistance to plant damage, *Crop Sci.,* 23, 1210, 1983.

Williams, W. P., Buckley, P. M. and Davis, F. M., Larval growth and behavior of the fall armyworm (Lepidoptera: Noctuidae) on callus initiated from susceptible and resistant corn hybrids, *J. Econ. Entomol.,* 78, 951, 1985.

Williams, W. P., Buckley, P. M. and Davis, F. M., Feeding response of corn earworm (Lepidoptera: Noctuidae) to callus and extracts from corn in the laboratory, *Environ. Entomol.,* 16, 532, 1987a.

Williams, W. P., Buckley, P. M. and Davis, F. M., Tissue culture and its use in investigations of insect resistance in maize, *Agric. Ecosyst. Environ.,* 18, 185, 1987b.

Williams, W. P., Buckley, P. M., Hedin, P. A. and Davis, F. M., Laboratory bioassay for resistance in corn to fall armyworm (Lepidoptera: Noctuidae) and southwestern corn borer (Lepidoptera: Pyralidae), *J. Econ. Entomol.,* 83, 1578, 1990.

Wilson, R. L. and Wissink, K. M., Laboratory method for screening corn for European corn borer (Lepidoptera: Pyralidae) resistance, *J. Econ. Entomol.,* 79, 274, 1986.

Wiseman, B. R. and Isenhour, D. J., Effects of resistant maize silks on corn earworm (Lepidoptera: Noctuidae) biology: a laboratory study, *J. Econ. Entomol.,* 83, 614, 1990.

Wiseman, B. R. and Isenhour, D. J., A microtechnique for antibiosis evaluations against the corn earworm, *J. Kans. Entomol. Soc.,* 64, 146, 1991.

Wiseman, B. R., Isenhour, D. J. and Bhagwat, V. R., Stadia, larval-pupal weight, and width of head capsules of corn earworm (Lepidoptera: Noctuidae) after feeding on varying resistance levels of maize silks, *J. Entomol. Sci.,* 26, 303, 1991.

Zhou, D., Guthrie, W. D. and Chen, C., A bioassay technique for screening inbred lines of maize for resistance to leaf feeding by the European corn borer, *Maydica,* 29, 69, 1984.

Index